# ENVIRONMENTAL PHILOSOPHY

# ENVIRONMENTAL PHILOSOPHY

## Critical Concepts in the Environment

*Edited by*
*J. Baird Callicott and*
*Clare Palmer*

## Volume V:
## History and Culture

Routledge
Taylor & Francis Group
LONDON AND NEW YORK

First published 2005
by Routledge
2 Park Square, Milton Park, Abingdon, Oxon, OX14 4RN

Simultaneously published in the USA and Canada
by Routledge
270 Madison Ave, New York, NY 10016

Reprinted 2006 (twice)

*Routledge is an imprint of the Taylor & Francis Groups, an informa business*



Typeset in Times New Roman by
Newgen Imaging Systems (P) Ltd, India
Printed and bound in Great Britain by
MPG Books Ltd, Bodmin



*British Library Cataloguing in Publication Data*
A catalogue record for this book is available from the British Library

*Library of Congress Cataloging in Publication Data*
A catalog record for this book has been requested

ISBN 10: 0–415–32646–X
ISBN 13: 9–780–0–415–32646–9
ISBN 10: 0–415–34145–0 (Volume V)
ISBN 13: 9–780–0–415–34145–5 (Volume V)

**Publisher's Note**

References within each chapter are as they appear in the original complete work

# CONTENTS

## VOLUME V: HISTORY AND CULTURE

# ACKNOWLEDGEMENTS

The publishers would like to thank the following for permission to reprint their material:

The American Association for the Advancement of Science for permission to reprint Lynn White, Jr., 'The historical roots of our ecological crisis', *Science*, 155 (1967), 1203–7. Copyright 1967 AAAS.

Blackwell Publishing for permission to reprint Yi-Fu Tuan, 'Discrepancies between environmental attitude and behaviour: examples from Europe and China', *Canadian Geographer*, 12 (1968), 176–91.

University of New Mexico Press for permission to reprint J. Donald Hughes, 'The ancient roots of our ecological crisis' in J. Donald Hughes (ed.), *Ecology in Ancient Civilizations*, Albuquerque: University of New Mexico Press, 1975, pp. 147–56.

The author for permission to reprint Arne Naess, 'Spinoza and ecology', *Philosophia*, 7 (1977), 45–54. http://www.tandf.co.uk

Taylor & Francis for permission to reprint Genevieve Lloyd, 'Spinoza's environmental ethics', *Inquiry*, 23 (1980), 293–311. http://www.tandf.co.uk.

The author for permission to reprint K. L. F. Houle, 'Spinoza and ecology revisited', *Environmental Ethics*, 19 (1997), 417–31.

The estate of Charles Hartshorne for permission to reprint Charles Hartshorne, 'The rights of the subhuman world', *Environmental Ethics*, 1 (1979), 49–60.

Brill Academic Publishers for permission to reprint Clare Palmer, 'Identity, community and the natural environment: some perspectives from process thinking', *Worldviews: Environment, Culture, Religion*, 2 (1998), 257–67.

The MIT Press for permission to reprint James E. Lovelock, 'Geophysiology—the science of Gaia' in Stephen Schneider and Penelope Boston (eds), *Scientists on Gaia*, Cambridge, MA: MIT Press, 1991, pp. 1–10. Copyright © 1991, The MIT Press.

The author for permission to reprint Anthony Weston, 'Forms of Gaian ethics', *Environmental Ethics*, 9 (1987), 217–30.

The MIT Press for permission to reprint David Abram,'The mechanical and the organic: on the impact of metaphor in science', in Stephen Schneider and Penelope Boston, *Scientists on Gaia*, Cambridge, MA: MIT Press, 1991, pp. 66–74. Copyright © 1991, The MIT Press.

Brill Academic Publishers for permission to reprint Freya Mathews, 'Letting the world grow old: an ethos of countermodernity', *Worldviews: Environment, Culture, Religion*, 3 (1999), 119–37.

The author for permission to reprint Jim Cheney, 'The neo-stoicism of radical environmentalism', *Environmental Ethics*, 11 (1989), 293–325.

Oxford University Press for permission to reprint Yuriko Saito, 'The Japanese appreciation of nature', *British Journal of Aesthetics*, 25 (1985), 239–49.

The author for permission to reprint Chung-ying Cheng, 'On the environmental ethics of the *Tao* and the *Ch'i*', *Environmental Ethics*, 8 (1986), 351–70.

The author for permission to reprint David Kalupahana,'Man and nature: toward a middle path of survival', *Environmental Ethics*, 8 (1986), 371–80.

Gerald Larson, "Conceptual resources", in South Asia for "Environmental Ethics", reprinted by permission from *Nature in Asian Traditions of Thought: Essays in Environmental Philosophy*, J. Baird Callicott and Roger T. Ames (eds), the State University of New York Press, © 1989. State University of New York. All rights reserved.

The authors for permission to reprint G. W. Burnett, and Kamuyu wa Kang'ethe, 'Wilderness and the Bantu mind', *Environmental Ethics*, 16 (1994), 145–60.

Taylor & Francis for permission to reprint Paul Faulstich, 'Mapping the mythological landscape: an aboriginal way of being-in-the-world', *Ethics, Place, and Environment*, 1 (1998), 97–221. http://www.tandf.co.uk.

The author for permission to reprint J. Baird Callicott, 'The metaphysical implications of ecology', *Environmental Ethics*, 8 (1986), 301–16.

The authors for permission to reprint Karen J. Warren and Jim Cheney, 'Ecosystem ecology and metaphysical ecology: a case study', *Environmental Ethics*, 15 (1993), 99–116.

S. T. A. Pickett and Richard S. Ostfeld, 'The shifting paradigm in ecology', pp. 261–78, from *A New Century for Natural Resources Management*, by Richard Knight and Sarah Bates (eds), Copyright © 1995 by Island Press. Reproduced by permission of Island Press, Washington, DC.

## Disclaimer

# INTRODUCTION TO VOLUME V

This volume contains a selection of papers exploring foundational questions in environmental philosophy, in particular concepts of nature, and the human place in and relationship to nature. These concepts are considered from a wide range of philosophical and metaphysical positions emerging out of different times and cultures. It might be argued, indeed, that this volume is the lynchpin for all the others, since all work in environmental philosophy must either explicitly consider, or rest on assumptions about, these more fundamental questions concerning nature, human nature, and the relationship between the two. Given the huge range of potentially relevant material and the restricted space available, this volume proved most difficult in terms of limiting the selection of papers. In the end, we confined our selection to papers that met one or more of the following conditions: (a) being primarily philosophical in content; (b) making important contributions to subsequent debates in environmental philosophy; and (c) representing *either* important schools of thought in environmental philosophy *or* environmental perspectives derived from philosophical schools of thought important for other reasons. These criteria led us—not without some debate—to the selection that appears here.

The volume begins with a section called *Historical Roots*, a direct reference to the title of the lead article by Lynn White, Jr. 'The historical roots of our ecologic crisis' must have been one of the most significant and controversial articles published on the environment in the second half of the twentieth century. Certainly it was among the most influential. The reverberations from White's claim that (for a variety of reasons) attitudes towards the natural world embedded in Western Christianity are to blame for the environmental crisis can still be felt. In particular, his claim that members of Western cultures with a Judaeo-Christian heritage see themselves as separate from, and dominant over, the rest of nature, has been widely accepted within environmental philosophy. An extensive literature of responses to White's specific claims now exist: some supportive, some partially supportive, and some harshly critical. We were able to include only two further chapters in this part, both of which respond to some of White's concerns whilst making interesting and important claims in their own right. Yi-Fu Tuan's chapter 'Discrepancies between environmental attitude and behaviour: examples from Europe and China' argues that the link White makes between belief and action does not obtain, looking at examples of cultures where prevailing beliefs suggest that the environment is of very high value, but where cultural practices indicate that environmental destruction is nonetheless systematic. J. Donald Hughes argues in 'The ancient roots of our ecological crisis' that the roots of the ecological crisis reach into the soil of Greco-Roman classical civilisations. All three chapters are concerned with the relationship between belief and action. This relationship is crucially significant to environmental philosophy, because so much work in the subject aims at changing beliefs, both about the nature of the natural world and the human place in it, as a necessary condition for changing human behaviour in regard to the environment.

The second part of the volume contains a sample of contributions to an extensive debate about whether the work of Benedict Spinoza, or some adaptation of it, might form an appropriate metaphysical underpinning for some forms of environmental philosophy. This in turn is but a sample of one approach to reconstructive environmental philosophy—the revival of some revered historical figure whose philosophy is arguably an alternative to what Richard Routley in his seminal paper (Volume I) called the 'dominant tradition' of Western philosophy. In addition to Spinoza, Heraclitus, among the ancient, and Heidegger, among contemporary philosophers have been dusted off and served up as remedies for what ails us environmentally. In the chapters included here, Arne Naess in 'Spinoza and ecology' opens the discussion by suggesting that Spinoza's *Deus sive Natura*, alive, creative and all-inclusive, may provide a vision to inspire environmental philosophy. In Spinoza's thought, according to Naess, humans are part of, not separate from, nature (thus contrasting with White's portrayal of Christianity) and all beings—not only human beings—strive to actualise or realise their *conatus*. Genevieve Lloyd, in 'Spinoza's environmental ethics' raises questions about Naess' positive account of Spinoza, arguing that whilst it may be the case that Spinoza rejects an anthropocentric perception of the world, this rejection is coupled with 'an equally strong affirmation of man-centred morality'. Spinoza's work, she suggests, cannot be used as the basis for a non-anthropocentric environmental ethic. The third chapter in this part, Chapter 86, 'Spinoza and ecology revisited' by K.L.F. Houle, reviews the wider debate about 'Spinoza as ecologist', taking into account the work of Naess, Lloyd and that of others, such as George Sessions. This chapter attempts to effect a reconciliation, of sorts, between the conflicting positions.

Discussion of Spinoza's pantheistic philosophy is followed, in the next part, by two chapters on process philosophy. Charles Hartshorne, a leading process philosopher and theologian, published the paper 'The rights of the subhuman world,' in the first edition of the journal *Environmental Ethics* in 1979. He outlines a panexperientialist and panentheistic position, in which all lives—indeed, every 'singular, active agent'—can contribute to and delight the divine life that generates intrinsic value. Clare Palmer's chapter, 'Identity, community and the natural environment: some perspectives from process thinking', Chapter 88, explores aspects of this panexperiential and panentheistic tradition further, outlining some difficulties that may flow from such a position. (As both lead author of this introduction and sole author of this chapter, Palmer would like to note that she has changed her views on a few points. However, because one rule of this collection is that papers should appear as originally published, she has not revised the paper to take account of these changes—and doubtless many other authors might like to offer similar caveats if given the chance.) Both pantheist and panentheist metaphysical positions, therefore, are represented and critically discussed in Parts 2 and 3 of this volume respectively.

The following part considers the idea of a living Earth, in the form of the Gaia Hypothesis (not completely unrelated either to pantheist or process thinking, a point that Anthony Weston makes in his chapter here). An extensive and disparate literature (ranging across biology, ecology, philosophy, and religion) on the Gaia Hypothesis has emerged since James Lovelock first published his book *Gaia: A New Look at Life on Earth*, Oxford: Oxford University Press, in 1979. Again, we could only include a few chapters here, each of which provides different entry points to thinking about the Gaia Hypothesis. The first chapter, Chapter 89, 'Geophysiology: the science of Gaia' by James Lovelock introduces and outlines the main contours of the Gaia Hypothesis. He suggests that it is helpful scientifically—as well as in other ways—to see the earth as a kind of *super-organism*. Lovelock's chapter is followed by Anthony Weston's chapter 'Forms of Gaian ethics'. (Although Lovelock's paper was published after Weston's, Chapter 90, in this part, we have included it first because it is a later summary of an earlier idea, on which the following two chapters are commenting.) Weston explores what the Gaia Hypothesis might mean in ethical terms, questioning the widely held view that the hypothesis necessarily implies some form of strong, nonanthropocentric environmental ethic. Weston points out that different readings of the hypothesis, resting on different interpretations of the place of humans on Earth, are possible. Humans might be understood to be just one more Gaian species, or, alternatively, they might be seen as the mind of the Earth organism. The third chapter here, Chapter 91, David Abram's 'The mechanical and the organic', reflects on the significance of the metaphorical language of *organism* rather than *mechanism* in understanding the world. Drawing on work in the phenomenology of perception, he argues that the metaphor of mechanism separates scientists from the world in which they are working, while the metaphor of organism—as fleshed out in the language of Gaia—encourages a more participatory and less exploitative perspective.

Freya Mathews' chapter, Chapter 92, 'Letting the world grow old: an ethos of countermodernity' begins the next part we have called *Cosmic*

*Awareness*. (This chapter could well be read alongside her chapter 'Conservation and self-realization: a Deep Ecology perspective', Chapter 27, in Volume II—and it should be noted that the chapter reproduced here is just the first half of a two-section paper; the second part was published in the journal *Worldviews* in 1999.) Mathews' broadly panentheistic approach to the cosmos here leads her to maintain a position of letting things be, letting the given stand, let natural processes work without the need for constant human change and modernisation. As with Mathews' chapter, Jim Cheney's chapter 'The neo-stoicism of radical environmentalism', Chapter 93, which follows, also could be usefully read with the *Deep Ecology* section of Volume II. However, we included it here rather than there because Cheney's interest is primarily in the metaphysical orientation of Deep Ecology. He suggests that deep ecological metaphysical positions have a tendency, similar in some respects to ancient Stoicism, towards a kind of cosmic veneration and 'submission to a modern form of the divine logos'. The effects of this tendency, he suggests, are worrying: they may lead to a kind of totalising and essentialising voice—one that ultimately colonises other people.

Having considered some—broadly-speaking—Western approaches to metaphysical questions in environmental philosophy, the following part contains chapters introducing Asian philosophical approaches. In a collection of this kind it was impossible to aim even for semi-comprehensive coverage of the great range of Asian philosophical concepts of nature and the human place in relation to nature. Thus we have been able to include only a few papers presenting different approaches to environmental philosophy from Japanese, Chinese, and Indian perspectives, as well as a paper presenting a note of warning about appropriating traditions of thought such as Hinduism, Taoism, and Buddhism outside their contexts of origin. These papers, all published in the 1980s, have made a contribution to subsequent debates in environmental philosophy, and are all still widely used in the classroom.

Yuriko Saito's chapter 'The Japanese appreciation of nature', argues that 'traditional Japanese love of nature is based upon the conceived identity between man and nature'—a view that strongly contrasts with White's portrayal of Christianity in the first chapter in this volume. This identification of humans and nature, she suggests, stems from a kind of emotional affinity and an appreciation of the transience of nature and of humans, and is expressed in traditional Japanese aesthetic culture. Chung-ying Cheng in 'On the environmental ethics of the *Tao* and the *Ch'i*' considers what is meant by the term 'environment' and outlines a Chinese understanding of it, stemming from Taoism and Confucianism, 'based on integrative dependence and harmony between man and world'. He argues that a Chinese view of the natural environment understands it to be a totality, both phenomenal and organismic. David Kalupahana's chapter, 'Man and nature: towards a middle path of survival', considers how one might best think about 'nature' from a Buddhist perspective. He brings his interpretation of early Buddhism together with the work of William James, presenting a pragmatic approach to environmental problems, and suggesting that the Western Pragmatist philosophical tradition might provide a way forward, congruent with—but without borrowing from—Eastern traditions.

Precisely this concern—about borrowing from Eastern traditions—animates the final, and controversial, chapter in this part, by Gerald Larson. Larson in '"Conceptual resources" in South Asia for "environmental ethics"' questions the practice—relatively common in environmental thinking in general, and in environmental philosophy in particular—of incorporating ideas (or in Larson's terms, appropriating conceptual resources) from Asian philosophical traditions in Western environmental ethics. Such a practice, he maintains, *inter alia*, parallels similar exploitative colonial practices with natural resources, disembeds ideas from their historical and cultural context, and fails to appreciate the radical newness of the environmental problems that face us. (Interestingly, some parallels can be noted between both Larson's and Cheney's concerns in this volume.)

Complementing Western and Asian philosophical approaches to understanding nature and the human place in it, the penultimate part of this volume offers contributions from an East African Bantu, and an indigenous Australian perspective. G. W. Burnett and Kamuyu wa Kang'ethe in 'Wilderness and the Bantu mind' explore the metaphysics of the Bantu, and concentrate on their conception of wilderness as a social space, albeit a frightening one. They suggest that, unlike the post-colonial American–Australian wilderness idea, for the Bantu the wilderness is not seen as a place for self-realisation or inspiration; and indeed wilderness experiences can be harmful. (This chapter could helpfully be read with the chapters on wilderness in Volume IV.) Faulstich's chapter 'Mapping the mythological landscape: an aboriginal way of being-in-the-world' explores the place-orientedness of the Warlpiri, a group of Australian aboriginal people of the Central Australian Desert. He considers the way in which the Warlpiri interpret and make sense of the landscape through which they move, creating a kind of environmental philosophy mapped on the land, a land which has character and personality in its own right.

The final section of this volume—on *Philosophy and Ecology*—returns to concerns that appeared in earlier volumes, but now addressed in a different, metaphysical context. The first chapter here is J. Baird Callicott's 'The metaphysical implications of ecology' (which originally appeared alongside Kalupahana and Cheng's paper in *Environmental Ethics*, the journal). Callicott proposes that the science of ecology may suggest metaphysical views such as 'a field ontology, the ontological subordination of matter to energy, internal relations and systemic . . . holism'. However, this view is roundly criticised in the following chapter 'Ecosystem ecology and metaphysical ecology: a case study' by Warren and Cheney. Whilst recognising the importance of an understanding of ecology for environmental ethics, they argue that Callicott's idea of 'metaphysical ecology' fails to reflect the findings of modern ecological hierarchy theory, and cannot provide a satisfactory basis on which to establish an environmental ethic. The final chapter, 'The shifting paradigm in ecology', by S. T. A. Pickett and Richard Ostfield, provides an overview of just how ecological theory has changed since the dawning of the environmental crisis. This chapter illustrates both the *importance* to environmental philosophy of an understanding of current ecological theory (in particular when using it as an inspiration for metaphysical claims) and also the *problems* this can entail, given the changing nature of ecological theory. In raising questions about the relationships between scientific ecology and environmental philosophy and between shifting concepts of nature and environmental values, this part seemed a fitting conclusion both to Volume V and to the set.

# Part 1

# HISTORICAL ROOTS

81

# THE HISTORICAL ROOTS OF OUR ECOLOGIC CRISIS

*Lynn White, Jr.*

Source: *Science*, 155:3767 (1967), 1203–7.

A conversation with Aldous Huxley not infrequently put one at the receiving end of an unforgettable monologue. About a year before his lamented death he was discoursing on favorite topic: Man's unnatural treatment of nature and its sad results. To illustrate his point he told how, during the previous summer, he had returned to a little valley in England where he had spent many happy months as a child. Once it had been composed of delightful grassy glades; now it was becoming overgrown with unsightly brush because the rabbits that formerly kept such growth under control had largely succumbed to a disease, myxomatosis, that was deliberately introduced by the local farmers to reduce the rabbits' destruction of crops. Being something of a Philistine, I could be silent no longer, even in the interests of great rhetoric. I interrupted to point out that the rabbit itself had been brought as a domestic animal to England in 1176, presumably to improve the protein diet of the peasantry.

All forms of life modify their contexts. The most spectacular and benign instance is doubtless the coral polyp. By serving its own ends, it has created a vast undersea world favorable to thousands of other kinds of animals and plants. Ever since man became a numerous species he has affected his environment notably. The hypothesis that his fire-drive method of hunting created the world's great grasslands and helped to exterminate the monster mammals of the Pleistocene from much of the globe is plausible, if not proved. For 6 millennia at least, the banks of the lower Nile have been a human artifact rather than the swampy African jungle which nature, apart from man, would have made it. The Aswan Dam, flooding 5000 square miles, is only the latest stage in a long process. In many regions terracing or irrigation, overgrazing, the cutting of forests by Romans to build ships to fight Carthaginians or by Crusaders to solve the logistics problems of their expeditions, have profoundly changed some ecologies. Observation that the French landscape falls into two basic types, the open fields of the north and the *bocage* of the south and west, inspired Marc Bloch to undertake his classic study of medieval agricultural methods. Quite unintentionally, changes in human ways often affect nonhuman nature. It has been noted, for example, that the advent of the automobile eliminated huge flocks of sparrows that once fed on the horse manure littering every street.

The history of ecologic change is still so rudimentary that we know little about what really happened, or what the results were. The extinction of the European aurochs as late as 1627 would seem to have been a simple case of overenthusiastic hunting. On more intricate matters it often is impossible to find solid information. For a thousand years or more the Frisians and Hollanders have been pushing back the North Sea, and the process is culminating in our own time in the reclamation of the Zuider Zee. What, if any, species of animals, birds, fish, shore life, or plants have died out in the process? In their epic combat with Neptune have the Netherlands overlooked ecological values in such a way that the quality of human life in the Netherlands has suffered? I cannot discover that the questions have ever been asked, much less answered.

People, then, have often been a dynamic element in their own environment, but in the present state of historical scholarship we usually do not know exactly when, where, or with what effects man-induced changes came. As we enter the last third of the 20th century, however, concern for the problem of ecologic backlash is mounting feverishly. Natural science, conceived as the effort to understand the nature of things, had flourished in several eras and among several peoples. Similarly there had been an age-old accumulation of technological skills, sometimes growing rapidly, sometimes slowly. But it was not until about four generations ago that Western Europe and North America arranged a marriage between science and technology, a union of the theoretical and the empirical approaches to our natural environment. The emergence in widespread practice of the Baconian creed that scientific knowledge means technological power over nature can scarcely be dated before about 1850, save in the chemical industries, where it is anticipated in the 18th century. Its acceptance as a normal pattern of action may mark the greatest event in human history since the invention of agriculture, and perhaps in nonhuman terrestrial history as well.

Almost at once the new situation forced the crystallization of the novel concept of ecology; indeed, the word *ecology* first appeared in the English language in 1873. Today, less than a century later, the impact of our race upon the environment has so increased if force that it has changed in essence. When the first cannons were fired, in the early 14th century, they affected ecology by sending workers scrambling to the forests and mountains for more potash, sulfur, iron ore, and charcoal, with some resulting erosion and deforestation. Hydrogen bombs are of a different order: a war fought with them might alter the genetics of all life on this planet. By 1285 London had a smog problem arising from the brining of soft coal, but our present combustion of fossil fuels threatens to change the chemistry of the globe's atmosphere as a whole, with consequences which we are only beginning to guess. With the population explosion, the carcinoma of planless urbanism, the now geological deposits of sewage and garbage, surely no creature other than man has ever managed to foul its nest in such short order.

There are many calls to action, but specific proposals, however worthy as individual items, seem too partial, palliative, negative: ban the bomb, tear down the billboards, give the Hindus contraceptives and tell them to eat their sacred cows. The simplest solution to any suspect change is, of course, to stop it, or, better yet, to revert to a romanticized past: make those ugly gasoline stations look like Anne Hathaway's cottage or (in the Far West) like ghost-town saloons. The "wilderness area" mentality invariably advocates deep-freezing an ecology, whether San Gimignano or the High Sierra, as it was before the first Kleenex was dropped. But neither atavism nor prettification will cope with the ecologic crisis of our time.

What shall we do? No one yet knows. Unless we think about fundamentals, our specific measures may produce new backlashes more serious than those they are designed to remedy.

As a beginning we should try to clarify our thinking by looking, in some historical depth, at the presuppositions that underlie modern technology and science. Science was traditionally aristocratic, speculative, intellectual in intent; technology was lowerclass, empirical, action-oriented. The quite sudden fusion of these two, towards the middle of the 19th century, is surely related to the slightly prior and contemporary democratic revolutions which, by reducing social barriers, tended to assert a functional unity of brain and hand. Our ecologic crisis is the product of an emerging, entirely novel, democratic culture. The issue is whether a democratized world can survive its own implications. Presumably we cannot unless we rethink our axioms.

## The Western traditions of technology and science

One thing is so certain that it seems stupid to verbalize it: both modern technology and modern science are distinctively *Occidental*. Our technology has absorbed elements from all over the world, notably from China; yet everywhere today, whether in Japan or in Nigeria, successful technology is Western. Our science is the heir to all the sciences of the past, especially perhaps to the work of the great Islamic scientists of the Middle Ages, who so often outdid the ancient Greeks in skill and perspicacity: al-Rāzī in medicine, for example; or ibnal-Haytham in optics; or Omar Khayyám in mathematics. Indeed, not a few works of such geniuses seem to have vanished in the original Arabic and to survive only in medieval Latin translations that helped to lay the foundations for later Western developments. Today, around the globe, all significant science is Western in style and method, whatever the pigmentation or language of the scientists.

A second pair of facts is less well recognized because they result from quite recent historical scholarship. The leadership of the West, both in technology and in science, is far older than the so-callled Scientific Revolution of the 17th century or the so-called Industrial Revolution of the 18th century. These terms are in fact outmoded and obscure the true nature of what they try to describe—significant stages in two long and separate developments. By A.D. 1000 at the latest—and perhaps, feebly, as much as 200 years earlier—the West began to apply water power to industrial processes other than milling grain. This was followed in the late 12th century by the harnessing of wind power. From simple beginnings, but with remarkable consistency of style, the West rapidly expanded its skills in the development of power machinery, labor-saving devices, and automation. Those who doubt should contemplate that most monumental achievement in the history of automation; the weight-driven mechanical clock, which appeared in two forms in the early 14th century. Not in craftsmanship but in basic technological capacity, the Latin West of the later Middle Ages far outstripped its elaborate, sophisticated, and esthetically magnificent sister cultures, Byzantium and Islam. In 1444 a great Greek ecclesiastic, Bessarion, who had gone to Italy, wrote a letter to a prince in Greece. He is amazed by the superiority of Western ships, arms, textiles, glass. But above all he is astonished by the spectacle of water-wheels sawing timbers and pumping the bellows of blast furnaces. Clearly, he had seen nothing of the sort in the Near East.

By the end of the 15th century the technological superiority of Europe was such that its small, mutually hostile nations could spill out over all the rest of the world, conquering, looting, and colonizing. The symbol of this technological superiority is the fact that Portugal, one of the weakest states of the Occident, was able to become, and to remain for a century, mistress of the East Indies. And we must remember that the technology of Vasco da Gama and Albuquerque was built by pure empiricism, drawing remarkably little support or inspiration from science.

In the present-day vernacular understanding, modern science is supposed to have begun in 1543, when both Copernicus and Vesalius published their great works. It is no derogation of their accomplishments, however, to point out that such structures as the *Fabrica* and the *De revolutionibus* do not appear overnight. The distinctive Western tradition of science, in fact, began in the late 11th century with a massive movement of translation of Arabic and Greek scientific works into Latin. A few notable books—Theophrastus, for example—escaped the West's avid new appetite for science, but within less than 200 years effectively the entire corpus of Greek and Muslim science was available in Latin, and was being eagerly read and criticized in the new European universities. Out of criticism arose new observation, speculation, and increasing distrust of ancient authorities. By the late 13th century Europe had seized global scientific leadership from the faltering hands of Islam. It would be as absurd to deny the profound originality of Newton, Galileo, or

Copernicus as to deny that of the 14th century scholastic scientists like Buridan or Oresme on whose work they built. Before the 11th century, science scarcely existed in the Latin West, even in Roman times. From the 11th century onward, the scientific sector of Occidental culture has increased in a steady creascendo.

Since both our technological and our scientific movements got their start, acquired their character, and achieved world dominance in the Middle Ages, it would seem that we cannot understand their nature or their present impact upon ecology without examining fundamental medieval assumptions and developments.

## Medieval view of man and nature

Until recently, agriculture has been the chief occupation even in, "advanced" societies; hence, any change in methods of tillage has much importance. Early plows, drawn by two oxen, did not normally turn the sod but merely scratched it. Thus, cross-plowing was needed and fields tended to be squarish. In the fairly light soils and semiarid climates of the Near East and Mediterranean, this worked well. But such a plow was inappropriate to the wet climate and often sticky soils of northern Europe. By the latter part of the 7th century after Christ, however, following obscure beginings, certain northern peasants were using an entirely new kind of plow, equipped with a vertical knife to cut the line of the furrow, a horizontal share to slice under the sod, and a moldboard to turn it over. The friction of this plow with the soil was so great that it normally required not two but eight oxen. It attacked the land with such violence that cross-plowing was not needed, and fields tended to be shaped in long strips.

In the days of the scratch-plow, fields were distributed generally in units capable of supporting a single family. Subsistence farming was the presupposition. But no peasant owned eight oxen: to use the new and more efficient plow, peasants pooled their oxen to form large plow-teams, originally receiving (it would appear) plowed strips in proportion to their contribution. Thus, distribution of land was based no longer on the needs of a family but, rather, on the capacity of a power machine to till the earth. Man's relation to the soil was profoundly changed. Formerly man had been part of nature; now he was the exploiter of nature. Nowhere else in the world did farmers develop any analogous agricultural implement. Is it coincidence that modern technology, with its ruthlessness toward nature, has so largely been produced by decendants of these peasants of northern Europe?

This same exploitive attitude appears slightly before A.D. 830 in Western illustrated calendars. In older calendars the months were shown as passive personifications. The new Frankish calendars, which set the style for the Middle Ages, are very different: they show men coercing the world around them—plowing, harvesting, chopping trees, butchering pigs. Man and nature are two things, and man is master.

These novelties seem to be in harmony with larger intellectual patterns. What people do about their ecology depends on what they think about themselves in relation to things around them. Human ecology is deeply conditioned by beliefs about our nature and destiny—that is, by religion. To Western eyes this is very evident in, say, India or Ceylon. It is equally true of ourselves and of our medieval ancestors.

The victory of Christianity over paganism was the greatest psychic revolution in the history of our culture. It has become fashionable today to say that, for better or worse, we live in "the post-Christian age." Certainly the forms of our thinking and language have largely ceased to be Christian, but to my eye the substance often remains amazingly akin to that of the past. Our daily habits of action, for example, are dominated by an implicit faith in perpetual progress which was unknown either to Greco-Roman antiquity or to the Orient. It is rooted in, and is indefensible apart from, Judeo-Christian teleology. The fact that Communists share it merely helps to show what can be demonstrated on many other grounds: that Marxism, like Islam, is a Judeo-Christian heresy. We continue today to live, as we have lived for about 1700 years, very largely in a context of Christian axioms.

What did Christianity tell people about their relations with the environment?

While many of the world's mythologies provide stories of creation, Greco-Roman mythology was singularly incoherent in this respect. Like Aristotle, the intellectuals of the ancient West denied that the visible world had had a beginning. Indeed, the idea of a beginning was impossible in the frame-work of their cyclical notion of time. In sharp contrast, Christianity inherited from Judaism not only a concept of time as nonrepetitive and linear but also a striking story of creation. By gradual stages a loving and all-powerful God had created light and darkness, the heavenly bodies, the earth and all its plants, animals, birds, and fishes. Finally, God had created Adam and, as an afterthought, Eve to keep man from being lonely. Man named all the animals, thus establishing his dominance over them. God planned all of this explicitly for man's benefit and rule: no item in the physical creation had any purpose save to serve man's purposes. And, although man's body is made of clay, he is not simple part of nature: he is made in God's image.

Especially in its Western form, Christianity is the most anthropocentric religion the world has seen. As early as the 2nd century both Tertullian and Saint Irenaeus of Lyons were insisting that when God shaped Adam he was foreshadowing the image of the incarnate Christ, the Second Adam. Man shares, in great measure, God's transcendence of nature. Christianity, in absolute contrast to ancient paganism and Asia's religions (except, perhaps, Zoroastrianism), not only established a dualism of man and nature but also insisted that it is God's will that man exploit nature for his proper ends.

At the level of the common people this worked out in an interesting way. In Antiquity every tree, every spring, every stream, every hill had its own *genius loci*, its guardian spirit. These spirits were accessible to men, but were very unlike men; centaurs, fauns, and mermaids show their ambivalence. Before one cut a tree, mined a mountain, or dammed a brook, it was important to placate the spirit in charge of that particular situation, and to keep it placated. By destroying pagan animism, Christianity made it possible to exploit nature in a mood of indifference to the feelings of natural objects.

It is often said that for animism the Church substituted the cult of saints. True; but the cult of saints is functionally quite different from animism. The saint is not *in* natural objects; he may have special shrines, but his citizenship is in heaven. Moreover, a saint is entirely a man; he can be approached in human terms. In addition to saints, Christianity of course also had angels and demons inherited from Judaism and perhaps, at one remove, from Zoroastrianism. But these were all as mobile as the saints themselves. The spirits *in* natural objects, which formerly had protected nature from man, evaporated. Man's effective monopoly on spirit in this world was confirmed, and the old inhibitions to the exploitation of nature crumbled.

When one speaks in such sweeping terms, a note of caution is in order. Christianity is a complex faith, and its consequences differ in differing contexts. What I have said may well apply to the medieval West, where in fact technology made spectacular advances. But the Greek East, a highly civilized realm of equal Christian devotion, seems to have produced no marked technological innovation after the late 7th century, when Greek fire was invented. The key to the contrast may perhaps be found in a difference in the tonality of piety and thought which students of comparative thelogy find between the Greek and the Latin Churches. the Greeks believed that sin was intellectual blindness, and that salvation was found in illumination, orthodoxy—that is, clear thinking. The Latins, on the other hand, felt that sin was moral evil, and that salvation was to be found in right conduct. Eastern theology has been intellectualist. Western thelogy has been volutarist. The Greek saint contemplates; the Western saint acts. The implications of Christianity for the conquest of nature would emerge more easily in the Western atmosphere.

The Christian dogma of creation, which is found in the first clause of all the Creeds, has another meaning for our comprehension of today's ecologic crisis. By revelation, God had given man get Bible, the Book of Scripture. But since God had made nature, nature also must reveal the divine mentality. The religious study of nature for the better understanding of God was known as natural theology. In the early Church, and always in the Greek East, nature was conceived primarily as a symbolic system through which God speaks to men: the ant is a sermon to sluggards; rising flames are the symbol of the soul's aspiration. This view of nature was essentially artistic rather than scientific. While Byzantium preserved and copied great numbers of ancient Greek scientific texts, science as we conceive it could scarcely flourish in such an ambience.

However, in the Latin West by the early 13th century natural theology was following a very different bent. It was ceasing to be the decoding of the physical symbols of God's communication with man and was becoming the effort to understand God's mind by discovering how his creation operates. The rainbow was no longer simple a symbol of hope first sent to Noah after the Deluge: Robert Grosseteste, Friar Roger Bacon, and Thedoric of Freiberg produced startlingly sophisticated work on the optics of rainbow, but they did it as a venture in religious understanding. From the 13th century onward, up to and including Leibnitz and Newton, every major scientist, in effect, explained his motivations in religious terms. Indeed, if Galileo had not been so expert an amateur theologian he would have got into far less trouble: the professionals resented his intrusion. And Newton seems to have regarded himself more as a theologian than as a scientist. It was not until the late 18th century that the hypothesis of God became unnecessary to many scientists.

It is often hard for the historian to judge, when men explain why they are doing what they want to do, whether they are offering real reasons or merely culturally acceptable reasons. The consistency with which scientists during the long formative centuries of Western science said that the task and the reward of the scientist was "to think God's thoughts after him" leads one to believe that this was their real motivation. If so, then modern Western science was cast in a matrix of Christian thelogy. The dynamism of religious devotion, shaped by the Judeo-Christian dogma of creation, gave it impetus.

## An alternative Christian view

We would seem to be headed toward conclusions unpalatable to many Christians. Since both *science* and *technology* are blessed words in our contemporary vocabulary, some may be happy at the notions, first, that, viewed historically, modern science is an extrapolation of natural theology and, second, that modern technology is at least partly to be explained as an Occidental, voluntarist realization of the Christian dogma of man's transcendence of, and rightful mastery over, nature. But, as we now recognize, somewhat over a century ago science and technology—hitherto quite separate activities—joined to give mankind powers which, to judge by many of the ecologic effects, are out of control. If so, Christianity bears a huge burden of guilt.

I personally doubt that disastrous ecologic backlash can be avoided simply by applying to our problems more science and more technology. Our science and technology have grown out of Christian attitudes toward man's relation to nature which are almost universally held not only by Christians and neo-Christians but also by those who fondly regard themselves as post-Christians. Despite Copernicus, all the cosmos rotates around our little globe. Despite Darwin, we are *not,* in our hearts, part of the natural process. We are superior to nature, contemptuous of it, willing to use it for our slightest whim. The newly elected Governor of California, like myself a churchman but less troubled than I, spoke for the Christian tradition when he said (as is alleged), "when you've seen one redwood tree, you've seen them all." To a Christian a tree can be no more than a physical fact. The whole concept of the sacred grove is alien to Christianity and to the ethos of the West. For nearly 2 millennia Christian missionaries have been chopping down sacred groves, which are idolatrous because they assume spirit in nature.

What we do about ecology depends on our ideas of the man-nature relationship. More science and more technology are not going to get us out of the present ecologic crisis until we find a new religion, or rethink our old one. The beatniks, who are the basic revolutionaries of our time, show a sound instinct in their affinity for Zen Buddhism, which conceives of the man-nature relationship as very nearly the mirror image of the Christian view. Zen, however, is as deeply conditioned by Asian history as Christianity is by the experience of the West, and I am dubious of its viability among us.

Possibly we should ponder the greatest radical in Christian history since Christ: Saint Francis of Assisi. The prime miracle of Saint Francis is the fact that he did not end at the stake, as many of his left-wing followers did. He was so clearly heretical that a General of the Franciscan Order, Saint Bonaventura, a great and perceptive Christian, tried to suppress the early accounts of Franciscanism. The key to an understanding of Francis is his belief in the virtue of humility—not merely for the individual but for man as a species. Francis tried to depose man from his monarchy over creation and set up a democracy of all God's creatures. With him the ant is no longer simply a homily for the lazy, flames a sign of the thrust of the soul toward union with God; now they are Brother Ant and Sister Fire, praising the Creator in their own ways as Brother Man does in his.

Later commentators have said that Francis preached to the birds as a rebuke to men who would not listen. The records do not read so: he urged the little birds to praise God, and in spiritual ecstasy they flapped their wings and chirped rejoicing. Legends of saints, especially the Irish saints, had long told of their dealings with animals but always, I believe, to show their human dominance over creatures. With Francis it is different. The land around Gubbio in the Apennines was being ravaged by a fierce wolf. Saint Francis, says the legend, talked to the wolf and persuaded him of the error of his ways. The wolf repented, died in the odor of sanctity, and was buried in consecrated ground.

What Sir Steven Ruciman calls "the Franciscan doctrine of the animal soul" was quickly stamped out. Quite possibly it was in part inspired, consciously or unconsciously, by the belief in reincarnation held by the Cathar heretics who at that time teemed in Italy and southern France, and who presumably had got it originally from India. It is significant that at just the same moment, about 1200, traces of metempsychosis are found also in western Judism, in the Provençal *Cabbala.* But Francis held neither to transmigration of souls nor to pantheism. His view of nature and of man rested on a unique sort of pan-psychism of all things animate and inanimate, designed for the glorification of their transcendent Creator, who, in the ultimate gesture of cosmic humility, assumed flesh, lay helpless in a manger, and hung dying on scaffold.

I am not suggesting that many contemporary Americans who are concerned about our ecologic crisis will be either able or willing to counsel with wolves or exhort birds. However, the present increasing disruption of the global environment is product of a dynamic technology and science which were originating in the Western medieval world against which Saint Francis was rebelling in so original a way. Their growth cannot be understood historically apart from distinctive attitudes toward nature which are deeply grounded in Christian dogma. The fact that most people do not think of these attitudes as Christian is irrelevant. No new set of basic values has been accepted in our society to displace those of Christianity. Hence we shall continue to have a worsening ecologic crisis until we reject the Christian axiom that nature has no reason for existence save to serve man.

The greatest spiritual revolutionary in Western history, Saint Francis, proposed what he thought was an alternative Christian view of nature and man's relation to it: he tried to substitute the idea of the equality of all creatures, including man, for the idea of man's limitless rule of creation. He failed. Both our present science and our present technology are so tinctured with orthodox Christian arrogance toward nature that no solution for our ecologic crisis can be expected from them alone. Since the roots of our trouble are so largely religious, the remedy must also be essentially religious, whether we call it that or not. We must rethink and refeel our nature and destiny. The profoundly religious, but heretical, sense of the primitive Franciscans for the spiritual autonomy of all parts of nature may point a direction. I propose Francis as a patron saint for ecologists.

82

# DISCREPANCIES BETWEEN ENVIRONMENTAL ATTITUDE AND BEHAVIOUR

## Examples from Europe and China*

*Yi-Fu Tuan*

Source: *Canadian Geographer*, 12:3 (1968), 176–91.

Discrepancy between stated ideal and reality is a worrisome fact of our daily experience: in the political field one learns to discriminate between an orator's fulsome profession and what he can or will, in fact, carry out. The history of environmental ideas, however, has been pursued as an academic discipline largely in detachment from the question of how—if at all—these ideas guide the course of action, or how they arise out of it. Needless to say, there are many paradigmatic views of nature, such as those of science, that have great explicatory power and may, once they are applied, affect the lives of many people; but in themselves they do not enjoin a specific course of action. In contrast, the acceptance of certain specific environmental ideas can have a definite effect on decision and on behaviour. If it is widely held, for example, that a dry and sunny climate is a great restorer of health, we may suppose that an appreciable number of people will seek out these areas for health. But what of less specific ideas? We may believe that a world-view which puts nature in subservience to man will lead to the exploitation of nature by man; and one that regards man as simply a component in nature will entail a modest view of his rights and capabilities, and so lead to the establishment of a harmonious relationship between man and his natural environment. But is this correct? And if essentially correct, how direct or tenuous is the link? These are some of the questions I wish to explore with the help of examples from Europe and China. The discrepancies are noted here; their resolution must await another occasion.

* The substance of this paper was read at the University of British Columbia, 29 November, 1967.

# I

To the question, what is a fundamental difference between the European and the Chinese attitude towards nature, most people with any opinion at all will probably make some such reply: that the European sees nature as subordinate to him whereas the Chinese sees himself as a part of nature. Taken as a broad generalization and with a grain of salt there is much truth in this distinction; a truth illustrated with diagrammatic force when one compares the formal European garden of the seventeenth century with the Chinese naturalistic garden. The geometric contrast reflects fundamental differences in environmental evaluation. The formal European garden in the style of the Le Nôtre was designed to produce a limited number of imposing prospects. It can be appreciated to the full only at a limited number of favoured spots where the onlooker is invited by the garden's design to gaze at distant vistas. Or, seen in another way, the European garden is a grandiose setting for man; in deference to him, nature is straitjacketed in court dress. The Chinese garden, on the other hand, is designed to produce almost constantly shifting scenes: there are no set prospects. The nature of the garden requires the perceiver to move along a winding path and to be more than visually involved with the landscape. It is not nature that is required to put on court dress in deference to man: rather, it is man who must lay aside his formalistic pretensions in order to enter nature.

This widely recognized distinction is valid and important. On the other hand, by the crude test of the total tonnage of earth removed there may not be so very much difference between the European formal and the Chinese naturalistic garden. Both are human artifacts. It is not widely known that some of the famous scenic areas of China are works of man rather than of geologic processes. The West Lake of Hang-chou, for example, was celebrated by T'ang and Sung poets and it remains to this day an adornment of China. To the casual visitor, the West Lake region may appear to be a striking illustration of how the works of man can blend modestly into the magistral context of nature. However, the pervasiveness of nature is largely an illusion induced by art. Some of the islands in the lake are man-made. Moreover, the lake itself is artificial and has to be maintained with care. In the thirteenth century, military patrols, under the command of specially appointed officials, looked after its policing and maintenance; it was forbidden, for example, to throw any rubbish into it or to plant in it lotuses or water-chestnuts. Peasants were recruited to clear and enlarge the lake, to keep it from being cluttered up by vegetation and silt.[1] Hang-chou's environs, then, owe much of their calm, harmonious beauty to human art and effort. The sense of open nature in Hang-chou is enhanced by its scale: the West Lake region is a cluster of public and semi-public parks. In the much smaller compass of the private garden the illusion of pervasive nature is far more difficult to achieve: nevertheless the aim of the Chinese gardener was to achieve it with cleverly placed, water-worn limestone whose jagged outlines denoted wildness, and by means of winding footpaths that give the stroller an illusion of depth and space. In this line the Oriental's ultimate triumph is symbolized by the miniature garden, where wild nature is reduced to the scale of a dwarf landscape that can be fitted into a bowl. Complete artifice reigns: in the narrow confines of a bowl, shrubs are tortured by human skill into imitating the shape and posture of pines, the limbs of which may have been deformed by winds that swept the China Seas.

## II

I have begun with a contrast and then proceeded to suggest that, from another perspective, the contrast is blurred. The publicized environmental ethos of a culture seldom covers more than a fraction of the total range of environmental behaviour. It is misleading to derive the one from the other. Simplifications that can mislead have at times been made. For example, Professor Lynn White has recently said: "What people do about their ecology depends on what they think about themselves in relation to things around them. Human ecology is deeply conditioned by beliefs about our nature and destiny—that is, by religion."[2] He goes on to say that the victory of Christianity over paganism was the greatest psychic revolution in Western culture. In his view, despite all the talk of "the post-Christian age" and despite evident changes in the forms of modern thinking, the substance often remains amazingly akin to that of the Christian past. The Western man's daily habits of action are dominated by an implicit faith in perpetual progress which was unknown either to Greco-Roman antiquity or to the Orient. It is rooted in, and is indefensible apart from, Judeo-Christian teleology. Peoples of the Western world continue to live, as they have lived for about 1700 years, very largely in a context of Christian beliefs. And what has Christianity told people about their relations with the environment? Essentially that man, as something made in God's image, is not simply a part of nature; that God has planned the universe for man's benefit and rule. According to White, Christianity is the most anthropocentric religion the world has seen. It has not only established a dualism of man and nature but has also insisted that it is God's will that man exploit nature for his proper ends.[3]

To press the theme further, it is said that Christianity has destroyed antiquity's feeling for the holiness of landscapes and of natural things. The Greek religious tradition regarded the land not as an object to be exploited, or even as a visually pleasing setting, but as a true force which physically embodied the powers that ruled the world. Vincent Scully, the architectural historian, has argued that not only were certain landscapes regarded by the ancient Greeks as holy and expressive of specific gods, but also that the temples and the subsidiary buildings of their sanctuaries were so formed in themselves and so placed in the landscapes and to each other as to enhance, develop, and complement the basic meaning of the landscape.[4]

Martin Heidegger, a modern philosopher whose insights have been greatly influenced by early Greek philosophy, characterized the Greek temple as disclosing the earth on which it stands. The whiteness of the temple discloses the darkness and the strength of the rock underneath; it reveals the height and blueness of the sky, the power of the storm and the vastness of the sea.[5] In the Christian tradition, on the other hand, holiness was invested not in landscapes but in man-made altars, shrines, churches, and basilicas that dominated the landscapes. Constantine and Helen are said to have built basilicas over caves in the Holy Land to celebrate the triumph of Christianity over the "cave cultus" of the pagan world. In the Christian view it was not emanation from the earth but ritual that consecrated the site; man not nature bore the image of God and man's work, the hallowed edifice, symbolized the Christian cosmos. In pagan antiquity, at the level of the common people, each facet of nature had its own guardian spirit. Before one ventured to cut a tree, mine a mountain, or dam a book, it was important to placate the spirit in charge of that particular situation, and to keep it placated. By destroying animistic beliefs, Christianity made it possible to exploit nature in a mood of indifference to the feeling of natural objects.

Much of this is now Western folklore and Lynn White is among the more recent writers to give it eloquent expression. The thesis, then, is that Christianity has introduced a fundamentally new way of evaluating the environment, and that this new evaluation has strongly affected Western man's traffic with the natural objects around him. The generalization is very useful, although one should take note of facts that appear to contradict it. As Clarence Glacken has demonstrated, in the ancient world there was no lack of interest in natural resources and their quick exploitation. Economic activities such as mining, the various ways of obtaining food, canal building, and drainage are clear proof of man's incessant restlessness in changing the earth about him.[6] Glacken points out that in Sophocles' *Antigone* there are lines which remind one of the eulogies of science in the eighteenth century, and of contemporary enthusiasm for man's control over nature. At one point in the play the chorus declares how the earth has felt man's ungentle touch:

> Oh, Earth is patient, and Earth is old,
> And a mother of Gods, but he breaketh her,
> To-ing, froing, with the plough teams going,
> Tearing the soil of her, year by year.[7]

The tearing of soil has led to erosion. In Plato's *Critias* there is the well-known passage in which he describes how the soils of Attica have been washed down to the sea. "And, just as happens in small islands, what now remains compared with what then existed is like the skeleton of a rich man, all the fat and soft earth have wasted away, and only the bare framework of the land being left." Plato then describes the former arable hills, fertile valleys, and forested mountains "of which there are visible signs even to this day." Mountains which today have food only for bees could, not so long ago, grow trees fit for the largest buildings. Cultivated trees provided pasturage for flocks, and the soil was well watered and the rain was "not lost to it, as now, by flowing from the bare land to the sea."[8] Plato's comments sound remarkably modern; they remind us almost of the lamentations of latter-day conservationists.

If there is evidence of man's awareness of his power to transform nature—even destructively—in the time of Sophocles and Plato, there is evidence of much greater awareness of the almost limitless capabilities of man in Hellenistic times. Agriculture and related occupations such as cattle-breeding were then the most important source of wealth in the ancient world. Land reclamation was not a haphazard affair but one based on the science of mechanics and on practical experience with canal-digging, irrigation, and swamp drainage. It was a time of faith in progress. But far more than the Greeks, the Romans have imposed their will on the natural environment (Figure 1). And perhaps the most dramatic example of the triumph of the human will over the irregular lineaments of nature is the Roman grid method of dividing up the land. As Bradford puts it, centuriation well displayed the arbitrary but methodical qualities in Roman government. With absolute self-assurance and great technical competence, the Romans have imposed the same formal pattern of land division on the well-watered alluvium of the

*Figure 1* The growing "second nature" in antiquity. In spite of a passive or adaptive attitude toward nature (suggested by the importance of the "environmental theory" in philosophy and by respect for the *genius loci* of natural object in folk religion), the ancient landscapes have been markedly transformed by man. Figure 1 shows how nature can even be dominated by man's engineering achievements. Reproduced from Clarence Glacken, *Traces on the Rhodian Shore*.

Po Valley as on the near-desert of Tunisia. Even today the forceful imprint of centuriation can be traced across some thousands of square miles on both sides of the central Mediterranean, and it can still stir the imagination by its scale and boldness.[9]

Against this background of the vast transformations of nature in the pagan world, the inroads made in the early centuries of the Christian era were relatively modest. Christianity teaches that man has dominion over nature. St. Benedict himself had cut down the sacred grove at Monte Cassino because it was a survival of pagan worship. And the story of how monks moved into the forested wilderness, and by a combination of work and prayer, had transformed them into cloistered "paradises" is a familiar one. But for a long time man's undisputed power over nature was more a tenet of faith than a fact of experience: to become a realized fact Europe had to wait for the growth of human numbers, for the achievement of greater administrative centralization and for the development and wide application of new technological skills. Fields that were cleared in heavy forests testified to the mediaeval farmer's great capacity for changing his environment: it was a change, however, that must continually be defended against the encroachments of nature (Figure 2). Farmsteads and arable lands multiplied through

*Figure 2* A fifteenth-century print showing a mediaeval scene. Foreground: ploughing team and field; mid-ground: water mill; background: castle. Technological innovations, such as the heavy plough and the water mill, appear to have given man a position of dominance vis-à-vis nature, from about the ninth century onward, that he lacked in the earlier Christian centuries.

the Middle Ages at the expense of forests and marshes, but these man-made features lacked the permanence, the geometric order, and the prideful assertion of the human will that one can detect more readily in the Roman road system, aqueducts, and centuriated landholdings. The victory of Christianity over paganism may well have been, as Lynn White says, the greatest psychic revolution in Western culture; but for lack of real, as distinct from theologically postulated, power the full impact of that revolution on ecology was postponed.

## III

As to China, Western humanists commonly show bias in favour of that country's Taoist and Buddhist traditions. They like to point out the virtues of the Oriental's quiescent and adaptive approach towards nature in contrast to the aggressive masculinity of Western man. Support for the quiescent view is easily found in the Taoist classics. The *Tao Tê Ching*, for a example, has a rather cryptic message of seven characters (*wei wu wei, tzu wu pu chih*) which James Legge has translated as: "When there is this abstinence from action, good order is universal." And Joseph Needham has recently interpreted it to mean: "Let there be no action (contrary to Nature), and there is nothing that will not be well regulated."[10] It is easy to see how these words might appeal to the modern man, who finds in his own environment the all-too-evident consequences of human action "contrary to nature." In another influential Taoist book of much later date (*T'ai shang kan ying p'ien*), one finds the belief that "even insects and crawling things, herbs and trees, may not be injured." These Taoist texts have been much translated into European languages; the latter, with its injunction against injuring even insects and crawling things, is believed to have had some influence on the thought of Albert Schweitzer.[11]

Another aspect of Chinese attitude towards nature, which has found favour among some Western humanists, is embodied in the concept of *feng-shui* or geomancy. This concept has been aptly defined as "the art of adapting the residences of the living and the dead so as to co-operate and harmonize with the local currents of the cosmic breadth."[12] If houses and tombs are not properly located, evil effects would injure the inhabitants and the descendants of those whose bodies lay in the tombs. On the other hand, good siting would favour wealth, health, and happiness. Good siting involves, above all, taking proper note of the forms of hills and directions of water courses since these are themselves the outcome of the moulding influences of winds and waters, that is, of *feng-shui*; but in addition one must also consider the heights and forms of buildings, the directions of roads and bridges. A general effect of the belief in *feng-shui* is to encourage a preference for natural curves—for winding paths and for structures that seem to fit into the landscape rather than to dominate it; and at the same time it promoted a distaste for straight lines and geometrical layouts. In this respect it is of interest to note the short life of China's first railway. This was built in 1876 and connected Shanghai with its port of Wu-sung. Although the venture was at first well received, the mood of the local people turned sour after a native was killed by the locomotive. The people in their hostility thought that the railway had offended the principle of *feng-shui*. On 20 October, 1877, the Chinese government closed the railway, and so a symbol of Western progress was temporarily sacrificed to the local currents of the cosmic breadth.[13]

An adaptive attitude towards nature has ancient roots in China. It is embodied in folklore, in the philosophical-ethical precepts of Taoism, and later, Buddhism, and it draws support from practical experience: the experience that uncontrolled exploitation of timber, for example, brings hurtful results. In ancient literature one finds here and there evidence of a recognition for the need to regulate the use of resources. Even as early as the Eastern Chou period (eighth century–third century B.C.), deforestation necessitated by the expansion of agriculture and the building of cities seems to have led to an appreciation of the value of trees. In that ancient compendium of songs the *Shi Ching*, we find the sentiment expressed in lines such as these:

On the hill were lovely trees,
Both chestnut-trees and plum trees.
Cruel brigands tore them up;
But no one knew of their crime.

Trees were regarded as a blessing. As another poem put it,

So thick grow those oaks
That the people never look for firewood.
Happiness to our lord!
May the spirits always have rewards for him.[14]

In the *Chou Li*—a work which was probably compiled in the third century B.C., but may well include earlier material—we find mentioned two classes of officials whose duties were concerned with conservation. One was the *Shan-yu*, inspector of mountains, and the other the *Lin-heng*, inspector of forests. The inspectors of mountains were charged with the care of forests in the mountains. They saw to it that certain species were preserved, and in other ways enforced conservation practices. Thus trees could only be cut by the common people at certain times; those on the south side in the middle of winter and those on the north side in the middle of summer. At other seasons the people were permitted to cut wood in times of urgent need, such as when coffins had to be made or dykes strengthened, but even then certain areas could not be touched. The inspectors of forests (in the *Lin-heng* office) had similar duties. Their authority covered the forests that lay below the mountains.[15] Another ancient literary reference to conservation practice was in the *Mencius*. The sage advised King Huai of Liang that he would not lack for wood if he allowed the people to cut trees only at the proper time.[16]

Through Chinese history perspicacious officials have from time to time warned against the dire consequences of deforestation. A scholar of the late Ming dynasty reported on Shan-hsi, a province in North China: "At the beginning of the reign of Chia-ching" (1522–66), he wrote, "people vied with each other to build houses, and wood from the southern mountains were cut without a year's rest. The natives took advantage of the barren mountain surface and converted it into farms. . . . If heaven sends down a torrent, there is nothing to obstruct the flow of water. In the morning it falls on the southern mountains; in the evening, when it reaches the plains, its angry waves swell in volume and break ambankments causing frequent changes in the course of the river."[17]

Deforestation was deplored by the late Ming scholars not only because of its effect on stream flow and on the quality of the soil in the lowlands, but also—interestingly enough—because of their belief that the forests on mountain ridges were effective in showing down the horse-riding barbarians. As one scholar put it, "I saw the fact that what the country relies on as strategically important is the mountain, and what the mountain relies on as a screen to prevent advance are the trees."[18] There was also recognition of the aesthetics of forested mountains. Wu-tai mountains in northern Shan-hsi, for example, were famous everywhere. But the question was asked: since they have become almost bare, what remained to keep them famous?

These brief notes suggest that there existed in China an old tradition of forest care. Officials encouraged the practice but the people engaged in it on their own initiative when it did not conflict with the urgent needs of the moment. Nearly forty years ago, the American conservationtist W. C. Lowdermilk noted how thousands of acres in An-hui and Ho-non were planted with pine from local nurseries, a practice he recognized as ancient and independent of the modern forestry movement. Lowdermilk also found that the North China plain "actually exports considerable quantities of logs of *Paulownia tomentosa* to Japan and poplar (*Populus tomentosa*) to match factories. It is true that no forests are to be found in this plain, but each village has its trees, which are grown according to a system."[19]

In Communist China trees are extensively planted to control soil erosion, in answer to pressing economic needs but also for aesthetic reasons. Roadside planting, a practice dating back to the Eastern Chou period, uses the "traditional" trees (*Populus simonii*, *Pinus tabulaeformis*, *Salix babylonica*, *S. matsudana*, *Aesculus chinesis*, *Ulmus parvifolia*), but in particular the poplars. Afforestation proceeds in villages, and most conspicuously, in cities, new suburbs, and industrial districts where the trees hide a great deal of the raw ugliness of new construction.[20]

## IV

Thus far I have sketched what may indeed be called the "official" line on Chinese attitude towards environment; it is widely publicized and commonly accepted. There is however another strain: the enlightened memorials to the emperor on the need for the conservation of resources are in themselves clear evidence of the follies that have already been committed. Unlike the Western man of letters the geographer is ususally aware of China's frequent mistreatment of nature. He perceives that country, not through the refined sentiments of Taoist philosophy, Neo-Confucianism, and Oswald Siren, but through the bleak reports of Mallory, Lowdermilk, and Thorp. Deforestation and erosion on the one hand, the building of cities and rice terraces on the other are the common foci of his attention rather than landscape painting or poetry contests in the cool precincts of garden. The two images of reality complement each other: in an obvious but not trite sense, civilization is the exercise of human power over nature, which in turn may lead to the aesthetic appreciation of nature. Philosophy, nature poetry, gardens, and orderly countryside are products of civilization, but so equally are the deforested mountains, the clogged streams, and, within the densely packed, walled cities, the political intrigue.

If animistic belief and Taoist nature philosophy lie at the back of an adaptive attitude towards nature, what conceptions and ideals—we may ask—have encouraged the Chinese, through their long history, to engage in gigantic transformation of environment—whether this be expressed positively in huge works of construction or negatively in deforested mountains? Several ancient beliefs and conceptions may be recognized and they, individually or together, have allowed the Chinese to express the "male" principle in human nature. Consider, for example, the fact that one of the greatest culture heroes of China was Yu, the legendary founder of the Hsia dynasty. He was famed primarily for his magnificent deeds: He "opened up the rivers of the Nine Provinces and fixed the outlets of the nine marshes"; he brought peace and order to the lands of Hsia and his achievements were of an enduring kind which benefited succeeding dynasties.[21] Chinese rulers were bidden to imitate the ancient culture heroes, and one way to imitate them was to ensure order and prosperity by large-scale engineering works. Another ancient idea of importance to the "male" principle of dominance was to see in the earthly environment a model of the cosmos. The regular motions of the stars were to be translated architecturally and ritually to space and time on earth. The walled city oriented to the cardinal directions, the positioning of the twelve city gates, the location of the royal compound and the alignment of the principal axial street were given a geometric pattern that reflected the order to be found in heaven. The key concept was built on the related notions of rectilinearity, order, and rectitude. This key concept acquired architectural and social forms which were then imposed on earth, for the earth itself lacked paradigms of perfect order. Indeed the experience of mountains and waters has led to such unaggressive prescriptions as the need to observe and placate the spirits of the earth, the need for man to understand the balance of forces in nature, to contemplate this harmony and to adapt himself to it. By contrast, the observation of the stars has inspired such masculine attitudes as geometric order, hierarchy, and authoritarian control over earth and men.

The two outlooks—celestial and terrestrial, masculine and feminine—are not easy to reconcile. Events in heaven affect events on earth but not in any obvious or dependable way: abnormal floods and droughts have traditionally been taken as warnings by those who derive their power from astronomy. Tension, if not contradiction, is also revealed when these two ideas find architectural and geographical substance. The construction of Ch'ang-an in the Sui and T'ang dynasties illustrates the triumph of the cosmic principle of order and rectilinearity over the earth principle of complex harmony and natural lines (Figure 3). Ch'ang-an was laid on new ground and on an unprecented scale. The site in the Wei Ho valley was chosen for functional reasons but also because of its great histroical links: the site received the sanction of the great men and deeds in the past. Geomantic properties of the site were studied; however, unlike villages and rural roads the topographical character of the region seems to have made little impact on the city's fundamental design. Astronomers had an important role in the laying out of the city: they measured the shadow of the noon sun on successive days and observed the North Star by night in order to arrive at accurate alignments of the city walls to the four directions.[22] In the course of building Ch'ang-an, which had an enclosed area of 31 square miles, villages were levelled and trees uprooted; broad straight avenues were laid out and then rows of trees planted. Thus, despite the geomantic gestures, in Ch'ang-an the superposition of man's and heaven's order on natural terrain was complete. Or rather not quite complete, if we accept the charming story of why one great old locust tree was not in line. It had been retained from the old landscape because the chief architect had sat under it as he supervised the construction, and a special order from the emperor in honour of his architect spared it from being felled.[23]

## V

The natural environment of both Mediterranean Europe and China has been vastly transformed by man: constructively in the building of cities and roads, in the extension of arable land and the introduction of new crops; destructively in deforestation and erosion. Of any long-settled, thoroughly civilized part of the world, we can draw up a list of forces and the motives for their use that would more or less account for the transformation of the biotic mantle. Such lists may well agree in fundamentals: fire is widely used to

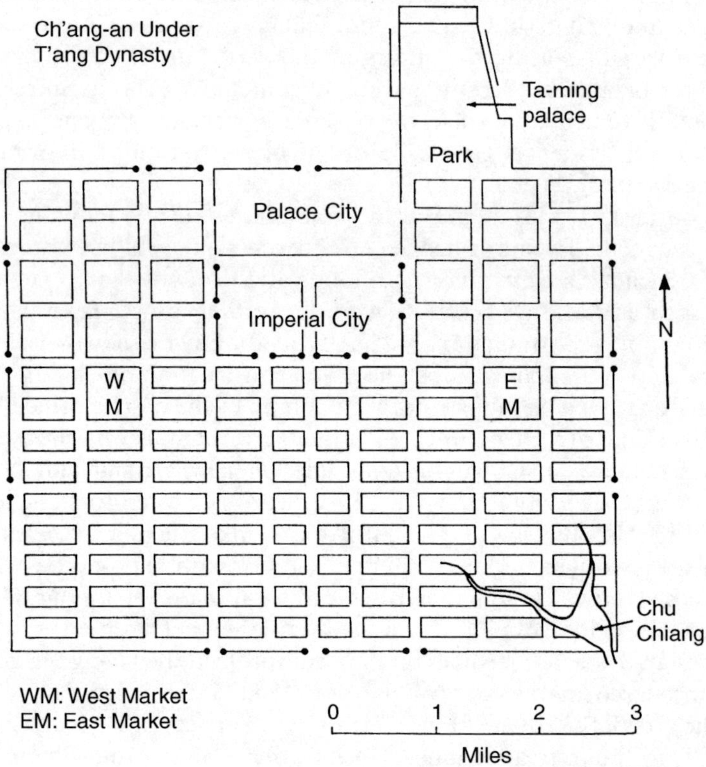


*Figure 3* The cosmic paradigm in Chinese city-building as exemplified by Ch'ang-an during the T'ang dynasty. The city grid reflects the superposition of a regular cosmic pattern on the irregularities of the earth. The curves of Taoist nature are represented mainly by Chü chiang park in the southeastern corner of the city and by the gardens of the wealthy and of monasteries but the rectilinear geometry of the city streets and enclosed wards dwarfs these manifestations of terrestrial nature. The imposing, rather lifeless, north-south avenues are more than 400 feet wide.

clear vegetation; the forest is cleared to create more grazing and arable land; timber is needed for the construction of palaces, houses, and ships, for domestic and industrial fuel, or as raw material for paper mills. Then again the forest is pushed back because it may shelter dangerous wild animals or provide hiding places for bandits. Naturally, the means at hand and the motives for using them vary from region to region: in contrast to the Mediterranean world, for example, China's vegetation suffered less from sheep and goats, and less from the enormous demands of shipbuilding which flourished with the Mediterranean maritime powers. China's forests, on the other hand, suffered more from the demands of city building and the need for domestic fuel.

To illustrate further the kinds of force that work against conservation practices in China, consider some of the causes of deforestation. One is the ancient practice of burning trees in order to deprive dangerous animals of their hiding places. There exists a passage in the *Mencius* of how in ancient times the luxuriant vegetation sheltered so many wild beasts that men were endangered. The great minister Shun of legendary repute ordered Yih to use fire, and "Yih set fire to, and consumed the forests and vegetation on the mountains and in the marshes, so that the birds and beasts fled away to hide themselves."[24] Even in the early decades of the twentieth century non-Chinese tribes in Kuang-hsi and Kuei-chou provinces are known to burn forests to drive away tigers and leopards; and in North China, in such long-settled areas as central Shen-hsi province, fires were ostensibly started by Chinese farmers for no other purpose. It is not always easy to establish the real reason for setting fire to forest. When asked, the farmers may say that it is to clear land for cultivation, although the extent of burning far exceeds the need for this purpose; or it is to leave fewer places in which bandits may hide; or to encourage the growth of small-sized sprouts in the burnt over area, which would then save the farmers the labour of splitting wood![25] The last reason tends to upset any residual illusion we may have of the Chinese farmer's benign attitude towards nature. A fire can of course also be started accidentally. A risk that is special to the Chinese is the forest fire caused by the burning of paper money at the grave mounds, which, in the rugged parts of the South, are commonly located beyond the fields and at the edge of the forested hills.

Forests in North China were depleted in the past for the making of charcoal as an industrial fuel. Robert Hartwell has shown how, from the tenth century onward the expanding metallic industries had swallowed up many hundreds of thousands of tons of charcoal each year, as did the manufacture of salt, alum, bricks, tiles, and liquour.[26] By the Sung dynasty (960–1279 A.D.) the demand for wood and charcoal as both household and industrial fuel had reached a level such that the timber resources of the country could no longer meet it; the result was the increasing substitution of coal for wood and charcoal.

An enormous amount of timber was needed in the construction of the old Chinese cities, probably more than that required in building Western cities of comparable size. One reason for this lies in the dependence of traditional Chinese architecture on timber as the basic structural material. Mountains may be stripped of their cover in the construction of a large palace.[27] And if a large palace required much timber, a whole city would require much more, especially if it were of the size of Ch'ang-an, capital of T'ang dynasty, and Hang-chou, capital of the southern Sung dynasty. Both had populations of more than a million people. The great expansion in the size of Hang-chou in the thirteenth century led to the deforestation of the neighbouring hills for construction timber. The demand for timber was such that some farmers gave up rice cultivation for forestry.[28] Cities in which houses were so largely made of wood ran the constant danger of demolition by fire; and this was especially true of the southern metropolises where the streets tended to be narrow. The necessity of rebuilding after fire put further strain on timber resources. But of even greater consequence than the accidental burning of parts of cities was the deliberate devastation of whole cities in times of upheaval, when rebels or nomadic invaders toppled a dynasty. The succeeding phase of reconstruction was normally achieved in quick time by armies of men who made ruthless inroads upon the forest.

## VI

The theme we have yet to trace is the involved interplay between environmental attitude and environmental behaviour, between the philosophy identified with a people and the actions that people may undertake. Besides the more glaring contradictions of professed ideal and actual practice, there exist also the unsuspected ironies: these derive from the fact that the benign institutions of a complex society, no less than the exploitative, are not always able to foresee all the consequences of their inherent character and action. For example, Buddhism in China is at least partly responsible for the preservation of trees around temple compounds, for the islands of green in an otherwise denuded landscape (Figure 4); on the other hand, Buddhism introduced to China the idea of the cremation of the dead; and from the tenth to the fourteenth century the practice of cremation was sufficiently

*Figure 4* Temple trees protected within the precincts of a village. This scene of northern Shan-hsi province in China shows a deforested and badly denuded landscape. Under pressure from rising population for more land, religion can do no more than save a few scattered groves. From Lowdermilk and Wickes.

common in the southeastern coastal provinces to have had an effect on the timber resources of that area.[29] The researches of E. H. Schafer provide us with another illustration of irony in Chinese life; for it would seem that the most civilized of arts was responsible for the deforestation of much of North China. The art was that of writing which required soot for the making of black ink. The soot came from burnt pine. And, as Schafer put it, "Even before T'ang times, the ancient pines of the mountains of Shan-tung had been reduced to carbon, and now the busy brushes of the vast T'ang bureaucracy were rapidly bringing baldness to the T'a-hang Mountains between Shansi and Hopei."[30]

## VII

I began by noting the contrast between the European formal garden and the Chinese naturalistic garden, and then suggested that these human achievements probably required comparable amounts of nature modification. To compare artworks and construction projects on the basis of the quantitative changes made on the environment is a useful exercise in so far as we wish to emphasize the role of man as a force for change along with other geophysical forces; but it is only the beginning in the interpretation of the meaning of these works and how they reflect cultural attitudes. It seems to me valid to see the European garden as an extension of the house: in the development of the European garden some of the formality and values of the house are taken outdoors in the form of courtyards, terraces, formal parterres, and avenues, and now the smooth, carpet-like lawn. The lawn displays the house; its sloping surfaces are a pedestal for the house. The Chinese garden, on the other hand, reflects a totally different philosophy from the orthogonal rectitude of the traditional Chinese house. In stepping through a circular gate, from the rectangular courtyard into the curvilinear forms of the garden, one enters a different world. Perhaps something of the difference in attitude towards outdoor spaces is retained to the present day. Simone de Beauvoir notes how a French family picnic is often an elaborate affair involving the transportation of a considerable portion of the household goods outdoors: it is not always a harmonious event for whatever tension that may exist in the house is carried to the less organized natural environment where it is exacerbated by entanglement with flies, fishing rods, and spilled strawberry jam. In Communist China, de Beauvoir spent an afternoon (1955) in the playgrounds of the Summer Palace outside Peking. She captures the peace of the scene with an anecdote: "In the middle of the lake I see a little boat: in it a young woman is lying down peacefully asleep while two youngsters are frisking about and playing with the oars. Our boatman cups his hands. 'Hey!' he calls. 'Look out for those kids!' The woman rubs her eyes, she smiles, picks up the oars, and shows the children how they work."[31]

## References


[1] GERNET, JACQUES, *Daily Life in China on the Eve of the Mongol Invasion 1250–1276* (London, 1962), pp. 51–52.

[2] WHITE, LYNN, "The Historical Roots of our Ecologic Crisis," *Science*, CLV (1967), 1205.

[3] *Ibid.*

[4] SCULLY, VINCENT, *The Earth, The Temple, and The Gods* (New Haven, 1962), p. 3.

[5] VYCINAS, VINCENT, *Earth and Gods: An Introduction to the Philosophy of Martin Heidegger* (The Hague, 1961), p. 13.

[6] GLACKEN, CLARENCE, *Traces on the Rhodian Shore* (Berkely and Los Angeles, 1967), p. 118.

[7] SOPHOCLES, *Antigone*, transl. by GILBERT MURRAY in ARNOLD TOYNBEE, *Greek Historical Thought* (New York, 1952), p. 128.

[8] PLATO, *Critias*, transl. by Arnold Toynbee in *Greek Historical Thought*, pp. 146–47.

[9] BRADFORD, JOHN, *Ancient Landscapes* (London, 1957), p. 145.

[10] NEEDHAM, JOSEPH, *Science and Civilisation in China*, vol. II (Cambridge, 1956), p. 69.

[11] SCHAFER, E. H., "The Conservation of Nature under the T'ang dynasty," *Journ. Econ. and Soc. Hist. of the Orient*, V (1962), 282.

[12] CHATLEY, H., "Feng shui" in *Encyclopaedia Sinica*, ed. by S. COULING (Shanghai, 1917), p. 175. See also ANDREW MARCH, "An Appreciation of Chinese Geomancy," *Journ. Asian Studies*, XXVII (1968), 253–67.

[13] *Encyclopaedia Sinica*, p. 470.

[14] *Shi Ching*, transl. by ARTHUR WALEY as *The Book of Songs* (New York, 1960), pp. 138, 213.

[15] *Chou Li*, transl. by E. BIOT as *Le Tcheou-li* (Paris, 1851), vol. I, 371–74.

[16] *Mencius*, Bk. I, pt. 1, 3:3.

[17] CHI, CH'AO-TING, *Key Economic Areas in Chinese History* (New York, 1963), p. 22.

[18] Gazetteer (1596) written by Chen Teng and translated by W. C. LOWDERMILK, and D. R. WICKES, *History of Soil Use in the Wu T'ai Shan Area*, Monog., Roy. Asiatic Soc., NCB, 1938, p. 8.

[19] LOWDERMILK, W. C., "Forestry in Denuded China," *Ann., Amer. Acad. Pol. Soc. Sci.*, CLII (1930), 137.

[20] RICHARDSON, S. D., *Forestry in Communist China* (Baltimore, 1966), pp. 152–53.

[21] SSU-MA CH'IEN, *Shi Chi*, chap. 29.

[22] WRIGHT, A. F., "Symbolism and Functions: Reflections on Changan and Other Great Cities," *Journ. Asian Studies*, XXIV (1965), 670.

[23] WU, N. I., *Chinese and Indian Architecture* (New York, 1963), p. 38.

[24] *Mencius*, Bk III, pt. 1, 4:7.

[25] Reported by A. N. STEWARD and S. Y. CHEO in "Geographical and ecological notes on botanical explorations in Kwangsi province, China," *Nanking Journ.*, V (1935), 174.

[26] HARTWELL, R., "A Revolution in the Chinese Iron and Coal Industries during the Northern Sung, 960–1126 A.D.," *Journ. Asian Studies*, XXI (1962), 159.

[27] See L. S. Yang, *Les aspects économiques des travaux publics dans la Chine impériale*, Collège de France, 1964, p. 37.
[28] Gernet, (n. 1), p. 114.
[29] Moule, A. C., *Quinsai* (Cambridge, 1957), p. 51.
[30] Schafer, (n. 11), pp. 299–300.
[31] De Beauvoir, Simone, *The Long March* (Cleveland, 1958), p. 77.

# 83

# THE ANCIENT ROOTS OF OUR ECOLOGICAL CRISIS

*J. Donald Hughes*

Source: J. Donald Hughes (ed.), *Ecology in Ancient Civilizations*, Albuquerque: Univeristy of New Mexico Press, 1975, pp. 147–56.

The damaging changes being suffered today by the natural environment are far more rapid and widespread than anything known in ancient times. Today deforestation proceeds on a worldwide scale, the atmosphere becomes more turbid and opaque every year, the oceans are being polluted on a massive scale, species of animals and plants are being wiped out at a rate unmatched in history, and the earth is being plundered in many other ways. But although the peoples of ancient civilizations were unfamiliar with such recent discoveries as radioactivity, insecticides, and the internal combustion engine, they faced problems sometimes analogous to those the modern world faces, and we may look to the ancients in order to see the beginnings of many of our modern difficulties with an environment which is decaying because of human misuse.

A human community determines its relationship to the natural environment in many ways. Among the most important are its members' attitudes toward nature, the knowledge of nature and the understanding of its balance and structure which they attain, the technology they are able to use, and the social control the community can exert over its members to direct their actions which affect the environment. The ancient world shows us the roots of our present problems in each of these areas.

In a well-known and often reprinted article, "The Historical Roots of Our Ecologic Crisis,"[1] Lynn White traced modern Western attitudes toward the natural world back to the Middle Ages. But both medieval and modern attitudes have ancient roots. Greece and Rome, as well as Judaism and Christianity, helped to form our habitual ways of thinking about nature. And it is evident that the modern ecological crisis is to a great extent the result of attitudes which see nature as something to be freely conquered, used, and dominated without calculation of the resultant cost to mankind and the earth.

These attitudes stem from similar ideas which were held by the ancient people who have most influenced us. Animism, which saw the natural world as sharing human qualities and treated things and events in nature as sacred objects of respect or worship, was the dominant attitude in early antiquity and persisted almost everywhere in the Mediterranean world, but it gradually gave way to other ways of thinking. In Israel, transcendent monotheism replaced animism's "world full of gods." Instead of being divine in itself, nature was seen as a lower order of creation, given as a trust to mankind with accountability to God. But in the later history of that idea, people tended to take the command to have dominion over the earth as blanket permission to do what they wished to the environment, conveniently forgetting the part about accountability to God, or else interpreting most human activities as improvements in nature and therefore pleasing to God.

Perhaps even more important in the history of human attitudes toward nature was the departure from animism made by the Greek philosophers. Rejecting traditional mythological and religious explanations of the natural world, they insisted on the ability of the human mind to discover the truth about nature through the use of reason. Instead of a place filled with spiritual being, or beings, a theater of the gods, the environment was to them an object of thought and rational analysis. Worship of nature became mere ritual, supposedly replaced with philosophical understanding. Since, in the words of Pythagoras, "man is the measure of all things,"[2] it followed that all things have usefulness to mankind as their reason for existence. This idea has persisted in Western thought in various forms until the present, for the belief that everything in nature must justify its existence by its purposeful relationship to mankind is firmly, though perhaps implicitly, held by most people.

What was for the Greeks a philosophical opinion became for the Romans a practical reality. Early Roman animism was overcome less by the ingestion of Greek ideas than by the Romans' own demonstrated ability to dominate and to turn most things to their own profit, but both Greek influence and Roman practicality helped the Romans to develop attitudes toward nature which are remarkably similar to those expressed and demonstrated today. The Romans treated the natural environment as if it were one of their conquered provinces. If they needed any justification of this beyond their own pragmatism and cupidity, they could find it in Greek philosophy, which reached them in a late, skeptical form that had removed the sacred from nature and made nature an object of manipulation in thought and, by extension, in action. Our Western attitudes can be traced most directly to the secular, businesslike Romans. Today the process of dominating the earth is seen not as a religious crusade following a biblical commandment but as a profitable venture seeking economic benefit. In this, we are closer to the Romans than to any other ancient people, and in this we demonstrate to a great extent our heritage from them.

Attitudes alone do not determine the way a human community will interact with the natural environment. People whose religion teaches them to treat the world as a sacred place may still manage to make their surroundings a scene of deforestation and erosion, because good intentions toward nature are not enough if they are not informed by accurate knowledge about nature and its workings.

The earlier civilizations of the Near East accumulated a vast amount of information about the world through trial and error, and that information was passed on through tradition. Some of what they thought they knew was correct and useful, and much was colorfully inaccurate, interwoven with myth and folk stories.

A few Greek thinkers were the first to approach the natural world in a consistently rational fashion, demanding that reasonable explanations be found for all natural phenomena. This enabled them to begin the process of gaining knowledge which eventually developed into what might be called the scientific method. Many of the Greek thinkers were also careful observers of nature and attempted to check their ideas against what could be observed, but all of them held rational thought to be superior to what could be seen in the world and assumed that the inner workings of the human mind are congruent with the outer workings of the universe. This assumption, along with the antipathy of Greek thinkers toward work done with the hands, limited the range of their discoveries and led them into some fallacious speculations. Nonetheless, the discoveries of the Greek philosophers and scientists are many and impressive.

Unfortunately, research and discovery in this field gradually diminished under the Romans, who were collectors of older bits of information about the world of nature rather than discoverers of new knowledge. A few Greek scientists continued to work under the Roman Empire, but the Romans themselves produced few creative thinkers in this field. With the advent of Christianity, the situation worsened. Living in a world which they believed to be temporary, early Christians seemed to regard study of the things of this world to be irrelevant, if not a positive barrier on the way to salvation. "The wisdom of this world is foolishness,"[3] said Paul. He spoke in a somewhat different connection, but the Christians of the later Roman Empire, with very few outstanding exceptions, tended to look at all scientific inquiry in the spirit of that statement.

The modern world, having revived the works of the ancient Greeks, has gone beyond them in developing a rigorous methodology for gaining knowledge about the natural environment. The extent and accuracy of the understanding of nature that is available today is truly impressive, but far from complete. Much remains to be discovered about the circulation of the earth's atmosphere, weather, and the effects of pollution of various kinds on climate, for example. The behavior of species of animals and the interaction of all forms of life in an ecosystem are only imperfectly understood.

Governments and institutions have not always seen the relevance of such knowledge, and support for research has been a sometime thing. At the same time, human activities are inexorably destroying the last few examples of relatively undisturbed ecosystems that remain on earth, so that soon they will no longer be available for study. Brazilians are proclaiming that the Amazon rain forest will be gone in thirty years, to give one example, and no one can accurately predict what effects that massive change will have on South America and the world. Careful study is still needed.

The speed, scope, and intensity of interaction with the natural environment are crucially determined by the level of technology available to a human community. Using human and animal motive power and the energy of water, wind, and fire with the relatively simple tools and machines that had been invented, the ancient peoples constructed huge monuments which still impress us, but their level of interaction with the natural environment was relatively low as compared with that of modern industrial society. The changes wrought in the environment by ancient civilizations are massive indeed, but involved centuries or millennia for their accomplishment. Today more significant changes take place in months or years—or even seconds, in the case of atomic explosions. While the real extent and nature of the impact of ancient technology must not be underestimated, what impresses us almost as much is the failure of the ancients to pursue inventions, and the slow rate of technological change that resulted. Of all ancient peoples, the Romans possessed the most highly developed technology, and in this respect they are closest to us. Their machines for war, construction, and industry foreshadowed some that are still in use today. The fact that ancient peoples absorbed and survived successive changes in the technology of war and peace cannot be of much comfort to us today, because the rapidity, size, and power of such changes today are of an entirely different order from anything experienced in ancient times.

Another factor determining the way a human community will interact with the natural environment is the degree of organization and social control the community possesses. This is true because environmental ends desired for the good of the community may involve sacrifices on the part of its individual members, sacrifices which they would not make without some degree of social encouragement or coercion. The early civilizations of the river valleys, for example, had to be able to call for large expenditures of human energy on the construction of canals which seemed to benefit the entire society. Ancient civilizations were able to exert a considerable degree of social control because the vast majority of ancient people regarded themselves primarily as parts of their societies, and only secondarily, if at all, as individuals. Each person had a place in the social hierarchy which was rigidly defined and rarely changed. This was true of the pharaonic autocracy of Egypt, perhaps the most marked example of social control in the ancient world. But Egypt suffered periodic breakdowns in social control, and no ancient civilization could have channeled the actions of its citizens with regard to the environment to the extent that is at least theoretically possible today.

All ancient societies depended to a large extent on slave labor, a fact which seems to indicate an extreme degree of social control, until it is remembered that the majority of slaves were owned by citizens, not by the state, and that citizens were to a surprising degree able to pursue their private goals, at least in Greece and Rome.

As indicated in the preceding chapters, Greek and Roman governments established policies in the fields of agriculture, forestry, mining, and commerce, but citizens were allowed a wide latitude of choice within certain guidelines. Greek citizens had carefully defined duties to the community, but the city-states are noted for the freedom they allowed. The later Roman Empire tried to interfere in and control the lives of its citizens to an unusual extent; the edicts of Diocletian attempted to stabilize occupations, regulate prices, and control religion, while his secret police kept him informed of activities dangerous to the state. But no ancient autocracy remotely approached the ability of a modern industrial state to keep informed about its citizens and see that they perform their social duties. Greece operated without imprisonment as a punishment, and the Roman Empire supported itself financially without an income tax. The degree of control that can be exercised in the modern world by governments with electronic surveillance, computers, chemical and psychological methods, bureaucracies, police, deportations, and prisons is unmatched by anything seen before in world history. In democracies, environmental policies can be established only when widespread public support for them exists. Over the last few decades, such policies have in part been established even over the opposition of powerful pressure groups. Some needed measures have been blocked by the same groups or by the tendency of the public to prefer short-term personal gains to long-term benefit for society. Totalitarian states such as the Soviet Union have also taken some steps to preserve the environment, but we have not yet seen a major government take all of the steps which seem called for in the present ecological dilemma. Neither in ancient times nor in modern times have human communities become fully aware of the role which their relationship to the natural environment plays in their long-term welfare and even survival.

One conclusion which seems clear to this author is that the modern ecological crisis grew out of roots which lie deep in the ancient world, particularly in Greece and Rome. The problems of human communities with the natural environment did not begin suddenly with the ecological awakening of the 1960s, nor indeed with the onset of the Industrial Revolution or the Christian Middle Ages. Mankind has been challenged to find a way of living with nature from the earliest times, and many of our habitual answers to that challenge received their first conscious formulation within ancient societies, especially the classical civilizations.

At this point, one might well ask whether this study of ancient civilizations has produced any insights which might be of use in meeting the present crisis. If our ecological crisis has ancient roots, it might also be possible to learn from some of the successes and failures of ancient civilizations as we look to the future.

First of all, it might be possible for people today to recover something like the attitude of respect for the earth and nature that was felt by many in ancient societies. This could come not as a renaissance of animism, or a revival of ancient religions which have lost their ability to infuse human minds, but as a new insight compatible with many religions and philosophies. Judaism and Christianity could expand their concepts of human stewardship to recapture the biblical inclusion of the whole created natural world within the responsibility of people to God. Islam has its own unique insights along similar lines. Other will be impressed by Albert Schweitzer's demonstration that the concept of reverence for life serves as a basis for philosophical ethics. Eastern philosophies, which have long contained attitudes toward nature which emphasize harmony, respect, and refusal to exploit, might find ways to realize their insights. Recent interest in the American Indian feeling for the land and its creatures reveals that the native Americans had ecological wisdom which can be studied and emulated. Better attitudes toward the natural environment will have to develop in a pluralistic human community, as people of varying traditions and points of view come to see the necessity of caring for the earth in order to preserve life itself and improve the quality of life.

Second, a concentrated effort to study the natural environment in all of its facets and interrelationships is needed. This is particularly crucial at the present moment, before much of the evidence about nature is altered, marred, or erased by human activities. These activities themselves and their effects upon the natural environment must also be investigated thoroughly. No wise environmental policy can be based on ignorance of the workings of nature. So that we may learn from what mankind has experienced through millennia of interaction with nature, more research is needed into the ecological relationships of past human societies, to correct and fill out the broad outline which is presented here.

Third, each human community must seek a viable relationship with the natural environment at the level made possible by the technology available to it. A study of ancient civilizations should demonstrate that a rejection of modern technology or an attempt to turn back the clock would not in itself assure a proper balance with nature or prevent environmental degradation. Rather, we should find ways to use our technological abilities in order to minimize the destructive impact of our civilization upon the natural environment and to enhance our relationship with nature in ways which are beneficial both to people and to the environment. This would no doubt mean that some possible avenues of technological development ought to be abandoned, and that human population ought to be stabilized at some optimum size. No level of technology could support an unlimited increase of human numbers without catastrophic damage to the natural world and resultant crisis for mankind.

Finally, as human beings, we must be willing to accept freely certain limitations on our actions which affect the earth. In democracies, these limitations can be based on public awareness of the magnitude of environmental problems and of the options which exist to meet them. The alternatives to freely chosen environmental policies, consistently administered, are probably few. History does not provide us with an example of an ecologically aware dictatorship, willing to coerce its people to take the courses of action which it deems necessary for survival in balance with nature, but such a government is certainly a future possibility somewhere in the world, unpleasant as it may be to contemplate. History does, however, provide us with many examples of ancient peoples who failed to adapt themselves to live in harmony with the ecosystems within which they found themselves, who depleted their environment, exhausted their resources, and exist today only as ruins within eroded and desiccated landscapes. That fate might also await our own civilization, but this time on a global scale. Ancient history is a warning and a challenge to our attitudes, our ability to understand, our technological competence, and our willingness to make far-reaching decisions. The challenge will not go away, and the response we will make is not yet clear.

## Notes

1 Lynn White, "The Historical Roots of Our Ecologic Crisis," *Science* 155 (1967): 1203–7.
2 Plato *Theaetetus* 160D (15).
3 21 Cor. 3:19.

# Part 2

# SPINOZA AS DEEP ECOLOGIST?

# 84

# SPINOZA AND ECOLOGY

*Arne Naess*

Source: *Philosophia*, 7 (1977), 45–54.

In what follows I do not try to prove anything. I invite the reader to consider a set of hypothetic connections between spinozist and ecological thought. Most of them seem obvious to me, but every one needs to be carefully scrutinized. They are (of course) built upon a set of definite interpretations of ecology and of the texts of Spinoza.

The industrial states adopt policies which to some extent limit pollution and conserve non-renewable resources. There is also a slight recognition of overpopulation in the sense of too great consumption per capita. In short, there is in the industrial states a shallow movement in favour of protection of the environment, or better, the ecosystems. But there is also a deeper, international movement which tries to modify attitudes towards nature and the whole conception of the relations of culture to nature. It has deep social and political implications.

Rachel Carson, who started this international movement fifteen years ago, found man's arrogance or indifference towards nature ethically unacceptable. The driving force of the movement was and is still philosophical and religious. The field bioecologists, who work in nature, are on the whole manifesting attitudes of love and respect that have made an impact upon millions of people.

There is a deep convergence in metaphysics, ethics and life styles among the people inspired by field ecological thinking. The issues of pollution, resource depletion and overpopulation are not neglected within the deep ecological movement, but they are integrated in a vastly more comprehensive frame of reference. This frame includes the study of non-industrial cultures, some of them showing a remarkable ecological equilibrium combined with affluence.[1] History is littered with the remains of cultures that lost the equilibrium. There is a growing understanding that even if we cannot and will not imitate any of the original affluent cultures, we shall have to establish post-industrial societies in equilibrium. Spinoza may turn out to be an important source of inspiration in this quest.

In what follows I accordingly invite friends of Spinoza to consider the many aspects of his philosophy that seem to accord with basic strivings within the deep ecological movement. I do not think it important to get to a final conclusion about exactly which concepts or aspects accord best with which concepts or aspects of that movement. There is room for different interpretations. I offer only one.

The interpretation made use of in this article is elaborated in more detail in A. Naess, *Freedom, Emotion and Self-subsistence* (Oslo: Universitetsforlaget 1975). Footnotes will refer mainly to that work. (Abbreviation: *FES*.)

1. The nature conceived by field ecologists is not the passive, dead, value-neutral nature of mechanistic science, but akin to the *Deus sive Natura* of Spinoza. All-inclusive, creative (as *natura naturans*), infinitely divers, and alive in the broad sense of panpsychism, but also manifesting a structure, the so-called laws of nature. There are always causes to be found, but extremely complex and difficult to unearth. The Nature with capital N is intuitively conceived as perfect in a sense that Spinoza and out-door ecologists have more or less in common: It is not a narrowly moral, utilitarian or aesthetic perfection. Nature is perfect 'in itself'.

*Perfection* can only mean *completeness* of some sort when applied in general, and not to specifically human achievements. In the latter case it means reaching what has been consciously intended.[2]

2. The value-dualism spirit/matter, soul/body does not hold in Spinoza nor is it of any use in field ecology. The two aspects of Nature, those of extension and thought (better: non-extension), are both complete aspects of one single reality, and *perfection characterizes both*.

In view of the tendency to look upon the body as something more crude than spirit, both field ecologists and Spinoza oppose most forms of idealism and spiritualism—and, of course, moralism. (I am not sure these "ism" words deserve to be used in this connection.)

3. Nature (with capital N) according to Spinoza, and the 'universe' of modern physics, are not in time. As an *absolutely* all-embracing reality, Nature has no purpose, aim or goal. If it had a purpose, it would have to be part of something still greater, e.g., a grand design. Times is only defineable *within* the network of relations of Nature, therefore Nature as a whole cannot have aims or goals which refer to time.

There is, in ecological thought, a marked reaction against facil finalism, especially in sophisticated research. The development of 'higher' forms of life does not make field ecologists less impressed with the 'lower' forms, some of which have flourished countless millions of years and still 'are going strong'. (Too strong, some will say, thinking of recent epidemics of flu!) There is no 'purpose' in time such that the bacteria do not have any function or value when 'higher' forms have developed.

4. There is no established moral world-order. Human justice is not a law of nature. There is, no the order hand, no natural laws limiting the endeavour to extend the realm of justice as conceived in a society of free human beings.

These spinozist thoughts are important for striking the balance between a submissive, a-moral attitude towards all kinds of life struggle, and a shallow moralistic and antagonistic attitude. Future societies in ecological equilibrium presuppose such a 'third way'.

5. Good and evil must be defined in relation to beings for which something is good or evil, useful or detrimental. The terms are meaningless, when not thus related.[3]

This accords well with the effort of field ecologists to understand each culture 'from within'. It contrasts with explaining or moralizing on the basis of a definite value code dominant within particular (mostly industrial) societies.

6. Every thing is connected with every other. There is a network of cause-effect relations connecting everything with everything. Nothing is causally inactive, nothing wholly without an essence which it expresses through being a cause.

The ecologist Barry Commoner has called 'All things are connected' the first principle of ecology. Intimate interconnectedness in the sense of internal rather than external relations characterizes ecological ontology.

7. Every being strives to preserve and develop its specific essence or nature. Every essence is manifestiation of God or Nature. There are infinite ways in which Nature thus expresses itself. And there are infinite kinds of beings expressing God or Nature.

The pervasive basic *striving* is not mere effort to adapt to stimuli from the outside. It is an active shaping of the environment. Successful acts create new wider units of organism/environment. The basic urge is to gain in extent and intensiveness of self-causing. The term 'self-realization' is therefore better than 'self-preservation', the first suggesting activeness and creativity, the latter a passive conservative or defensive attitude.[4]

8. Another name for the ability to act out one's nature or essence is 'power', *potentia*, then substantivation of the verb 'to be able', *posse*. It is not the same as to coerce others.

The power of each thing is part of God's power. God or Nature has no other power than ours. "Each and every existing thing expresses God's power. God or Nature has no other power than ours. "Each and every existing thing expresses God's nature or essence in a certain determinate way... that is,... each and every thing expresses God's power..." (Part I, Proof of Proposition 36.)

All beings strive to maintain *and gain* power. This need not be a striving to dominate, subdue or terrorize. The establishment of symbiosis, 'living together', rather than cut-throat competition marks and gain in power. At higher levels of self-realization, the self encompasses others in a state of increasing intensity and extension of 'symbiosis'.[5] The freedom of the individual ultimately requires that of the collectivity.

9. If one insists upon using the term "rights", every being may be said to have the right to do what is in its power. It is a "right" to express its own

nature as clearly. and extensively as natural conditions permit. "That right which they (the animals) have in relation to us, we have in relation to them." (Part 4, First Scholium to Proposition 37.) Rights as part of a separate moral world order is a fiction.[6]

Field ecologists tend to accept a general 'right to live and blossom'. Humans have no special right to kill and injure, Nature does not belong to them.

10. There is nothing in human nature or essence, according to Spinoza, which can *only* manifest or express itself through injury of others. That is, the striving for expression of own nature does not inevitably imply an attitude of hostile domination over other beings, human or non-human. Violence in the sense of violent activity is not the same as violence as injury to others.

The human attitude of violence and hostility towards some species of animals have made is impossible to study realistically their life and function within the whole. The field ecologist who deeply identify with the species studied is able to live peacefully with any kind of "wild" animal. This attitude harmonizes with the view of Spinoza concerning free man (*homo liber*). Spinoza's doctrine about the development of affects (Part 3 and 4 of the Ethics) makes the field ecologists symbiotic attitude inevitable if the development proceeds far enough.

In what follows other spinozist thoughts are mentioned which harmonize with those of field ecologists even if the latter do not often develop them consciously.

11. Every being has its unique direction of self-realization, its particular essence, but "the greatest good" is the "understanding realization of the union (*cognitio unionis*) of our mind with whole Nature". (*On the improvement of the understanding*, § 13–14.)

12. The realization of union with the whole Nature is made through the understanding of the particular things as a manifold of expressions or manifestations of Nature (God). But Nature or God is nothing apart from the manifestations.

Spinoza rejects the kind of *unio mystica* which results in a turning away from particulars and from nature. "The more we understand the particular things, *res singulares*, the more we understand God". (Part 5, Proposition 24.) Spinoza's concept of Nature and its manifestations lack the features which makes nature (in the more common connotations) something inferior to spirit, or to God.

Ecological thinking presumes an identification with particulars in their internal relations to others. The identification process leads deeper into Nature as a whole, but also deeper into unique features of particular beings. It does not lead away from the singular and finite. It does not lend itself to abstract thinking or contemplation, but to *conscious*, *intuitive*, *intimate interaction*.

13. 'Rationality' is wise conduct maximizing self-realization. It cannot be separated from perfection, virtue and freedom. "Since reason does not demand anything contrary to Nature, it demands that everyone loves himself, look for what is useful, . . . , and that he strives to obtain all which really leads man to greater perfection . . ." (Part 4, Proposition 18, Scholium.) Since self-realization implies acts of understanding with increasing perspective, rationality and virtue increases with the development of understanding. The maximum is 'an understanding love of Nature', *amor intellectualis Dei*. This implies acts of understanding performed with the maximum perspective possible, or loving immersion and interaction in Nature.[7]

14. Interacting with things and understanding things cannot be separated. The units of understanding are not propositions, but acts. To the content of ideas in the "attribute of non-extension" there corresponds an act in the "attribute of extension". Ultimately these attributes are attributes of the same, but the human way of understanding is such that we have to treat them separately.

Increase of rationality and freedom is proportional to increase of activeness, each action having the aspects of understanding and of a behaviour or inter-action. Not all acts need be overt.

15. Since a gain in understanding expresses itself as an act, it is in its totality a process within the extended aspect of Nature and can be studied as such.

This point is of prime importance to the methodology of ecology: The 'world' of a living being is investigated through study of its manifest ("molar" not "molecular") behaviour. Spinoza furnishes ecology a frame of reference completely devoid of the kind of "mentalism" and "introspectionism" that often has obstructed the study of cognition in animals and men.

The framework of Spinoza and general etology is also well suited to counteract the tendency to conceive human knowledge as something existing independent of acts of particular human beings in particular situations—and stored wholesale in libraries.

The formulation of Spinoza does not point to any definite form of "behaviourism". We are free to inspect critically any contemporary version. There is no reason to identify the concepts of 'behaviour' with that of Watson or Skinner.

16. Most of the basic concepts used in the *Ethics* when characterizing the human predicament are such as can be used whatever the cultural context. They are furthermore adapted to general characterizations covering smaller or greater parts of the animal, plant and mineral kingdoms. Some of these concepts have already been mentioned.

Spinoza rarely touches upon questions concerning animals, but where he does, he shows that his main concepts are not only intended to apply to humans.[8] He warns, however, against thinking that the joys of insects are the same as those of humans. Each kind of living being is content with and delights in what corresponds to its nature or essence.

Among the important concepts which have a wider application than to the human species one may not the following:

Perfection (cf. point 2)
good and evil (cf. point 4 and 5)
striving to express one's nature or essence (cf. point 7 and 8)
self-preservation, self-realization (cf. points 7, 11 and 13)
power (cf. points 8, 9 and 10)
rationality (cf. points 13 and 14)
virtue (cf. points 13, cp. the expression '*potentia seuvirtus*')
freedom (cf. points 13 and 14)
understanding (cf. points 14 and 15) feeling
emotion (The passive ones are confused *ideas*.)
confused idea

For all these terms it holds that Spinoza's definitions are *open as regards their exact range*.[9] Some are clearly intended to be applicable at least to a major part of the kingdom of animals. Because of equivalences holding between many of them the range of all of the can without doing violence to Spinoza's texts be made as large as suitable within ecology and theory of evolution.

The wide applicability of Spinoza's *concepts* does not imply uncritical *statements* about similarities between humans and other living beings. It ensures a broad *continuity* of outlook, and the possibility of fighting human haughtiness and cruelty.

Ecological thought typical of active field ecologists is not entirely homogeneous. And Spinoza's texts are of course open to various interpretations. In spite of this my conclusion is positive: No great philosopher has so much to offer in the way of clarification and articulation of basic ecological attitudes as Baruch Spinoza.

## Notes

1 Highly readable: Sahlins, Marshall, *Stone Age Economics*, London: Tavistock, 1974.

2 'Perfection is not a term which is introduced in the *Ethics* by means of separate definition, and it is something admitting of degrees.

Joy is an emotion through which mind is said to *become more* perfect. (Part 3, Scholium to Proposition 11.) Whatever its connotation, 'more perfect' cannot be separated in denotation from 'more powerful'. Compare Part 4, proof of Proposition 4): "Joy . . . is the emotion through which the power of the body [and therefore also of the mind] to act, increases or is furthered". The relation to action, and therefore to understanding, is intimate. There more perfect, the more active and the less passive, according to Proposition 40, Part 5. In short, more perfect than' cannot in denotation be separated from a number of other basic relations. The application of the term to Nature or God clearly is on par with the application of terms like 'love' (amor), 'rationality', 'mind', that is, it cannot be taken in any precise sense known from phenomena *in* Nature.

3 The occurrences of the words *bonus* and *malus* in the *Ethics* admit of various conceptualizations. According to Definition 1, Part 4, in the *Ethics*, '*x* is good for *y*' does not mean more than '*x* is useful for *y*' or '*x* is known by *y* to be useful for *y*'. The expression '*x* is useful for *y*' is equivalent to *x* causes an increase in *y*'s power', '*x* causes an increase of *y*'s freedom' and '*x* causes an increase of *y*'s perfection.' More about these equivalences in *FES*, pp. 107–109.

4 According to Part 3, Proposition 6, *every* thing, as far as it is in, itself, strives to preserve in its being. The term *perseverare* I take to mean something much more active than just to survive. Therefore I accept as equivalent '*x* increases in power' and '*x* increases in level of self-preservation'. (*FES*, p. 97.)

5 Good relations to others are obtained, according to Part 4, Propositions 46, 50 (Scholium), 72, a.o., through generosity and other forms of non-injury (*ahimsa*). According to Part 4, Definition 1, it can never be useful to us. Therefore it cannot cause increase in power or understanding.

6 It must conceded that Spinoza holds that we cannot be friends of animals or include them in our society. Only humans can be friends of humans and be members of our societies. (See Part 4, Appendix, Chapter 26). And because we are more powerful than animals, we have more rights. We may *use* animals as we see fit, and one cannot issue laws against molesting them. (Cp. Part 4, Scholium 1 to Proposition 37, and Appendix, Chapter 26.)

7 The basic position of 'understanding', *intelligere*, in Spinoza's system is seen from its intimate relation to 'causing'. It something is caused adequately through something else, it is adequately understood through this something, *and* vice versa. (*FES*, p. 40, cp. Part 3, first part of first definition.) Activeness is internally related to understanding because the specific activity of the mind is understanding. It is also related to power and freedom in so far as increased activeness is internally related to increase in power and freedom. In this way not only intuitive understanding of highest (third) kind but also the understanding of laws of nature is promoting power, freedom, joy and perfection.

8 The panpsychism of Spinoza is expressed in the Scholium to Proposition 13 in Part 2 (of the *Ethics*). Other individuals than humans are animated, *animata*, but in different degrees, *diversis gradibus*. He even (in the proof of Proposition 1, Part 3) uses the expression "the minds of other things," (*aliarum rerum mentes*). About the difference in appetities and joys between various kinds of animals, see Scholium to Proposition 57, Part 3.

9 Spinoza does not directly say so, but I think he would deny rationality of any kind to other beings than humans. He speaks, however, about "virtue or power" of animals, and he more or less identifies virtue and rationality. '... to act virtuously is nothing else than to act according to reason; ...' (Part 4, Proof of Proposition 56).

10 Spinoza may evidently be interpreted in various ways as regards the relation of animals to man. We have been interested in the main trend of his reasoning and the main features of his terminology.

# 85

# SPINOZA'S ENVIRONMENTAL ETHICS

*Genevieve Lloyd*

Source: *Inquiry*, 23 (1980), 293–311.

The paper explores an apparent tension in Spinoza's thought between his treatment of man as part of nature, with no specially privileged position within it; and his treatment of morality as circumscribed by what is good for human beings. These two themes, it is argued, are in fact interconnected in Spinoza's thought. The paper goes on to consider some possible responses, from a contemporary standpoint, to Spinoza's rejection of animal rights. Finally, it is argued that the apparent tension in Spinoza's thought becomes more understandable in the light of his treatment of the importance of truth to human beings; and that this throws into relief some puzzling features of some contemporary formulations of the theme of man as part of nature.

Some contemporary philosophers concerned with ethical issues related to the environment are looking to Spinoza in the hope of finding a firm metaphysical basis for environmental ethics.[1] Such a hope is by no means entirely misplaced. Spinoza, perhaps more than any other modern philosopher, is concerned with the integration of metaphysics and ethics, with the metaphysical bases of ethical positions. A very dominant theme in his thought, moreover, is the cultivation of what can only be described as an attitude of reverence for nature. Spinoza extends to the world as 'Deus sive Natura' the attitude of reverence, love, and humility which religious orthodoxy had given to a transcendent God. He himself calls this attitude the 'intellectual love of God'. Related to this theme is Spinoza's constant insistence that man has no privileged position in nature. The cultivation of Reason involves coming to grasp one's own status as 'part of a whole', subject to the same necessities that govern the rest of nature. This persistent theme of 'man as part of nature' does, on the face of it, readily lend itself to the development of an environmental ethic. And it is reinforced by his treatment of individuality as essentially demanding an integration between the individual and its surrounding totality of being.[2] Spinoza lays constant stress on the wider systems that comprise the sustaining framework for individual existence. His metaphysics of individuality, if it can be extricated from the assumptions of seventeenth century dynamics, may well seem to flow naturally into an ethic of concern for the non-human constituents of reality. Finally, there is, of course, a certain ironic appeal in the prospect of being able to enlist a seventeenth-century arch-rationalist into the ranks of the environmental movement.

Despite all this, it would, I think, be quite misplaced to claim Spinoza as patron philosopher of the environmental movement. This is partly for obvious reasons. Spinoza's system is an extremely complex network of interconnected theses. And the whole is too abstruse and, in some crucial respects, too alien to modern thought for us to seriously envisage environmental education for the future fruitfully including a solid dose of Spinoza's repudiation of individual substance. But there is, in any case, another cluster of themes in Spinoza's thought which seems in fact to be strongly inimical to the possibility of an environmental ethic. Spinoza claims that morality is entirely circumscribed by the human species; that it is inextricably grounded in drives for *human* self-preservation; that other species can be ruthlessly exploited for human ends. And these are not, I think, themes that can be readily removed from Spionza's system while leaving intact the theses that cluster around the theme of man as part of nature. Both spring from the same metaphysical presuppositions, and Spinoza himself certainly saw them as interconnected. It is precisely because Spinoza recognizes no privileged position fro man with respect to nature that he is, in the contemporary phrase, a 'speciesist'.[3] The very recognition of man as part of nature that looks, at first sight, so promising as a basis for an environmental ethic, for Spinoza goes hand in hand with a reasoned acquiescence in the exploitation of other species. Anyone who looks to the *Ethics* for a viable, coherent metaphysical system to ground a belief in the rights of the non-human will look in vain.

Is the relevance of Spinoza for environmental ethics then restricted to the possibility of extracting a few purple passages stimulating to the preception of man as part of nature? I think not. From our perspective, there is an extraordinary tension in Spinoza's juxtaposition of the theme of man as part of nature with a strongly anthropocentric morality. And this is itself of considerable philosophical interest. Spinoza sets himself firmly against a man-centred perception of the universe; against the view that man has a privileged place in the universe. But from this very insight he derives a moral position that seems unequivocally 'speciesist'. The apparent strangeness of this juxtaposition throws into relief some major differences between the kinds of view that can be plausibly combined in seventeenth-century thought and in our own. Such contrasts serve, at the very least, to illuminate the structures of our own thought. But the apparent tension in Spinoza's thought also throws light, I think, on some confusing aspects of current debate on environmental ethics. Spinoza manages to combine a strong rejection of *anthropocentric perception* with an equally strong affirmation of a man-centred *morality*. That he manages to do so throws into question, I think, the assumption of many contemporary thinkers on environmental ethics that condemning man's 'exploitation' of his environment involves extending the 'moral community' to include the non-human.

To bring out the apparent tension in Spinoza's thought on man's relationship to his environment I will begin from a passage in the *Ethics* where he speaks on the rights of animals, and try to explicate some of the background assumptions of the stand he takes there. I will then look at some possible responses to Spinoza's rejection of animal rights from a contemporary standpoint. Finally, I will suggest that this tension becomes more intelligible in the light of Spinoza's treatment of the importance of truth of human beings; and that this throws into relief a puzzling tension in at least some contemporary formulations of the theme of man as part of nature.

## I The rights of animals

Spinoza addresses himself to the question of animal rights in the following passage:

> [I]t is plain that the law against the slaughtering of animals is founded rather on vain superstition and womanish pity than on sound reason. The rational quest of what is useful to us further teaches us the necessity of associating ourselves with our fellowmen, but not with beasts, or things, whose nature is different from our own; we have the same rights in respect to them as they have in respect to us. Nay, as everyone's right is defined by his virtue, or power, men have far greater rights over beasts than beasts have over men. Still I do not deny that beasts feel: what I deny is, that we may not consult our own advantage and use them as we please, treating them in the way which best suits us; for their nature is not like ours, and their emotions are naturally different from human emotions. . . .[4]

Spinoza thinks, contrary to Descartes, that animals are sentient beings. This follows from his account of minds as ideas of bodies. Any idea of a body with a requisite degree of complexity will be for him a sentient mind. But being of a different nature from ourselves – being, that is, ideas of bodies of a different structure – their emotions are different from ours:

> Horse and man are alike carried away by the desire of procreation; but the desire of the former is equine, the desire of the latter is human. So also the lusts and appetites of insects, fishes, and birds must needs vary according to the several natures.[5]

Other species have their own joys, their own lusts. And this being so, they cannot enter into collaboration with us in the shared pursuit of self-preservation, of the enhancement of our own activity, which is the sole basis of morality.

> No individual thing, which is entirely different from our own nature, can help or check our power of activity, and absolutely nothing can do us good or harm, unless it has something in common with our nature.[6]

> . . . that which is entirely different from our nature can neither bc to us good nor bad.[7]

Everything that is incapable of entering into collaboration with us, Spinoza sees as morally indifferent. This is a consequence of his grounding morality in a Hobbesian drive towards self-preservation.

> To act absolutely in obedience to virtue is in us the same thing as to act, to live, or to preserve one's being (these three terms are identical in meaning) in accordance with the dictates of reason on the basis of seeking what is useful to one's self.[8]

This end is served by collaboration with other men. The rational pursuit of intra-species collaboration rests, for Spinoza, not on any Aristotelian conception of man as by nature a social animal. He sees it rather as a natural extension of the Hobbesian drive towards individual self-preservation. The Hobbesian principle that every man is bound to seek what is useful to himself Spinoza see not, as did some of his contemporaries, as the foundation of impiety but rather that of piety and virtue.[9] Man's self-preservation is fostered by collaboration with others of the same species. An active fostering of the good of other men is a natural outcome of the rational man's pursuit of his own individuality.

> As every man seeks most that which is useful to him, so are men most useful one to another. For the more a man seeks what is useful to him and endeavours to preserve himself, the more is he endowed with virtue (IV. xx.), or, what is the same thing (IV. Def. viii.), the more is he endowed with power to act according to the laws of his own nature, that is to live in obedience to reason. But men are most in natural harmony, when they live in obedience to reason . . . therefore . . . men will be most useful one to another, when each seeks most that which is useful to him.[10]

> Let satirists then laugh their fill at human affairs, let theologians rail, and let misanthropes praise to their utmost the life of untutored rusticity, let them heap contempt on men and praises on beasts; when

> all is said, they will find that men can provide for their wants much more easily by mutual help, and that only by uniting their forces can they escape from the dangers that on every side beset them. . . .[11]

The pursuit of self-preservation, which for Spinoza is what the life of reason really amounts to, would naturally issue, if left to itself, in willing collaboration without any need for external restraints. It is the passions that are the source of conflict and division in the pursuit of self-interest. The shared cultivation of reason draws men together, enhancing their own powers, their own activity, their own perfection. But men for the most part do not live in accordance with reason. And, given this, the Hobbesian picture of man emerging from the state of nature through recognition of the necessity of external restraints is an accurate one. Men come to avoid inflicting injury through fear of incurring a greater injury themselves.[12] The harmful, destructive passions are inhibited by the stronger emotion, fear. In an ideal situation this role of curbing the harmful passions would be played by reason. The knowledge of good and evil – the understanding of the passions – would be at all times a stronger emotion curbing the destructive impulse to harm others. But in the lack of reason's dominance it must be replaced by fear.

The contrast between the 'state of nature' and the moral community which replaces it when men contract to accept a common authority is crucial to the understanding of Spinoza's position on animal rights and its relationship with the theme of man as part of nature. This contrast is reflected in a distinction between two kinds of 'right'. There are, first, the rights that apply to men in the state of nature along with everything else. And, secondly, there are rights that apply to men as citizens who have contracted into a state, agreeing to collaborate for the sake of thereby avoiding greater evil.

Animals have rights only in the first sense. And, for Spionza, unlike Locke, 'natural right' has no moral connotation. That animals have rights in this sense does not make them in any way members of a moral community. It is, for Spinoza, trivially true that the stronger – those who can do more, those with more power – have greater rights than the weaker. And it is in this sense that he says we have more rights in respect of animals than they have in respect of us. This is just because human bodies are superior to animal bodies; they are capable of a greater range of activity. We have more rights with respect to animals just in virtue of the fact that we have more power than them; that is to say, just in virtue of the fact that we can do more.

In the *Tractatus Theologico-Politicus*, Spinoza is more explicit about the amorality of the state of nature. Like Hobbes, he rejects any conception of 'natural law' or 'natural rights' as a basis of morality. Moral laws and rights are consequent on the emergence from the state of nature.

> By the right and ordinance of nature, I merely mean those natural laws wherewith we conceive every individual to be conditioned by

> nature, so as to live and act in a given way. For instance, fishes are naturally conditioned for swimming, and the greater for devouring the less; therefore fishes enjoy the water, and the greater devour the less by sovereign natural right.[13]

In the state of nature every individual

> has sovereign right to do all that he can; in other words, the rights of an individual extend to the utmost limits of his power as it has been conditioned. Now it is the sovereign law and right of nature that each individual should endeavour to preserve itself as it is, without regard to anything but itself; therefore this sovereign law and right belongs to every individual, namely, to exist and act according to its natural conditions.[14]

This emphasis on the amorality of the state of nature is associated with Spionza's rejection of a prevalent conception of Reason. Like Hobbes he refuses to see the laws of reason as governing the whole of nature, or man as separated out from the rest of nature by having an innate grasp of those laws. Men in the state of nature are 'no more bound to live by the dictates of an enlightened mind than a cat is bound to live by the laws of the nature of a lion'.[15]

> [T]he right and ordinance of nature, under which all men are born, and under which they mostly live, only prohibits such things as no one desires, and one can an attain: it does not forbid strife, not hatred, nor anger, nor deceit, nor, indeed, any of the means suggested by desire.
>
> This we need not wonder at, for nature is not bounded by the laws of human reason, which aims only at man's true benefit and preservation; her limits are infinitely wider, and have reference to the eternal order of nature, wherein man is but a speck. . . .[16]

Here we see the interconnection between Spinoza's anthropocentric morality and his view of man as part of nature. It is a fundamental illusion, Spinoza thinks, to think of everything being arranged according to the dictates of human reason. Reason is a means for achieving what is good for us. It has no bearing on what is good for the rest of reality. It is of importance only for that small speck of reality which consists in human life. But, having thus put Reason in its place, Spinoza goes on to insist on the crucial importance of the dictates of reason for human good.

> Nevertheless, no one can doubt that it is much better for us to live according to the laws and assured dictates of reason, for . . . they have

> men's true good for their object. . . . When we reflect that men without mutual help, or the aid of reason, must needs live most miserably . . . we shall plainly see that men must necessarily come to an agreement to live together as securely and well as possible if they are to enjoy as a whole the rights which naturally belong to them as individuals, and their life should be no more conditioned by the force and desire of individuals, but by the power and will of the whole body.[17]

Spinoza comes back to the theme of the irrelevance of the dictates of reason to the rest of reality in the Appendix to Part One of the *Ethics*, in a virulent attack on those 'prejudices and superstitions' which have taken root in the human mind around the theme of final causes:

> [A]s they find in themselves and outside themselves many means which assist them not a little in their search for what is useful, for instance, eyes for seeing, teeth for chewing, herbs and animals for yielding food, the sun for giving light, the sea for breeding fish, etc., they come to look on the whole of nature as a means for obtaining such conveniences. Now as they are aware, that they found these conveniences and did not make them, they think they have cause for believing, that some other being has made them for their use.[18]

They come then to believe in some 'ruler or rulers of the universe endowed with human freedom, who have arranged and adapted everything for human use', who direct the whole course of nature for the satisfaction of men's 'blind cupidity and insatiable avarice'.[19]

> Thus the prejudice developed into superstition, and took deep root in the human mind; and for this reason everyone strove most zealously to understand and explain the final causes of things; but in their endeavour to show that nature does nothing in vain, i.e., nothing which is useless to man, they only seem to have demonstrated that nature, the gods, and men are all mad together.[20]

> [S]uch persons firmly believe that there is an *order* in things, being really ignorant both of things and their own nature.[21]

The basic form of these superstitions is mistaking man's position in nature, in thinking that it is in reference to him that things are to be judged more or less perfect. But such reasoners, Spinoza says in concluding the Appendix, are

> easily confuted, for the perfection of things is to be reckoned only from their own nature and power; things are not more of less

> perfect, according as they delight or offend human senses, or according as they are serviceable of repugnant to mankind.[22]

The doctrine of man's privileged place in the universe was of course prevalent in the Judaeo-Christian tradition. It was, no doubt, part of popular consciousness in the seventeenth century. The targets of Spinoza's attack here were probably some fairly crude theological statements of the doctrine. But there is no lack of evidence of the same idea occurring in sober philosophical treatises in the seventeenth century. Thus Locke, in his discussion of man in the state of nature in the early sections of the second Treatise says

> there cannot be supposed any such subordination among us that may authorise us to destroy one another, as if we were made for one another's uses, as the inferior ranks of creatures are for ours.[23]

The doctrine occurs, in a rather more sophisticated from, also in Leibniz. '[I]f we think that God has made the world only for us, it is a great blunder . . . ', he says in the *Discourse on Metaphysics*.[24] But he does think nonetheless that all the other creatures should serve intelligent souls, the 'citizens of the republic of the universe whose monarch is God'.[25]

> The difference between intelligent substances and those which are not intelligent is quite as great as between a mirror and one who sees.[26]
>
> [T]he spirits with which (God) can, so speak, enter into conversation and even into social relations by communicating to them in particular ways his feelings and his will so that they are able to know and love their benefactor, must be much nearer to him than the rest of created things which may be regarded as the instruments of spirits.[27]

But for Leibniz the role of other creatures as 'instruments of spirits' is not confined to such mundane purposes as providing food:

> A single spirit is worth a whole world, because it not only expresses the whole world, but it also knows it and governs itself as does God. In this way we may say that though every substances expresses the whole universe, yet the other substance express the world rather than God, while spirits express God rather than the world. This nature of spirits, so noble that it enables them to approach divinity as much as is possible for created things, has as a result that God derives infinitely more glory from them than from the other beings, or rather the other beings furnish to spirits the material for glorifying him.[28]

Spinoza's acquiescence in the exploitation of other species has, we have seen, a very different basis from this anthropocentric perception of the world which he categorically condemns. But this rejection of anthropocentric perception, we have also seen, is linked with a thoroughly anthropocentric morality. We are more perfect than the animals; that is, we are capable of a wider range of activities. It is not that he sees this as a *reason* or a justification for our having 'more rights over them than they should have over us'. To be more perfect is just to be more active, to have more power. And virtue is just doing what our activity.[29]

## II The requirements of an 'environmental' ethic

Animals, then, have rights only in the sense of 'natural right'. And for Spinoza – like Hobbes, and unlike Locke – natural right is not a moral matter. Moral rights arise only within the framework of a moral community and this is species-relative. It depends on a shared nature. What is good for us is not at all commensurable with the goods of other species. There is, I think, no way for spinoza to even express the view that our goods ought to be curtailed for the sake of the goods of other animals. The 'ought' will not straddle the gap between our moral community and the 'state of nature' where the other animals reside. To use Passmore's distinction, man is not a 'depot of nature', in the way suggested by the perception of other species as existing for human good; but nor is he a 'steward of nature', responsible for the good of other species less perfect than himself.[30] It seems, then, that we cannot extract from Spinoza's thought even a basis for the responsible concern for the environment which Passmore expresses in the picture of man as 'steward of nature'; an approach which many contemporary writers find too conservative to provide a viable basis for a genuine environmental ethic.

Spinoza's system does, of course, provide a basis for so-called 'instrumental' arguments, grounded in human interest, for responsibility towards nature, for concern with man's habitat. But this, of course, is not what many environmental philosophers are looking for. The demand is for an ethic which will give other species a claim on human concern in cases where this seems not to be demanded by what is good for human beings. 'Human interests and preferences', Richard Routley claims, 'are far too parochial to provide a satisfactory basis for deciding on what is environmentally desirable'.[31] And Rolston, in an article dissecting what requirements would have to be met by an ethic that was 'genuinely ecological', says:

> If Leopold's preserving the ecosystem is merely ancillary to human interests, the veiled antecedent ought is still that we ought to maximise human good. Were we so to maximise the ecosystem we should have a corporate anthropological egoism, 'human chauvinism', not a planetary altruism. The optimum ecosystem would be

> but a prudential means to human welfare, and our antecedent ought would no longer be primarily ecological, but as before, simply a familiar one, incidentally ecological in its prudence.[32]

It is not clear that such an 'incidentally ecological' ethic stands to gain much from the provision of a metaphysical basis. We do not, after all, need metaphysical arguments to get men to pursue what can be shown to be in their own self-interest. Spinoza himself has a very strong statement of the overridingness of self-interest:

> [E]veryone will, of two goods, choose that which he thinks the greatest; and of two evils, that which he thinks the least. . . . This law is so deeply implanted in the human mind that it ought to be counted among eternal truths and axioms.[33]

What environmental philosophers seem to be looking for, in contrast, is a basis for the judgment that some activities that are regarded as good for human beings should be curtailed *because* they are bad for other parts of nature. And we seem to have drawn a blank in looking for this in Spinoza's thought.

Can we patch up Spinoza's thought to make its central insights more amenable to reconstruction as a genuinely 'ecological' ethic? Much of what Spinoza says, we might think, reflects the circumstances and knowledge of the time rather than what we might claim as the timeless truth of the ethical system itself. Can it be brought into line with the contemporary scene in ways that would enable us to salvage it as a basis for an environmental ethic? For a start, we might try to break down the rigidity of Spinoza's conception of what does and does not 'share a common nature with us'. Is there not something arbitrary about the setting of boundaries between natures? Can we perhaps recognize, as he could not, a common nature between ourselves and at any rate some other species? But in so far as it rests on scientific knowledge, Spinoza's insistence on a difference in 'nature' between us and other animal species rests ultimately on difference in bodily structure. The availability of more information about bodily structure might uncover unsuspected continuities. But it could also be expected to reinforce Spinoza's perception of the differences between human beings and other animals. The concept of 'sharing a nature' is too imprecise here to allow of any confidence that advance in scientific knowledge would allow for an extension of the privilege of 'sharing our nature' to other kinds of animal.

There is, in any case, another aspect of Spinoza's talk of common natures which seems to go against any attempt to extend his notion of sharing our nature to non-humans. There is a nationalist strand in his thought which does not readily lend itself to the abstracting of common natures. Men's common nature that underlies their creation of a moral community is not something statically present to be discerned by reason. The morally significant common nature between men seems rather to be something that is *achieved* through the cultivation of reason. There seems little prospect of extending the relevant kind of nature-sharing to the non-human without departing completely from Spinoza's thought.

Stanley Benn has suggested that the range of 'right-bearer' might be extended to include animals capable of engaging in projects.[34] But on Spinoza's concept of a moral right the mere capacity to engage in projects would not suffice for membership of 'the moral community'. The moral significance of the shared 'nature' of human beings derives from their being able to collaborate in the pursuit of reason. Those who share our nature in the relevant way have to be not merely capable of engaging in projects; they have to be capable of realizing the advantages to them of accepting a common authority. Thus even the current generation of apes who have been taught sign language would not be moral 'right bearers' for Spinoza unless they were able, for example, to realize the advantages to them of being brought up like human children. The mere fact that they can talk to humans would not, I think, shake Spinoza's speciesism.

It is not clear that we can salvage Spinoza's thought here by pointing to advances in knowledge. We might, alternatively, try to reconstruct it by extricating it from its surrounding social circumstances. Spinoza in a contemporary setting, we might say, would surely see man's well-being as more bound up with that of other species than it was possible to foresee in the seventeenth century. It was at that stage possible to see the well-being of animals as quite independent of that of man. And it was plausible, therefore, to see concern for their well-being as anti-human, as expressing a preference for beasts; or else as expressing an exaggerated sense of our own importance in the universe. But the impact of human beings on the environment has of course accelerated since then. The capacity of animals to get on with 'enjoying their own lusts' depends to a greater extent than Spinoza would have thought possible on our decisions. And this, we might say, *makes* valid the perception of human beings as responsible for the rest of nature. Man's activity in the world has made it true. It is no longer, then, a matter of how we *ought* to perceive our position in the universe but of acknowledging what we have done.

Whatever the intrinsic merits of these points I do not think they can be utilized in a reconstruction of Spinoza's thought. The suggestion amounts to an adaptation of Spinoza's insights to take account of the 'humanization' of the world. I do not think it can withstand the intrusion of that perception. Spinoza would in fact, I think, reject this way of describing what has happened to man's relationship to the rest of nature. The supposed humanization of the world he would see as an illusion, as a product of the very anthropocentric perception that he castigates in the *Ethics*. What has happened, he would say, is a falsification of reality. And, whatever the remedy, it does not lie in treating as a fact the humanization of the world. To pursue this approach to the problem would, I think, amount not to an extension of Spinoza's system but to abandoning it.

There is, however, another way of approaching the challenge to reconcile Spinoza's thought with contemporary ethical insights about the environment. Spinoza's ethical system is centered in self-preservation, in the survival of human beings. He sees the pursuit of human activities as an extension and enrichment of this basic pursuit of self-preservation – the maintenance and enrichment of human existence. But he has a very definite conception of the kind of behaviour that in fact enhances human activity and hence human perfection. It is, basically, the cultivation of reason, the pursuit of truth. And the ultimate expression of this human endeavour is the intellectual love of God, a detached, selfless perception of things as parts of wholes and as modes of substance. These are the ends Spinoza sees as the achievement of human good, in relation to which concern with the goods of other species is an irrelevant, anthropomorphic distraction. But the contemporary human activities and ends that prompt the concern of the environmentalists are, of course, very far removed from these rationalist ideals.

We can, Spinoza says, use animals in any way useful to us. But his conception of what is useful to human beings would certainly exclude many of the projects taken in the contemporary scene to justify the use of animals. Some current goals which involve as means the exploitation of other species are very far removed from Spinoza's conception of the 'intellectual love of God'. Many of them could readily be brought under his criticism of men's distracted preoccupation with short-term goals pursued in the blindness of passion against their long-term interests.[35] Using animals in ways that are useful for us means, for him, using them in ways that enhance *our* activity, *our* perfection. But this human perfection characteristically issues in an attitude of reverence for the whole of nature. Spinoza's remark that we can use animals in any way we see fit has to be taken in the context of a much more limited range of exploitative acts than that since made possible by advanced technology. It may be that we could extract from a Spinozistic ethic a condemnation of *some* forms of human exploitation of other species. But again, this would be based on assessment of the activities involved from the point of view of human well-being – on the irrelevance, or, in some cases, the inconsistency of the end with Spinoza's paradigm of virtuous activity – the detached appreciation of oneself as part of nature, the 'intellectual love of God'. All that this yields is some restriction on what use of other species a contemporary Spinoza would recognize as enhancing human good.

We are here still within the limits of an ethic grounded in human interests. What has emerged is that, given Spinoza's conception of what are ultimate human goods, his claim that we can use animals in any way useful to us is less exploitative than might at first appear. But, it might be replied, all this presupposes that we are prepared to accept Spinoza's valuation of the contemplative life. And this, after all, is, in our contemporary scene, a restricted taste. Intellectual contemplation is not a basic human interest of a kind we could expect to impress at all those individuals and institutions whose current laying waste of the environment is a cause for concern.

It now seems, at best, that we can salvage for Spinoza some basis for the condemnation of the use of other species in pursuing trivial human ends, or ends that can be shown to be in conflict with the acquisition of what he calls 'intellectual love of God'. We can perhaps reconcile some of his heavier remarks about the use of animals with contemporary sensitivity to the exploitation of other species. But this is a far cry from being able to find in the *Ethics* a basis for concern with the *well-being* of the rest of nature.

Can we extract more than this from Spinoza? Can we utilize his thought in articulating what Rolston calls a 'genuinely ecological' ethic, which treats the non-human as in themselves proper objects of human moral concern?[36] Or are we to say that Spinoza's thought cannot be reconciled with the most important moral insights of the environmental movement? Environmental ethics wants to be able to express the moral judgment that human goods should in some circumstances be curtailed for the sake of what is good for the rest of the world; that the goods of other things should be taken into account in our moral judgments. If Spinoza's ethical system, making morality entirely species-relative, makes it impossible to say such things, then does not that in itself show that system to be defective as a vehicle for the articulation of contemporary moral consciousness?

But what exactly is it that Spinoza cannot express? And how crucial is it to a viable environmental ethic? Spinoza's system does make it impossible to extend the moral community to include other animals. And it also makes it impossible to conceive of ourselves as 'stewards of nature'; as morally answerable for its well-being. These ways of construing the claims of the non-human on our moral concern seem to be ruled out by Spinoza's rejection of the view of reason as relevant to the whole of reality and of man's rationality as conferring on him a privileged position in the scheme of things. He would regard them as unacceptably anthropocentric in their perception of man's position in the universe. But this emphasis on the point of view, as it were, of the non-human should perhaps itself be seen as a rather dubious element in contemporary debate on environmental ethics.

It is common for contemporary writers on environmental ethics to run together being an object of moral concern and being included in a moral community. The proper locus of moral questions, we are told, should shift from the interests of the human individual or even the species to the ecosystem. As Aldo Leopold puts it:

> The land ethic simply enlarges the boundaries of the community to include soils, waters, plants, and animals, or collectively, the land.[37]

> A thing is right when it tends to preserve the integrity, stability, and beauty of the biotic community. It is wrong when it tends otherwise.[38]

Studying Spinoza's version of the theme that man is 'part of nature' is of value here in casting doubt on the assumption that we need to extend the moral community, to treat the non-human as bearers of rights, etc. in order to treat the non-human as morally relevant. The non-human for Spinoza have no rights in any moral sense. What is not human is not part of our moral 'community'. It is quite wrong to suggest, as George Sessions has done, that Leopold and Spinoza share the view that the total system of nature should be seen as ethically fundamental.[39] Non-human things can be objects of moral concern. But this does not involve, as Leopold suggests, their *goods* entering into our moral judgments.

Let us look again at what is involved in Spinoza's repudiation of man's supposed privileged position in nature. This involves seeing reason as a 'speck' in the universe. But that speck is nonetheless absolutely crucial for the good of man. What this means is that truth becomes for Spinoza an absolute good, an 'intrinsic value'; although this value of course remains circumscribed as what is good *for man*. In the concern for truth, which plays so central a role in Spinoza's system, we have a 'human interest' with a difference. Spinoza's thought here straddles the gap that is assumed in much current debate on environmental ethics between 'instrumental', 'man-centred' values and 'intrinsic values'. Spinoza cannot say: 'Things (such as butterflies, whales, rain forests) are good, or have value, or have rights, independently of man.' But he can say: 'It is *good for man* to perceive things as independent of himself. It is good for man, that is, to perceive things as they really are. It is good for man to perceive things truly.'

## III Conclusion

We began with the apparent strangeness of Spinoza's juxtaposition of a morality circumscribed by human interests and repudiation of an anthropocentric perception of the world. By now, I hope, this juxtaposition should appear less strange. And, at the same time, the juxtaposition of stances characteristic of much current debate on environmental ethics can be seen as itself containing some tensions. A rejection of anthropocentricity in morals is quite often juxtaposed with what seems nonetheless a thoroughly anthropocentric way of perceiving the non-human. And this seems strange. We are, it seems, supposed to render morality less anthropocentric by humanizing the non-human. We are recommended to extend to animals the status of 'preference-havers' or 'project-initiators' and, on the basis of this, to regard them as 'right-bearers'. Again, we are urged to regard non-human things as bearers of 'intrinsic value', as things whose existence and preservation we ought to see as somehow good in themselves, independently of us, demanding that our own good be curtailed accordingly.

Consideration of Spinoza's *Ethics* casts light on what seems incoherent about this talk of non-human intrinsic values, and, at the same time, on what it is about this notion that captures something we very much want to be able to say. According to Spinoza, we cannot recognize any non-human intrinsic values. But we do not have to think that there are such values in order to be able to see it as a *human* good to perceive things as independent of ourselves: the contrast here is the one I adverted to earlier between (1) things are good, are of value, independently of human beings; and (2) it is good for human beings to perceive things as independent of them, that is, to perceive them as they really are.

Spinoza's approach captures also, I think, something of what is attractive in the suggestion that concern for the ecosystem should replace concern for the self construed as set over against and competing with its environment. Rolston makes the point in a picturesque way:

> Ecology does not know an encapsulated ego over against his environment. . . . The self metabolically, if metaphorically, interpenetrates the ecosystem. The world is my body.[40]

Spinoza's system gives metaphysical content to this fanciful sounding idea of perceiving the world as my body. But it is a matter of transcending a distorted perception of our true position in nature; not of treating the non-human as bearers of rights or values in relation to which human rights and values should be curtailed. In a letter, Spinoza compares our fragmented, distorted perception of the world to that of a worm in the blood, perceiving parts without grasping their relation to wider wholes:

> Let us now, if you please, imagine that a small worm lives in the blood, whose sight is keen enough to distinguish the particles of blood, lymph, etc., and his reason to observe how each part on collision with another either rebounds, or communicates a part of its own motion, etc. That worm would live in this blood as we live in this part of the universe, and he would consider each particle of blood to be a whole, and not a part. And he could not know how all the parts are controlled by the universal nature of blood, and are forced, as the universal nature of blood demands, to adapt themselves to one another, so as to harmonise with one another in a certain way.[41]

Human individuals, for Spinoza, are located in wider systems to which they stand in the same kind of relationship as that in which, in his example, the particles of the blood stand to the blood. And the cultivation of reason involves the mind's coming to grasp the body of which it is the idea, and hence itself, as parts of a whole.

The concern for truth is here not without moral relevance. Such a change in perception can be expected to have some influence on behaviour. Children educated to regard themselves as 'but a part of nature' would, for the most part, surely, orientate themselves differently towards other species from those who are explicitly taught that man holds a privileged position in the universe. At least some of our exploitative responses to the non-human rest on the implicit belief that the rest of nature exists for us; and can be expected to wither away if this implicit belief is brought into the open and rejected.

Spinoza, I have argued, would endorse the condemnation of any exploitation of the rest of nature that rests on an implicit belief in man as the centre of the universe. But what would concern him about it is precisely that it does express a false belief about man's place in nature; that it expresses an ignorant, distorted attitude. As Stuart Hampshire puts it, such behaviour manifests a false philosophy of dominance, 'as if men were situated in the world as in their own garden'.[42]

Morality, on the Spinozistic approach, remains circumscribed by what is good for the human species. But it is good for human beings to perceive themselves as parts of wider systems, as parts of wholes; to perceive themselves as they really are.

Such an ethic remains man-centred. Spinoza would not endorse Session's condemnation of the failure of what he sees as the 'humanistic approach to Nature' to put the stability and well-being of the total system before the survival and well-being of human beings:

> [A]n ostensibly humanitarian rationale, untempered by a knowledge of population dynamics, provided the impetus for the technological assistance which led to the dramatic declines in death rates in the 'underdeveloped' countries after WWII, thus contributing significantly to their unparalleled population explosions. And now when this added population is starving in great numbers, a humanistic attitude dictates trying to feed all of these people together with their exponentially increasing progeny, again blissfully ignoring the ecological consequences of this course of action....[43]

A Spinozistic ethic, in contrast, would treat the survival and well-being of human beings as the proper locus of morality, regarding any attempt to shift the locus of morality beyond human concerns to the well-being of the total system as itself expressing an anthropocentric perception of man's place in the universe.

We may still feel that Spinoza's ethical system can be judged as deficient from the standpoint of contemporary concern with man as part of his environment. Spinoza sees human beings as parts of wider systems. But his rationalism precludes not only anthropocentric perception of man as the centre of the universe, but also the recognition of any bonds of sympathy between ourselves and the rest of nature. Again, without his account of the emergence of the state, his dichotomy between the human and the non-human parts of the total system of nature may well strike us as altogether too sharp. But, from the other direction, we can also see, I think, from the standpoint of Spinoza's version of the theme of man as 'part of nature', some gaps in contemporary argumentation for extending the moral horizons to take account of the non-human. Rejecting any privileged position for man in the universe does not commit us to regarding the non-human as having moral rights; or to introducing any new, epistemologically puzzling, non-human intrinsic values. It can be seen as just a reaffirmation of a pretty old intrinsic value – the concern for truth. Whether Spinoza's own explanation of exactly why truth is good for human beings can be salvaged in a form that would be relevant to contemporary moral consciousness is, of course, another matter.

## Notes

1 See esp. George Sessions, 'Anthropocentrism and the Environmental Crisis', in *Humboldt Journal of Social Relations*, Vol. 2, Fall/Winter, 1974; 'Panpsychism versus Modern Materialism: Some Implications for an Ecological Ethics', expanded version of paper delivered at the conference on 'The Rights of Non-Human Nature' held at Pitzer College, Claremont, California, April 1974; and 'Spinoza and Jeffers on Man in Nature', *Inquiry*, Vol. 20 (1977). The implications of Spinozism for ecology are also discussed by Arne Naess in 'The Shallow and the Deep, Long-Range Ecology Movement: A Summary', *Inquiry*, Vol. 16 (1973) and 'Spinoza and Ecology', *Philosophia*, Vol. 7 (1977).

2 See esp. the treatment of individual bodies in the digression after Prop. XIII of Pt. II of the *Ethics*.

3 Sessions notes the existence of this opposed theme in Spinoza's thought in his informal newsletter 'Ecophilosophy', No. 1, April 1976; and in 'Spinoza and Jeffers on Man in Nature', op. cit., p. 507. But he does not recognize its interconnection with the theme of man as part of nature.

4 Note I to Prop. XXXVII, Pt. IV of the *Ethics*; in R. H. M. Elwes (Trans.), *The Chief Works of Benedict de Spinoza*, Vol. II, Dover, New York 1955, p. 213.

5 Note to Prop. LVII, Pt. III; Elwes, op. cit., p. 170.

6 Prop. XXIX, Pt. IV; Elwes, op. cit., p. 206.

7 Proof to Prop. XXIX, Pt. IV; Elwes, loc. cit.

8 Prop. XXIV, Pt. IV; Elwes, op. cit., p. 204.

9 Note to Prop. XVIII, Pt. IV; Elwes, op. cit., p. 202.

10 Corollary 2 to Prop. XXXV, Pt. IV; Elwes, op. cit., p. 210.

11 Note to Prop. XXXV, Pt. IV; Elwes, loc. cit.

12 Note 2 to Prop. XXXVII, Pt. IV; Elwes, op. cit., p. 214.

13 *A Theologico-Political Treatise*, in R. H. M. Elwes (Trans.), *The Chief Works of Benedict de Spinoza*, Vol. I, Dover, New York 1951, p. 200.

14 Ibid., pp. 200–1.

15 Ibid., p. 201.

16 Ibid., p. 202.

17 Ibid., pp. 202–3.
18 *Ethics*, App. to Pt. 1; Elwes, op. cit., pp. 75–76.
19 Ibid., p. 76.
20 Loc. cit.
21 Ibid., p. 79.
22 Ibid., p. 81.
23 Locke, *An Essay Concerning the True Original, Extent and End of Civil Government*, Ch. II, Everyman, London 1924, p. 120.
24 Leibniz, *Discourse on Metaphysics*, Ch. 19; in G. R. Montgomery (Trans.), *Leibniz: Basic Writings*, Open Court, La Salle 1962, p. 34.
25 Ibid., Ch. 12, p. 19.
26 Ibid., Ch. 35, p. 59.
27 Ibid., Ch. 35, p. 60.
28 Ibid., Ch. 36, p. 61.
29 *Ethics*, Prop. XXIV, Pt. IV; Elwes, op. cit., p. 204.
30 J. Passmore, *Man's Responsibility for Nature*, Duckworth, London 1974.
31 R. Routley, 'Is There a Need for a New, an Environmental Ethic?', *Proceedings of the XVth World Congress of Philosophy*, Varna 1973, p. 210.
32 Holmes Rolston III, 'Is There an Ecological Ethic?', *Ethics*, Vol. 85 (1975), p. 103.
33 *A Theologico-Political Treatise*, Ch. XVI, Elwes, op. cit., p. 203.
34 S. Benn, 'Personal Freedom and Environmental Ethics: the Moral Inequality of Species', paper to the World Congress on Philosophy of Law and Social Philosophy, St. Louis, Missouri, Aug. 1975.
35 *Ethics*, Prop. LXVI, Pt. IV; Elwes, op. cit., p. 231.
36 Rolston, op. cit.
37 A. Leopold, 'The Land Ethic', in *A Sand County Almanac*, Oxford University Press, New York 1949, p. 219.
38 Ibid., p. 240.
39 George Sessions, 'Panpsychism versus Modern Materialism', op. cit., p. 24; 'Anthropocentrism and the Environmental Crisis', op. cit., p. 10.
40 Rolston, op. cit., p. 104.
41 Spinoza, Letter to Oldenburg (XXXII); in A. Wolf(Ed.), *The Correspondence of Spinoza*, Cass, London 1966, pp. 210–11.
42 S. Hampshire, *Two Theories of Morality*, Oxford University Press, Oxford 1977, p. 91.
43 George Sessions, 'Anthropocentrism and the Environmental Crisis', op. cit., p. 10.

86

# SPINOZA AND ECOLOGY REVISITED

*K. L. F. Houle*

Source: *Environmental Ethics*, 19 (1997), 417–31.

Spinoza has been appropriated as a philosophical forefather of Deep Ecology. I identify what I take to be the relevant components of Spinoza's metaphysics, which, at face value, appear to be harmonious with Deep Ecology's commitments. However, there are central aspects of his moral philosophy which do not appear to be "environmentally friendly," in particular the sentiments expressed in the *Ethics* IV35CI and IV37SI. I describe environmental ethics' treatment of these passages and then indicate what I take to be a more satisfactory route toward "ecologizing Spinoza."

## I Spinoza and environmental philosophy

Given the relative "greenness" of environmental philosophy, it is no surprise that a good deal of initial work in the discipline has been devoted to exploring the history of philosophy for "precursors" whose metaphysical or ethical ideas seem to share the spirit of contemporary ecosophy.[1] From these historical projects, an ecological "perennial philosophy" is mapped from past to present, sometimes with the aim of affording credibility to a new philosophical area, sometimes with the aim of celebrating the immense illumination that certain philosophers can shed on those new places.

Spinoza has a central part in that history.[2] Arne Naess declares that "No great philosopher has so much to offer in the way of clarification and articulation of basic ecological attitudes as Baruch Spinoza."[3] Among the extraordinary plurality of perspectives into which Spinoza's thought has been smuggled, it also makes an appearance in ecological thought. The appeal of "Spinoza as ecologist" largely proceeds from elements of Spinoza's metaphysics. In this paper, I describe which of those elements have been seized upon by the contemporary environmental philosophy movement and then explore the validity of this appropriation given some apparent "anti-environmental" difficulties present in his moral philosophy, particularly *Ethics* IV35C1 and IV37S1.[4] I critically describe the treatment of these difficulties by environmental philosophers. I end by proposing an alternate route through Spinoza's *Ethics* to ecology, a route which I consider to be more faithful to the sentiments of Spinoza's moral philosophy. My chief motive here is *also* to afford credibility to contemporary environmental philosophy.

## II Spinoza as ecologist: the appeal

The contemporary environmental philosophy movement shares the conviction that the natural environment is, in fact, in a state of crisis. It also shares the conviction that an alternative future path must be taken in order to avoid the devastation of the entire system of nature. This alternative path, such as is followed by the Green movement, reorients the relationship of humankind to the Earth as members within, rather than above, the community of beings. Central to achieving this alternative path is the rejection of the traditional "anthropocentric perspective"; that is, valuations of aspects of nature which are solely a function of human wants and needs. This anthropocentric value system is built upon the premise that humankind deserves privilege and bears exclusive value by virtue of its exclusive insight, intellect, or any other characteristic. Even the "stewardship" paradigm, which attempts to reorient human compassion and responsibility to non-human nature is rejected by deep ecologists as unduly anthropocentric in that it continues to operate on the assumption that humans can and do know what is good for other species, and thus it retains for humankind a privileged position among the Earth's stakeholders or community members.[5] They argue that what is required is an ecocentric perspective and a radical "holistic/preservationist" paradigm which at least purports to accommodate the needs and values of *all* living beings.[6]

There are three aspects of Spinozism which immediately appear to be, at least on the face of it, amenable to the general commitments of Deep Ecology. These are his metaphysical monism, his pantheism, and the democracy of virtue/power in nature.

First, Spinoza collapses the substantive dualism of mind and body by declaring all things to be of one substance, which is God (*Ethics* I:14). Given that mind or rational thought has been heralded as the quintessential human attribute justifying anthropocentrism, a metaphysical monism like Spinoza's is welcomed both for its extension of the community of rational beings and for its ontological parity of mind and matter. What environmental philosophy celebrates in Spinoza's ontology "is his constant insistence that man has no privileged position in nature."[7] Environmental philosophy strives to overturn that attitude toward nature, often associated with Bacon and Descartes, which allows and even to some extent compels, unrestricted use for our own advancement. An ontology such as Spinoza's, which appears to embed and thus constrain human activity, is therefore considered a significantly more ecological one, given that the environmental crisis is conceived as a byproduct of the history of this rationalized "unrestricted use."

Second, Spinoza's God "comes down" from his transcendent Judeo-Christian position and sheds any anthropomorphism. Although will and intellect may pertain to humankind, "neither will nor intellect pertain to the nature of [Spinoza's] God" (*Ethics* I:17, sch). He is impersonal and thus infinitely more ambivalent to humankind than he is within traditional Christian doctrine. Spinoza identifies that traditional Judeo-Christian "human God" as a species of human prejudice, equipped to "direct the whole of Nature so as to serve his blind cupidity and insatiable greed" (*Ethics* I:app.). Spinoza's God, *Deus Sive Natura is* all things and thus impartial to all things. Spinoza's God/ Nature concept has been variously named mysticism, materialism, monotheism and most often pantheism.[8] One thing is definite: it revolves around God's immanence in and coequivalence with nature. This is considered crucial to the holistic paradigm. The resacralization of the natural world—the recognition and love of trees and sky as God rather than as creations of an absent God—is seen as a key route to the respect and protection of those elements of nature.[9] Also, Spinoza's God of infinite attributes adds a humbling note. Under Spinoza's scheme, the universe remains to us infinitely mysterious and ungraspable in its completeness. This is sharply contrasted by the idea of nature implicit in rationalism: that nature is somehow complete or completable. In such a universe, exhausted by mind and/or matter, humankind, which possesses those two attributes, tends to believe that, in principle, it can arrive at an exhasutive understanding of the whole: it can complete the project of nature. The ignorance of humankind (indeed all beings) within Spinoza's endless, evolving universe is, on the other hand, a necessary fact. As Hampshire reports:

> Spinoza thought of humans as greatly limited in their powers to grasp and survey the natural order which must outrun their powers of perception and understanding. His philosophy is by implication a polemic against anthropocentrism. He does represent human intelligence as a not unnatural . . . elaboration of structures found elsewhere in nature; and he always insists that our perceptual apparatus and our intelligence cannot exhaust the infinite variety and extent of nature.[10]

Through Spinoza's ontology we arrive at an epistemological constraint which calls into question the validity of the "resource conservation and development" and "stewardship" paradigms, whose operating assumption is that there is an adequate human knowledge of nature.

Another part of Spinoza's philosophy which find currency in Deep Ecology involves his notion of "conatus." To Spinoza, "Everything, insofar as it is in itself, endeavours to persevere in its own being" (*Ethics* III:6). As far as environmental philosophy is concerned, what is significant here is that Spinoza apparently does not cleave off animalium from humankind with respect to the potential for self-empowerment; rather he appears to be applying mental predicates to the entire biological field. These include the potential for active and passive emotions, understanding, and degrees of genuine freedom. Arne Naess suggests that Spinoza's "conatus" *can* "meaningfully be applied to all living beings"[11] and the importance of doing so is partly indicated by Hampshire:

> The notion of conatus, of which there is no equivalent in Descartes or in purely mechanical and atomistic cosmologies, is exactly the concept which biologists have often demanded as essential to the understanding of organic and living systems.[12]

What is celebrated about conatus is that, like pantheism, it is perceived as a means for setting back within nature a matrix of badly needed restraints. My recognition of my own desire to arrive at self-realization, my understanding of the good I experience with that, combined with the understanding that like myself, "nothing by its own nature desires to destroy or harm itself" (*Ethics* III:4) amounts to my recognizing and respecting the conatus of other beings. This respect may manifest itself in my not harming them.

A closely related idea is that of intrinsic value, a prized but highly problematic concept in environmental philosophy. Sessions writes that "Spinoza makes it clear that all beings exist for their own sakes, for their own particular and individual forms of completion and self-realization, and not for the sake of anything else."[13] The undesirable "for the sake of anything else" which Naess and Sessions are referring to is the Baconian/Cartesian divisions of mind and body, man and nature, and the instrumental rationality which accompanies these divisions. Instrumental value and the rationality of use are generally considered as inadequate schemes for protecting the natural environment.

What is applauded by deep ecologists in these three aspects of Spinozism, which ultimately are seen to be a solidarity in a "more objective non-anthropocentrism,"[14] is the fact that, like themselves, "Spinoza sets himself firmly against a man-centred perception of the universe; against the view that man has any privileged place in the universe."[15]

## III Spinoza as ecologist? Some difficulties

When we examine Spinoza's moral philosophy, however, the transition from seventeenth-century rationalism to the aims of contemporary environmental ethics is significantly less smooth. As Lloyd (correctly) points out: "Spinoza manages to combine a strong rejection of anthropocentric perception with an equally strong affirmation of a man-centred morality."[16] There are two particular passages in Spinoza's *Ethics* which could be taken as fairly conclusive evidence of his anthropocentric anti-environmental morality, and these are precisely the ones that environmental philosophy must come to terms with in order to honestly carry out the project of ecologizing Spinoza:

> It is plain that the law against the slaughtering of animals is founded rather on vain superstition and womanish pity than on sound reason. The rational quest of what is useful to us further teaches us the necessity of associating ourselves with our fellowmen, but not with beasts, or things, whose nature is different from our own; we have the same rights in respect to them as they have in respect to us. Nay, as everyone's right is defined by his virtue, or power, men have far greater rights over beasts than beasts have over men. Still I do not deny that beasts feel: what I deny is, that we may not consult our own advantage and use them as we please, treating them in the way which best suits us; for their nature is not like ours, and their emotions are naturally different from human emotions. . . .
>
> (*Ethics* IV:37S1)

Because rational obedience to virtue requires that we seek what is useful to ourselves, and, since nothing which "is entirely different from our own nature can help or check our power of activity" (*Ethics* IV:29), the achievement of virtue likewise follows our engagement with essences most like our own: "There is no individual thing in the universe more advantageous to man than a man who lives by the guidance of reason" (*Ethics* IV:35, cor. 1). These statements seem to run directly against the grain of the environmentalist sympathies outlined above. Here Spinoza seems to be limiting the moral community entirely to our own kind—those who can live under the guidance of reason.

Now, the chief strategy in the environmental philosophy literature aimed at bringing about a reconciliation between these and the more "ecological" thoughts of Spinoza, seems to involve reinterpretations of Spinoza's terms: "nature," "useful," "persevering in one's being," and "laws of nature"—reinterpretations that are generally true to the meanings that these terms already have within the ecological paradigm. A "nonanthropocentric unity" between Spinoza's metaphysics and moral philosophy is then recoverable.

Bill Devall and Georga Sessions seem particularly guilty of misrepresenting Spinoza's ideas in their own terms. Sessions' lament in "Spinoza and Jeffers on Man in Nature" is that a philosophy which focuses attention on "man" does so at the expense of "nature." Western subjectivism, he argues, has led us to see "nature" as an aspect of our own experience, rather then what it *really is*: "an independent resisting structure . . . which we tamper with at our peril."[17] Deep Ecology, while applauding Spinoza's "ontological unity of all things," still operates with a human/nature dualism: Nature is something other than ourselves; something for which we could cultivate a "detached appreciation,"[18] something "out there" which we harm, or go to. Deep Ecology has consistently identified certain culprits of environmental degradation: reason/transcendental religiosity/egoism/urbanization/instrumental rationality/patriarchy, etc. The challenge for "man" is to redeem himself by shedding those nasty characteristics and breaking rank to join nature and the other "goods" like wilderness and solitude, where he will attain enlightenment by identifying with other beings. Spinoza is read as equipping humankind with the means to do just that:

> Spinozistic understanding involves a way of being and living with *the other*. Of course, Taoists and Buddhists have long held that to overcome society-produced ego and find true Self, one must at least temporarily get away from other humans and align one's Self with the Tao in the wilderness.[19]

Neither this dualism nor any other are features of Spinoza's own thought. The cardinal point in Spinoza's philosophy *is* nature as a unity. Nature is ourselves and everything else. But as such, it is neither good nor bad: to say so is to reveal the ignorant and prejudiced products of our own desires. Deep ecology's value system is a product of its prejudiced assumptions about good and bad and its recommendations (such as that the human community is a pollutant of sorts on our ecological hearts) arise directly from them: they are not discoverable as such in Spinoza's writings.

Sessions' solution to the "difficulties" that Lloyd identifies is to construct a Spinozist unity of all things solely under the attribute of biological, organic extension:

> Spinozism was intended to be an open-ended system in the sense that it delineates the necessary structure of the cosmos; the specific content under the attribute of extension is spelled out by an increasingly adequate theoretical science.[20]

Sessions says that Spinoza "insisted that man is a biological organism driven exclusively by the desire for self-preservation."[21] Here he sees biology as the basis of our true community, rather than one built around reason/body boundaries. This point is the central message of ecology. Arne Naess in *Freedom, Emotion and Self-subsistence* also draws a boundary around the community of biological beings. He writes, "... self-preservation and power are applied not only to humans, but to animals and perhaps all living things... [except] beings of the mineral kingdom."[22] Again, because Spinoza's nature is all things and the ides of things and because the beings of the mineral kingdom are knowable under the attributes of thought and extension, in principle they too are parts of nature to whom the predicates of power, virtue, persistence in their own essence, joy, and sorrow belong, if only by a lesser degree. Naess' division, then, does not seem justified.

Furthermore, this narrow reading of Spinoza's nature and its corollary—the understanding of humankind as organic modes—is a misinterpretation of the Spinozistic terms *self-preservation* and that which is *useful* to us in rather Hobbesian terms. Lloyd views Spinoza's ethical system as "centred in self-preservation, in the survival of human beings."[23] Case in purely naturalistic terms, Sessions, Devall and Lloyd are able to excuse the problematic anthropocentrism as a function of the "abysmal biological ignorance of the seventeenth-century Western society"[24] rather than a function of Spinoza's true belief. In other words, the "necessity of nature" to which Spinoza refers is none other than the laws of [extended] nature which science and more recently, ecology has enunciated.

Because empowerment or self-preservation consists in understanding ourselves and having adequate ides and, since having adequate ideas consists in understanding either the essence of a thing (intuitively) or its proximate cause (*Ethics* II:1S2), knowing these scientific laws of nature, which supply explanations of the proximate causes of extended things, is, under Naess and Sessions' readings, a means to empowerment. Direct intuitive understanding, Spinoza's third level of knowledge, is seen as too "rare, noncommunicable" and vague, mired in such questionable notions as God and substance rather than such organic tangibles as biomass and symbiosis.[25] Thus, the scientific approach practically guarantees access to knowledge of nature and thus to the kind of enlightenment which Spinoza was recommending.[26]

It is overtly suggested that had Spinoza himself had access to biological knowledge, he would not have made the unsightly anthropocentric comments noted above. Sessions laments:

> Christian man, intent upon the destiny of his own soul, and increasingly urban, drew back from too great an intimacy with the natural, and . . . this is a particularly apt description of Spinoza in that, for all his prescience and innovativeness, he remained a typically "civilized" urbanized western man; as a result, his ultimate insight suffered accordingly. At once removed from the biological world . . . he was the very antithesis of the Taoist sage, seeking enlightenment in the solitude of mounting wilderness; reading the unity of God/Nature/Man directly from the living organic world.[27]

Had he left the company of his fellow men for grassland and the company of wolves, he would have understood that other beings are as "useful" to us as our own kind: they make the oxygen we breathe, they clean the rivers we swim in, they pollinate our roses. . . . He might have recognized the extent of "common nature" in nature: the cycles of carbon and nitrogen and oxygen we all participate in; the forces of electron attraction which affect us all. . . . He would have had a truer understanding of nature.

> Non-human nature is not "useful" to us merely as a source of food, shelter, and the like, as Spinoza seemed to think. And the "state of nature" is not strictly one of competition for survival, but rather one of symbiosis. It is clearly to our "advantage" as individuals and as a species that the delicate equilibriums of ecosystem functioning, upon which our lives literally depend, remain viable.[28]

From these comments, we can see that the strategy of upgrading Spinoza's morality into an ecological ethic has been to argue that he was simply mistaken about the extensive organic relationship of man to nature. But it is also clear, first of all, that the rapprochement is founded upon a non-Spinozistic notion of nature (especially attending to the aspect of extension). Spinoza would have surely known that nonhuman nature is "useful to us as a source of food, shelter and the like." Nevertheless, his statement is *Ethics* IV:35c1 that there is nothing more useful to man than reasoned man repells a simple Hobbesian interpretation of "useful," *even* with respect to the basic human wants and needs which are secured for us by the association of citizens.

Rather, what is really "useful" or "good" for an individual is that which causes an increase in one's "perfection," "freedom," or what Naess calls "self-realization"—that by which one is active, rather than passive—and this increase *must* be understood in terms of the highest possible sense of apprehending and understanding the love of God, which is all things.

Naess does not make the common mistake of supposing Spinoza to mean by conatus something like mere physical survival of the individual. He notes that striving under the reason of nature, understood in the broadest Spinozist way, may demand compromising physical acts of a being which amount to their physical self-limitation or even destruction.[29] However, this demand too eventually amounts to little more than a less Hobbesian, less egoistic rewriting of how one is to proceed as an individual *among* the community of biological beings in order to preserve that community, or in order to enable the majority to fulfill their own self-empowerment. Naess' statements that "the establishment of symbiosis . . . marks a gain in power" and that "the freedom of the individual ultimately requires that of the collectivity"[30] are certainly more holistic, less individualistic models of politics and community, and are well informed by recent ecological insight. I do not deny that this insight could be accommodated into Spinoza's general doctrine of *social justice*. After all, Spinoza's *Ethics* and *Theologico-Political Treatise* do not solely address spiritual reform. One could not make sense of the entire Book IV of the *Ethics* if one did not take seriously Spinoza's commitment to practical matters like the welfare of humankind.[31] There certainly is something like the common Hobbesian notion of "useful" in Spinoza. However, there *is* an important difference between this "useful" and Spinoza's special sense of it which implies a very rich instrumental relationship between social welfare and spiritual reform.

Recall the three levels of virtue in Spinoza: "nature virtue" is the unreasoned action of the body found in the state of nature whereby, determined in its reactions by external constraints like thirst and fatigue, the being is a slave.[32] Human action in accordance with reason Spinoza calls "social virtue." This second level of virtue drives the formation of the state and is based both in the active desire for friendship, peace and concord and in the reasoned means to secure relief from the enslavement of the state of nature. The individual here abides by social custom, and is willing to subject himself to the kind of habits and proscriptions which Spinoza recommends in Book IV. He does so because, "the free man desires that other be free too."[33] Thus, the chief characteristic of the "man who is led by reason" is fortitude, and it has *two* facets which Hallet suggests are key to making sense of Spinoza's commitment to social justice and spiritual emancipation:

> Fortitude, understood as strength of mind . . . is marked by the active attitudes of temperance and sobriety [which] are conducive to the preservation of one's own being. . . . Fortitude, understood as nobility of mind . . . is marked by the active attitudes of moderation and mercy . . . and the desire, under reason, to help others in friendship.[34]

Under the guidance of reason, the individual actively engages with the human community and seeks to bring about its good as part of his own.[35] The highest level of virtue in Spinoza, "understanding or free virtue," is that which originates in reason. The individual here lives according to the laws of his own nature. As Bidney remarks, "In this sense, the free man transcends social morality (social virtue) and goes beyond the common notions of good and evil."[36] To this end, "man's greatest instrument is the state."[37] Taken at face value, this seems to be saying something rather obvious, like: citizenship frees individuals from constraints by providing for their basic wants and needs (and certain "higher" needs like fidelity and fraternity) and as such facilitates their self-perseverance.

But in Spinoza there seems to be something less obvious. And, here is where we can reconnect the dangling end of the miscast "Spinoza as ecologist." I have described how Spinoza has been taken to task for not appreciating the extent to which non-citizenry ("nature") is party to the fulfillment of the basic wants and needs of human beings. In addition, there is the suggestion that citizenship (as opposed to wildernesship?) actually thwarts its members' self-fulfilment.[38] The obvious solution is simply to extend "the state" to include "the state of nature." Because "the state" is "man's greatest instrument," social virtue must embrace a number of constraints with respect to this broader realm just as Spinoza's overlay of a social justice already applies to the realm of state. Human beings under the guidance of reason, and marked by the character of fortitude, ought willingly to conform to habits and prohibitions within that expanded "state" such as are necessary to ensure that what is "useful" to themselves will be accessible to their fellowkind. It is good both in the ecological sense of absolutely necessary for human survival, and possibly in the Spinozistic sense of "conducive to enlightenment." Yet even this ecologized Spinoza is still embedded in instrumental conception of human values—albeit a richer versions than, say, hedonist calculus and a less self-assured version than, say, rationalism. The position resembles an expanded resource and conservation paradigm, *not* a "nonanthropocentric morality" or an "intrinsic value" scheme. It doesn't involve the discovery of a deep ecologist *in Spinoza*, at least insofar as I understand what a deep ecologist is.

## IV Through the human community to ecology?

Enlightenment is "the knowledge of the union existing between the mind and the whole of Nature" (*Ethics* IV:26). That empowering understanding amounts to the recognition and love of God. Melamed suggest that this enlightenment, which in Spinoza's "intuitive knowledge," consists in "visualizing the particular and the general simultaneously"[39] in a gestalt-like moment which is not mediated by reason. I take it that direct experience of the particular is a crucial aspect of this kind of knowing.[40] Such knowing is a very joyful revelation. The particulars which cram our lives are, in principle, conducive to our perfection, rather than being inhibitive.[41]

Several things follow which can be brought to bear on the Spinoza and ecology question. First, the particulars in nonhuman nature are potential sites of our empowerment. Nevertheless, they must be experienced directly. In Spinoza one might read an exhortation to actually engage oneself with nature, which, if understood in Deep Ecology's terms as wilderness, could provide an argument for a basic preservationist/conservationist ethics whereby what one ought to do is ensure that the variety of species and particulars in nonhuman nature are protected.[42] It seems to me that the anti-human, zero-access wilderness preserves which have been successfully staked out in North America would not be supported here. However, I'm not entirely sure whether one could differentiate between, say, in situ or museum experience.

What's more, what one experience in these particulars of nature is not "the laws of nature" as described by the science of ecology. I do not understand (in Spinoza's sense of the word) magnetic fields and interspecies competition. I do not even experience "species." I experience particulars, such as a brown oak leaf, a spotted mule deer, or Mount Yoho, at least the parts of them that are under my hiking boots. It is their proper names, Curely suggests, and not the laws which they instantiate, which come the closest to Spinoza's notion of intuitive understanding to capturing the adequate idea of a thing. The laws of ecology which one can learn completely outside the experience of biosystems "out there" are general, rather than adequate ideas. Spinoza would say that insofar as they are not adequate ideas, they are confused. They are about "suchness" rather than "thisness." This confusion raises the question of whether or not scientific knowledge, gained under experimental control, is conducive to intuitive knowledge or inhibitive to it. Curley reads Spinoza as suggesting the former, he even suggests that such general ideas as "mass" are necessary before we can understand the essence of a thing.[43] If we accept Melamed's characterization of the dual nature of understanding, then perhaps scientific knowing "primes" us to see the universal in the particular. But, depending on the degree of one's commitment to the particular, depending on the motive behind approaching the particular, it might also get in the way. For example, is this bit of vegetation a sample of *Monarda augustifolia* which you might use interchangeably with the next in order to arrive at a general thesis about this *kind* of herb or an adequate pot of tisane, or, is it a unique individual which you eventually recall is a member of the general idea *Monarda augustifolia*? Operating with particulars under the first kind of commitment, as one might discover in the field ecologist, seems to me to habituate one against the discovery of the uniqueness of each thing, pressured as she is by the activities of categorization and induction.[44]

Even if one is not approaching nonhuman nature as a scientist, the nature of nonhuman nature is such that it appears to us less diverse than it actually is.[45] Thus, looking at an oak tree one sees a collection of what one takes to be roughly identical particulars. Even one's direct experience of each and every one of those individuals, even a genome analysis, may not bring one to seeing their uniqueness, as the relationship between the leaf and the human being is quite simple; less passionately reciprocal.

Finally, in Spinoza one discovers an exhortation to engage oneself with nature understood in the second, Spinozistic sense of "all there is." More specifically, Spinoza urges our engagement with our own kind. Here we arrive again at that anthropocentric *Note 1*. But we simply have to make sense of the special role that humans engaging humans as particulars have in the Spinozist scheme of enlightenment. I do not think we are free to discard it as an empty vestige of the history of egoism which impedes human empowerment as nonhuman nature does not. To do so is to miss a crucial point of Spinoza entirely.

In Spinoza, I understand the overriding virtuous stance as a human being to be one of respect, of acceptance, of reconciliation, of *Gelassenheit*. Respect is the active refusal to come to a judgment or a conclusion about a circumstance or an individual—both to realize the vanity of one's assessment of good or bad and constantly to recognize that one is never in the sort of privileged position that allows one to understand the whole situation or the whole individual.[46] This respect is the conviction behind Spinoza's refusal to set down his system of universal norms; rather, he suggests the cultivation of a particular attitude which is conducive to loving everything, loving God. Curley calls this as "hesitating from committing the is/ought fallacy."[47] It does not require any one concrete action, nor forbid any. Rather, it requires "due consideration . . . to the facts . . . and to the response of the other in determining what it is precisely that reason dictates."[48]

In this sense, Spinoza's moral theory is relativistic, or, as I prefer to call it, an "ethical empiricism." The fully active ethical individual is neither an egoist nor a principled "objectivist." Spinoza's objectivism, which one might call "the moral truth" is one and the same as being in state of perfection, of loving and understanding God. One arrives at this state through the love of particulars, in the most common case, through the active respect for each human being. Living among human beings teaches one, unlike living among oak leaves, the precariousness of constantly arriving at judgment in the face of indeterminacy. On the other hand, if one consciously and constantly cultivates the "affect of the mind" which Iris Murdoch calls "active attention to detail,"[49] then one *can* become more ethical, in Spinoza's terms, more perfect. Among humankind that task is experience as inexhaustible. Individual human beings present to other individual human beings a limitless moral challenge, a very tangible limitless understanding which resists closure the way a snippet of *Monarda augustifolia* does not. That is because we are not entirely mired in the mode of identification and classification. Because among our own kind we are largely freed from the task of classification,[50] we can become engaged in the deeper moral challenge of understanding each individual essence and the essence of humankind. Because much of our diverse training as human beings, both biological and social, amounts to enabling us to intuit universals, it is crucial that we also are "trained" to intuit particulars as particulars. The need for this trained intuitive ability explains why "no man is more useful to man than man." It is for this reason that what I take to be the Spinozistic moral sense, this "anthropocentrism," is a fundamental part of the enabling of the human species to be loving ecologists, and furthermore, is not untrue to the Spinozistic "nonanthropocentric" ontology.

## Notes

1 Among them, most notably are the "process philosophers": Heraclitus, Whitehead, and Spinoza. See Susan Armstrong-Buck, "Whitehead's Metaphysical System as a Foundation for Environmental Ethics," *Environmental Ethics* 8 (1986): 241–59; Charles Birch and John B. Cobb, Jr., *The Liberation of Life: From the Cell to the Community* (Cambridge: Cambridge University Press, 1983); Bill Devall and George Sessions, *Deep Ecology* (Salt Lake City: Gibbs M. Smith, 1985), pp. 236–42.

2 The leading spokesperson for this merger of the history of environmental thought and Spinozism has without question been Arne Naess. See "The Shallow and the Deep, Long-Range Ecology Movement, A Summary," *Inquiry* 16 (1973): 95–100; "Spinoza and Ecology," Sigfried Hessing, ed., *Speculum Spinozanum (1677–1977)* (Boston: Routledge and Kegan Paul, 1977), pp. 418–25; "Environmental Ethics and Spinoza's Ethics: Comments on Genevieve Lloyd's Article," *Inquiry* 23 (1980): 313–25. See also George Sessions, "Spinoza and Jeffers on Man in Nature," *Inquiry* 20 (1977): 481–528; Genevieve Lloyd, "Spinoza's Environmental Ethics," *Inquiry* 23 (1980): 293–311; Stuart Hampshire, *Spinoza* (London: Penguin, 1951); Errol E. Harris, *Salvation from Despair: A Reappraisal of Spinoza's Philosophy* (The Hague: Martinus Nijhoff, 1973).

3 Devall and Sessions, *Deep Ecology*, pp. 237–38. They claim to be quoting Arne Naess, "Through Spinoza to Mahayana Buddhism or through Mahayana Buddhism to Spinoza," in Jan Welesen, ed., *Spinoza's Philosophy of Man* (Oslo: Universitetslorlaget, 1978): 136–58, but this quotation is a misquote. Errol E. Harris has even gone so far as to claim that "implicit in Spinoza's thought is the essential *solution* to the issues distracting our era [among them that] the extinction of the race within the foreseeable future seems threatened from every quarter, whether by the exhaustion of the resources of the earth, or by the pollution of the sea and its life-giving waters, or by the destruction of the ecological systems in which living species cooperate to maintain themselves and one another." Harris, *Salvation from Despair*, p. 3 (emphasis added).

4 Benedictus de Spinoza, *The Ethics and Selected Letters*, trans. Samuel Shirley, ed. Seymour Feldman (Indianapolis: Hackett, 1982).

5 For instance, Bill Devall, criticizing Wendell Berry's *The Unsettling of America* (San Francisco: Sierra Club Books, 1977), writes that "the main shortcoming of this otherwise powerful book is that it is based on the 'stewardship model' and thus Berry fails to reach ecological consciousness." Devall and Sessions, *Deep Ecology*, p. 259.

6 Even within this holistic perspective, many are not content with "biocentric egalitarianism" at all and call instead for some even more radical version of ecological retributive justice whereby humankind, having historically enjoyed extensive privilege at the expense of nonhuman nature, must now forfeit it. Deep ecology endorses what one might call ecological affirmative action. As Genevieve Lloyd points out, "what environmental philosophers seem to be looking for is a basis for the judgement that some activities that are now regarded as good for human beings should be curtailed because they are bad for other parts of nature." Lloyd, "Spinoza's Ethics," p. 301. My purpose in this essay is not to provide a critique deep ecology, but rather to locate Spinoza in contemporary ecosophy.

7 Lloyd, "Spinoza's Environmental Ethics," p. 293. What is prized by deep ecologists is that this is taken as placing restrictions on the power of humankind: since it is a part of nature, its ultimate good is bound up with[in] the laws of nature.

8 Samuel Max Melamed, *Spinoza and Buddlha: Visions of a Dead God* (Chicago: University of Chicago Press, 1933), p. 153.

9 For extended discussion of the ecological merits of a pantheistic approach, see Eugene C. Hargrove, *Foundations of Environmental Ethics* (Englewood Cliffs, N.J.: Prentice-Hall, 1989); Eugene C. Hargrove, ed., *Religion and Environmental Crisis* (Athens: University of Georgia Press, 1989); J. Donald Hughes, "Ecology in Ancient Greece," *Inquiry* 18 (1975): 122–24; and implicitly, in James Lovelock, *Gaia: A New Look at Life on Earth* (Oxford: Oxford University Press, 1979). Note, however, especially in Hughes, that a response of recognition, reverence, and noninterference to the natural world as God is a function of a reverent individual: one to whom the notion of "the sacred" already makes powerful sense. The project of resacralizing the natural world under pantheism, if it is to have the wide-ranging reformist response that deep ecology claims it will have, will only be achieved in conjunction with a program of spiritual reawakening or social religiosity in general. Note also that deep ecologists tend not to include axes or environmental mutagens in their discussions of the identification of God with nature.

10 Stuart Hampshire, quoted in Sessions, "Spinoza and Jeffers," p. 524.

11 Arne Naess, *Freedom, Emotion and Self-subsistence* (Oslo: Universitetsforlage, 1975), p. 119. He writes that "there is nothing in the *Ethics* to stop us in our attribution of all the introduced predicates to living being down to insects.... Living beings very different from us have correspondingly different joys, drives,

etc. There is no hierarcy. . . . There is an ontological egalitarianism—of which ecology makes us more tolerant today."

12 Hampshire, *Spinoza*, p. 71. One notable place where this concept appears in environmental philosophy is in Lovelock's *Gaia*, where he attributes to the entire biosphere a homeostatic mechanism implying an awareness. For more discussion, see David Abram, "The Perceptual Implications of Gaia," *The Ecologist* 15 (1985): 96–103.

13 Session, "Spinoza and Jeffers," p. 507. Here, Sessions (and also Naess, "Comments on Lloyd," p. 321) seems to discover in Spinoza the notion of intrinsic value which is generally understood as a system of value which exists independent of our recognition or construction of it. This is difficult to square with Spinoza's own insistence that what *we call* "good" or "bad" are products of human affections and aversions (Appendix to Part I, *Ethics*). As I have described it here, in light of Spinoza's conatus, is that value begins with the intending human being who empathetically or sympathetically extends what he calls the good or bad to others: it is relative to one's own desires. There *is* the "*verum bonnum*" in Spinoza—that which is conducive to one's understanding (*Ethics* IV:18 and 27)—and the "*summum bonnum*" in Spinoza—that condition of Joy or Happiness which is one with complete understanding of and conformity to the laws of our own proper nature (*Ethics* IV:24). I suspect that Sessions' "intrinsic value" might be part of the "verum bonnum," but this value is never divorced from *our* desire or *our* understanding.

14 Devall and Sessions, *Deep Ecology*, p. 238.

15 Lloyd, "Spinoza's Environmental Ethics," p. 294.

16 Ibid., p. 295.

17 Sessions, "Spinoza and Jeffers," p. 491.

18 Lloyd, "Spinoza's Environmental Ethics," p. 305.

19 Devall and Sessions, *Deep Ecology*, p. 241 (emphasis added).

20 Sessions, "Spinoza and Jeffers," p. 508.

21 Ibid., p. 505.

22 Arne Naess, *Freedom*, p. 118.

23 Lloyd, "Spinoza's Environmental Ethics," p. 303.

24 Sessions, "Spinoza and Jeffers," p. 507. In the footnote to these comments he also takes interpretive license to discredit them: "But these passages do not reflect a major theme in Spinoza's outlook and are tucked away in notes and appendices" (p. 526). Naess, "Comments on Lloyd," p. 318, follows this lead on this same difficulty: "I attach more weight to the propositions than to the notes, which are largely personal in nature." He continues: "I reject Spinoza's formulation in his famous *Note*. I do so convinced that the structure of his system admits a considerable softening of the harsh words in that personal note."

25 Ibid, p. 501.

26 But, the more that scientific laws are taken as *fully* exhausting Spinoza's notion of a determined "eternal chain of casusation" the more that (biological) humankind is subject to that determination, and thus the further we move from being able to rescue moral responsibility, moral will and meaningful normative ethical postulates from Spinoza. If there is nothing but "biomass and energy," then there is nothing but curved space, energy flux and, depending on one's location in the universe, the laws of gravity. Humans can't *help* but to "act in accordance with the law of gravity." Ethical exhortations become vacuous. Environmental philosophy doesn't seem to have spied this difficulty yet, at least with respect to Spinoza. Hu-Shih, "Spinoza and Chuang-Tzu," *Speculum Spinozanum(1677–1977)*, pp. 330–32, delights in the likeness between Spinoza

and Chuang-Tzu's thought, the enlightenment achieved upon surrendering one's ego ad will to the Tao, the inevitable, the Way. This ethos becomes a meditative, acquiescent "non-striving" which does not seem to fulfill the external revisionist goals of deep ecology. There are, however, a number of environmental philosophers grappling with the difficulty presented by trying to mesh Taoism ("the way of nature") with environmentalism ("the how-to with nature") while still separating off one stream of nature ("harmony," "health," "diversity," etc.) as that to which we must be more acquiescent than another stream of nature ("nuclear war," "rationalism," etc.). See Po-keung Ip, "Taoism and the Foundations of Environmental Ethics," *Religion and Environmental Crisis*, pp. 94–106. It is precisely the same model of the challenge that is found in Spinoza's moral philosophy. R. P. Peerenboom, "Beyond Naturalism: A Reconstruction of Daoist Environmental Ethics," *Environmental Ethics* 13 (1991): 3–22, presents what is, to my mind, the most meaningful and satisfying solution to this paradox in that he equates, for humankind, "acting naturally" to "acting intelligently," which contains within it a degree of choice. Here, he recalls Hampshire's insistence that "Spinoza's metaphysics and dependent theory of knowledge are designed to show man's place in nature as a thinking being," *Spinoza*, p. 115. I think the "solution" in Spinoza, which has largely been overlooked, is that as we are determined under the attribute of extension (this is what Hu-Shih and Sessions are on to), so too are we determined under the attribute of thought. Within these alternatives, as Naess' *Freedom* suggests, there may be fields of contingency under a locus of determination; that is, enough flexibility or probability that these "chains of causation" are loosened are not always accurate or in concert. Humans can both understand and bring about better or worse alignment.

27 Sessions, "Spinoza and Jeffers," pp. 508–09.

28 Ibid., p. 508.

29 Naess, "Spinoza and Ecology," p. 424.

30 ibid., p. 423.

31 Melamed, *Spinoza and Buddha*, p. 14. Attempting to east Spinozism in Buddhistic terms, Melamed reads his mysticism as internal and thus "indifferent to the lot of the lower castes" and "not concerned with economic, political or social justice."

32 Although the action of the body may be self-sufficient, that is, accompanied by a mental action.

33 David Bidney,*The Psychology and Ethics of Spinoza* (New York: Russell and Russel, 1962), p. 300. Examples of these prohibitions which "beget harmony and are attributable to justice, equity, and honorable living" are found in the appendix to Part IV:xv–xxxii.

34 H. F. Hallet, *Creation Emendation and Salavation: A Spinozistic Study* (The Hague: Martinus Nighoff, 1962), p. 205.

35 Ibid., p. 332.

36 Bidney, *The Psychology*, p. 296 (reference to *Ethics* IV:68).

37 Ibid., p. 298.

38 Ibid. See notes 29 and 39 for this sentiment. Also, see Holmes Rolston, III, *Philosophy Gone Wild* (Buffalo: Prometheus Books, 1986); "Valuing Wildlands," *Environmental Ethics* 7 (1985): 23–48.

39 Melamed, *Spinoza and Buddha*, p. 181.

40 E. M. Curley, "Experience in Spinoza's Theory of Knowledge," in Maejorr Grene, ed., *Spinoza: A Collection of Critical Essays* (Notre Dame: University of Indiana Press. 1979), p. 28. Curley includes "number" in the list of particulars with which one can have direct knowledge.

41 I strongly disagree with Melamed who speaks of Spinoza's union with a "dead God" as man's final goal such that "[l]ife becomes superfluous, a burden and a liability which must be overcome. *Spinoza and Buddha*, p. 355.

42 I don't deny that under Spinoza's fullest sense of nature this would also mean preserving varieties of garbage, mutant bacteria and hideous architecture, to name a few.

43 Curley, "Experience," pp. 57–58.

44 As an aside, there is a Canadian photographer named Cheryl Sourkes who, in a recent exhibition at Toronto's "A" Space Gallery, juxtaposed signs of the genus and species of herbs with close-up monochromatic photos of the herbs themselves. She seemed intent on making the point that labels—that language or text—gets in the way of our "real" experience of these things. I beg to differ. It probably depends upon the kind of relation that exists between one's experience of some thing, and one's naming of it, or use of its name. Having had what I might characterize as an affectionate relationship with the *Basilicum opulata* plants in my garden last summer, hearing or writing that name evokes a definite fondness, especially in mid-December.

45 This difficulty with appearance may also be true of the other realms of Spinozist nature that are not leaves, mountains, canoes, etc. In urban settings, the diversity of things may be even less appreciable, the tendency to simplify sensory data even stronger. In fact, I think this "tendency" is hardwired, based in human survival and thus, very rarely, overcome. Nevertheless, it must be possible.

46 Following Curley, "Experience," p. 358.

47 Ibid., p. 370.

48 Hallet, *Creation*, p. 206. And here is where Spinoza might not accept the rules of deep ecology or the kinds of reforms it advocates because they seem based less on circumstance than on principle.

49 Iris Murdoch, *The Sovereignty of Good* (London: Routledge and Kegan Paul, 1970).

50 Or again, hard-wired to identify with our own kind, which frees us to notice the differences among us.

# Part 3

# PROCESS THINKING AND ENVIRONMENTAL PHILOSOPHY

# 87

# THE RIGHTS OF THE SUBHUMAN WORLD*

*Charles Hartshorne*

Source: *Environmental Ethics*, 1 (1979), 49–60.

Religion, philosophy, and science are best interpreted as supporting the ieda that nonhuman forms of existence have intrinsic as well as instrumental values that we are ethically obligated to try to safeguard as best we can.

Legal rights are those which, in accordance with laws and constitutions, magistrates and police will defend. Moral rights are those acknowledged by the conscience of the community. By "ethical" rights we can perhaps denote those which sufficiently enlightened, reflective, disinterested, and knowledgeable individuals would support. I shall concentrate on rights in this third or ethical meaning. Do the subhuman creatures have them?

Three great Asiatic religions teach the duty of respecting life in general, at least in its animal forms, and derive the ethical superiority of a vegetarian diet from this duty. Insofar as these religions assume the possibility that the "souls" of subhuman animals may be in some sense idential with those of past of possible future human beings, their ethical justification of vegetarianism is to this extent questionable. (Certain other reasons for vegetarianism, especially economic ones, seem to me stronger.) However, an ethics of reverence for life can be supported and often has been (e.g., by Schweitzer) without reference to the transmigration theory. An ethical position common to Jain, Hindu, and Buddhist sages, Western humane societies, Albert Schweitzer, and Mr. Dodgson (author of *Alice in Wonderland* as well as a textbook on symbolic logic) must, it seems, represent a considerable degree of the "enlightenment" spoken of above. Let us consider what may be the ultimate basis for this position.

* An earlier version of this paper appeared as "Rechte—nicht nur für Menschen," in *Zeitschrift fur evangelische Ethik* 22 (1978): 3–14.

An ethics can appeal to religious convictions, to scientific facts of principles, to a philosophical system, or to some two or all three of these possible rationales. Ideally, the religious, scientific, and philosophical aspects should form a coherent whole. To exhibit such a whole will be my goal and hope.

My religious faith is in a God who takes delight in the creation. Moreover, I hold that the ultimate value of human life, or of anything else, consists *entirely* in the contribution it makes to the divine life. Whatever importance we, and those we can help or harm, have is without residue measured by and consists in the delight God takes in our existence. Is it likely that God takes no delight whatever in the more than a million other living forms on this planet, yet does delight in, derive value from contemplating, the one human species lately emergent on the planet? If such an idea is not sheer anthropomorphic bias, what would be such bias?

Perhaps you say that God takes satisfaction in the other creatures only because they serve the needs of his darling, humanity. The other creatures are good, but instrumentally so only, whereas we are good intrinsically. Immanuel Kant so believed. I say that it took great genius to be so fantastically partial, and yet so knowledgeable, as Kant was. His God was an extremely moralistic one, interested directly only in creatures capable of "rational freedom." We, who are such creatures, find the nonrational animals more or less fascinating objects of our spontaneous or cultivated sympathy; but God is supposed to look upon them somewhat as an engineer might view parts of an intricate and useful machine of the engineer's own devising. I submit that such a theology is a museum piece, not a doctrine we can take seriously as an option for our own time or the future. God surpasses us not by the narrower but by the incomparably wider scope of his love or sympathy. Our human, somewhat-qualified anthropomorphism in God. Why think of God at all if he is but a more extreme version of our own limitation, our own preoccupation with our sort of animal?

Let us look at Kant's argument a moment.[1] He says that only a rational will that acts according to its rationality is intrinsically or unqualified good. But he also admits that human beings so act only imperfectly or incompletely and that only God always and entirely conforms to rational requirements in his actions. Consequently, only God is unqualifiedly and intrinsically good. If we add to this, as Kant would have refused to do, but I hold he ought to have done, the proposition that even the value of a rational animal is entirely contained in the value its existence contributes to the sole unqualifiedly rational or divine life, what we come to is that intrinsic value consists exclusively in what an existent does for God. Kant tried to locate the Summum Bonum, meaning the intrinsic value that measures all their values, in rational animals who are both good, that is, rational in their choices, and happy. I say that the Summum Bonum is located in God, who is to be loved by creatures "with all their minds and all their hearts and all their strength," And I say that, while rational animals make a special contribution to the

Summum Bonum, every creature makes some contribution, however humble it may be.

Not only are adult human beings imperfectly rational, human infants are not rational at all (they consult no general principles in their choices). the Kantian value of an infant is purely potential or instrumental: it may lead eventually to the existence of a rational animal. The line between the human and the subhuman is crossed in the life of each one of us, if by *human* one means actually rational. I think that a person needs to have something more than Kantian ethics to be a good parent.

She or he needs an ethics of sympathy broader than sympathy for life on the rational level alone. And I think that Deity should be supposed vastly less inhibited in sympathies than are human beings. *Life sharing*, delighting in the lives of others, this is my analysis of the most fundamental aspect of "love," whether human or divine. What makes the difference between the two forms of love is the presence or absence of limitations in the scope and quality of the sharing. Only God sympathizes with and takes delight in all the forms of life. As an old mystic had it, "All the forms of being are dear to the infinite being himself," So much for the religious aspect of our problem. That "not a sparrow falls from heaven but your Heavenly Father knoweth" I take to be true and to mean that there is divine participation, sympathetic delight, in the life of each sparrow. Its importance to God is greater than zero. How much greater? That very difficult question I shall consider later.

Turn now to the scientific aspect of our question. For contemporary biology our species is one very recent branch of a tree of life billions of years old. Innumerable steps led from the ancient beginnings to our kind of nervous system and brain. We have much in common in neural structure (and that is what counts) with other primates, also with whales and porpoises, and a good deal in common even with wolves, horses, and elephants. Indeed, we have something significant in common with all animals, and even with one-called plants as well. To ask scientists to suppose that all this development is, in terms of value, mere preparation for ourselves and our human ancestors is to ask them to exchange a sublime and coherent vision for a childishly arbitrary one. Scientists, if religious, will want to think of God as more comprehensive than themselves in appreciation for the beauty and intrinsic value of the entire vast creative process. It not religious, they will probably have some vague feeling for nature as a wondrous whole to the beauty of which every species makes its contribution, including humanity as (on this planet) supreme but far from sole example.

Up to this point I have been accusing my opponents, the deniers of rights to the lower animals, of anthropomorphism. In developing my own philosophy I may perhaps seem to be guilty of that weakness myself. I do hold that a thinking animal, in trying to understand things, is bound to employ analogies between itself and other parts of nature. For, if there are truly general principles in terms of which understanding is possible, the thinking animal itself must illustrate these principles. Charles Peirce spoke of an inevitable "zoomorphic" aspect of all our conceptions. I hold that Plato was well-advised to employ the analogy of the "ideal animal" in trying to conceive the created cosmos as a whole. And I believe also that the most significant model we can have even of the simplest parts of the universe, say molecules, atoms, and particles, is that they are the simplest, most primitive cases of that which our own natures illustrate in vastly more complex and highly evolved forms.[2]

There are several considerations by which this approach may be justified. First of all, whereas we know ourselves in two basic ways, we know most of nature in but one of these ways. We know ourselves by being ourselves, by direct feeling or memory of what it has been like to be human animals. We know other things, at least so far as they are outside our bodies, only by visual, auditory, olfactory, or tactual observation. By such observation we can also examine ourselves and our human fellows. Thus, if we can know what any sample of natural forces is like, a fortiori we can know those samples that we ourselves are.

There is a second reason, noted by Whitehead, for taking this closest-to-home sample seriously. It is that in ourselves the positive characteristics of animals generally, and for all we know of creatures at large, are present in highest degree, and therefore in most unmistakable form. For example, if adaptability is a characteristic of organisms at large, it is maximal and most obvious in our species. So are flexibility, versatility, originality of behavior, capacity to learn and remember, differentiation and control of numerous subordinate organs of parts, and finally, degree of consciousness and reasoning power. If we ask an atom whether there is such a thing as adaptability, creativity, or consciousness, we shall perhaps get no clear answer, unless this: either none at all or not very much. But the difference between zero and finite positive quantity makes *all* the difference when we are seeking the general principles of reality. So we had better start at our own end where attributes are likely to be present in sufficient magnitude or intensity to be unmistakable.

For the two reasons just specified I hold, as did Peirce and Whitehead (also Heidegger), that a cautiously positive form of anthropomorphism—that which attributes to other creatures neither the duplication, nor the total absence, but lesser degrees and more primitive forms, of those properties exhibited in high degree, and more refined or complex forms, of those in us—is the only rational initial hypothesis for us to form.[3] Primitive hylozoism was perfectly reasonable as a starting point for speculation. Indeed, it was more reasonable than the much later and seemingly so sophisticated model, which still fascinates many scientists, of a world machine whose parts are submachines. Machines occur when animals make them; animals are not given birth to by machines. Moreover,we simply do not, and could not, know that there is such a thing as a "mere machine." In Popper's brilliant metaphor, we do not know that clouds are really clocks, that is, purely mechanical in the behavior of their molecules or atoms, and indeed we do not know that even clocks are purely mechanical in this sense.[4] The natural interpretation of quantum physics implies (as Pierce and others has surmised) that there is no such thing as a mere machine. Quantum mechanics itself is not very mechanical. A quantum system behaves more like a society than like a clock, as Margenau, Čapek, and other have noted.

Although we do not know that there is such a thing as a mere machine, we do know that there are animals, creatures whose actions are at least altogether as *if* they were influenced by feelings, desires, hopes fears, likes and dislikes, memories and expectations. We know, too, that in animals the principle of spontaneous motion, what Plato called the "self-motion" of soul,[5] is normally apparent, whereas in some machines (as in a resting automobile) all the motion or change seems to be confined to the microscale. Since time itself is but an aspect of change or becoming, that model of reality which offers the best chance of understanding change is to be taken seriously, and that model which change a mere riddle is not to be taken seriously. It is an open secret of early Greek materialism that the atom of the materialists had, in itself, no principle of change. It was only a bit of stuff (*being* was the word used) in otherwise empty space (*nonbeing* or *the void*). Why should such a bit of stuff alter its position in the void? No reason was given. Today most thinkers realize that this will not do. So they take four-dimensional process or becoming, instead of being or substance, as the primitive concept. But what is it that becomes? Energy states? But what are these, apart from the geometrical and numerical patterns which the formulae of physics specify? Shapes of what? Velocities of what?

I will not prolong the argument. My view is that every singular active agent (there are no singular inactive ones, the seemingly inactive being composites whose constituents are active) resembles an animal in having some initiative or freedom in its activity, and that this spontaneous movement and change has, as in animals, inner aspects of feeling, memory, and expectation, though the effective scope of the latter functions may in many cases be extremely narrow, perhaps confined to a millionth of a second or less. In short, I hold, with Leibniz, Peirce, Whitehead, and many others, including both philosophers and scientists, that all process is at least sentient, and with some degree, however slight or nonconscious, of memory and expectation. I hold also that where there is feeling there is value in a more than instrumental sense. All nature involves self-enjoyment and contributes thereby to the participatory enjoyment of deity, which contribution is the absolute and final measure of its degree of importance.

To say, all process is at least sentient, as I have just done, might be taken to mean that trees and even rivers or mountains feel, enjoy, and perhaps suffer. But this is to forget the distinction between active singulars and composites. Science seems to show that the activity in a tree is in its cells not in the tree as a whole, and that the active agents in rivers or mountains are such entities as molecules, atoms, and still smaller things So we need not give up the common-sense view that vegetables, rivers, mountains, and other visible objects other than animals are insentient. It is only their invisible small constituents that are to be understood finally a remote analogy to our own inner life and activity.

The still unresolved question obviously is, "Which creatures are more and which less important?"—for certainly importance is a matter of degree. And it is clear that no creature can live without treating at least some other creatures as less important, in its view, than itself and its own king. Either our use of other animals and of plants as food is unethical, or our ethics assumes the greater importance of an individual of our own species that at least some of the nonhuman individuals in nature. Ability to interpret this is a critical test for ethical systems.

A logical requirement of any value system is that it should clarify the idea of no value, or the value zero. I hold that, as value diminishes, its limit of zero is to be sought not in a form of existence without value, but in total nonexistence. The zero of feeling, or of intrinsic value, and of actuality are one and the same. Take the criterion of integrity our unity, as in the aesthetic maxim, "beauty is unity in variety": with no such unity, there is no experience, no value, and no actuality. But there can be expertises with imperfect unity, as in all experience of conflict, suffering, bewilderment, confusion. Some experiences are better integrated, internally more harmonious, and so better than others. But variety, too, has a threshold below which experience fails. We experience contrasts, not mere homogeneties. When contrast falls below a certain minimum, we also consciousness, perhaps go into a dreamless sleep.

There is a subtle relationship between contrast and intensity, and between both of these and value. Failure to achieve any great degree of value may be because there is too little contrast for much intensity of too little intensity of any very strong contrasts; but it may instead be for the quite different reason that the intense contrasts are insufficiently integrated, unified, harmonized. There are cases of unified where the agree of intensity and level of complexity in the kinds and instance of variety is so low that the resulting beauty is rather trivial. We often use the work *pretty* for such a case. At the other extreme, where the complexity and intensity are great, we are likely to use the adjective *sublime* rather than *beautiful*. Where the variety is insufficiently unified, we may, if the intensity and complexity are not too great, experience the situation as *ridiculous*, *funny*; while, if intensity and complexity are great, the lack of unity or harmony may be felt as *tragic*. Thus, all the aesthetic values are accommodated.[6]

So far I have been using aesthetic criteria of value. What about moral criteria? I hold that these are either (1) special aspects or dimensions of the aesthetic, or (2) they specify means indirectly favoring the former.

The morally good individual is one who wills to optimize the harmony and intensity of living for all those lives he or she is in a position to affect. (1) In living out such a will to optimal aesthetic value for all, including self, the individual inwardly enjoys certain harmonies from the very fact of this form of volition. Insofar, moral value is also directly, or in itself, an aesthetic value. And there is much evidence that disinterested spectators experience a sense of beauty in observing a genuinely good person. But (2) the moral will is also a potential, probable, and intended source of future aesthetic values for those influenced by the actions. Where this is not the case the goodness is bogus, as indeed, alas, it not seldom is. The allegedly "good" people who make life general uglier, less harmonious, or more tedious, less zestful and intense, for themselves and others, are insofar not really good. It is a defect of our culture that this is so little realized.

In the light of the foregoing, what is the value of the human species compared to the others? Only in this species (in this solar system) is there any appreciable development of the moral dimension of value. Moreover, the purely aesthetic dimensions are not in human nature proportionately reduced to make room for this addition. Quite the contrary. Most birds sing, but their music, while good in its simple kind, is not comparable to human music in complexity and, we may presume, in intensity. All higher animals "communicate," but how extremely limited is such communication compared to that effected by human speech and writing.

The importance of a creature depends partly upon the effects it can have upon other creatures. In instrumental value, for ill, alas, as well as good, no other animal remotely compares with our species. By this test the incomparable importance of humanity is manifest.

John Cobb makes a helpful suggestion.[7] We tend to think that humanity is important because of the values in each individual person; but we tend to think that one nightingale or one hermit thrush is significant chiefly as a specimen of its species. For one thing, such an animal lives but a few years, if it is lucky enough even to survive its first year; for another, individual differences are rather slight in birds; finally, what consciousness does a bird have of its own identity? It surely has no concept of self. It makes no definite conscious plans for the future beyond a few seconds or minutes at most. The death of a bird is in all these respects very different from the death of a man. The animals know fear, but hardly fear of death. This requires a concept.

The significance of the term *murder* depends upon much more than the mere idea of killing a living individual. Predatory animals as such are not murderers. They are merely food gatherers whose food is other animals. But when a human killer strikes he knows about death, he has the idea of his species as one among many, and of fellow human beings with whom he shares that unique communication system called language; he realizes that he too might be killed; he knows that, if he kills another person, his act affects the meaning of existence for an entire social group and conflicts with principles without the observance of which by most people he himself would never have existed. Thus, murder is one thing; killing by nonhuman predators is a very different thing; and even human killing of nonhuman animals is far from the same thing. True, there is something in common, and those who condemn hunting can perhaps make a case. But the word *murder* is not to used to make this case.

For similar reasons, use of the same word for abortion is an intolerable abuse. It changes the whole quality of life for human adults, or even for children, if they know that there are murderers about. It does nothing to an embryo to know, for it could not know, that there are abortionists about. And what is destroyed in abortion is not a "rational animal" actually functioning as such. Any rational behavior attributable to an embryo is far surpassed by the more nearly rational behavior of a chimpanzee or porpoise. The human value of an embryo is essentially potential and future, not actual and present. Such animal functioning as is going on in it is nothing especially exalted, as such things go on this planet.

Of course, even the potentiality of a functioning rational animal has its value. But this has to be weighed against the values of the numerous actually functioning ones and the harm that allowing this potentiality to mature may do to these actual values. Suppose the unwanted and badly brought up child turns into a genuine murderer, as may rather easily happen. (There is, to be sure, the option of bearing the child but offering it for adoption.) Pearl Buck argued that an aborted embryo might have turned out a genius. Yes, it might, and with perhaps more probability it might have interfered with the maturing of some other person of genius, and might itself have become a criminal, or at best a mediocrity. And the world is getting rather too full of people for comfort or even for reasonable safety. In any case the moral question of abortion is one thing and the legal question quite another.

How far is the human species entitled, at its convenience, not only to subordinate the needs of other species to its own needs, but also, as it now threatens to do, actually exterminate hundreds, and soon thousands, of species? This is a hard question. If the tiny minority of adults who have the sense, shared between small children and adult nature lovers, of the fascination and beauty of all the forms of life could only be turned into a majority, the question would tend to answer itself. For to a nature lover it takes a lot of convenience to balance a radical diminution of natural beauty. Suppose the more than seven hundred species of birds in this country were reduced to one hundred, and in any one state from two or three hundred to two or three dozen, would life remain as interesting? Not for some of us. The lovely prairie grass is gone. Every lawn begins to look about like every other. How much monotony do we want? Only a few species of trees flourish in our cities. Our parks lack the rich undergrowth that many species of animals require, and so they are dull places, scarcely better than the tiny backyards of the urban poor.

At least we may be thankful, we nature lovers, that pollution sets prudential limits to the destruction of nature to make room for our artificially expanding populations. (Birth control and scientific hygiene are alike artificial, and until we effectively combine them there can be no genuine "natural law" in the case. What is natural without technology is one thing; what is natural with it is another.)

To say that the human species is more important than other species suggests the question, how *much* more? Can we, even if only in the vaguest way, quantify the answer? One human person is "of more value than many sparrows," but is one person of more value than an entire species of bird? Our intuitions seem less responsive to this question. There are not far from nine thousand species of birds. Which would be a greater loss to the universe, the disappearance of all birds, or premature deaths of nine thousand human beings? I could perhaps seriously consider giving up the remainder of my life if it would definitely save a threatened species for millennia. But it would be merely silly to risk one's life for a single individual bird, whose life expectancy anyway is probably less than three years. Perhaps the best we can do with this question of the relative values of ourselves compared to the other animals is the following. We can try to view man and the rest of nature as one ecosystem in which our species is, so far as possible, complementary, rather than competitive, with the other creatures in the system. To this end we can take zero population growth, or even, eventually and for a time, population decrease, as desirable goals. We can put a burden of proof upon each proposed destruction of wild nature. We can weigh seriously the need for luxuries which use large amounts of energy, such as wasteful modes of transportation that require more and ever more acres of concrete for superhighways and parking lots. We can ask if lawns and parks need be so bare of natural undergrowth or bushes as they now are. We can consider nourishing ourselves more largely by direct use of vegetable food, rather than by the very wasteful method of giving this food to animals whose flesh we ultimately eat. Complete vegetarianism need not be and probably is not the most appropriate solution. But the proportion of the grains, cereals, legumes, and other protein sources now going through the meat industry should not be expected to continue. Present high prices of meat may in the end do some good in educating the American people concerning the basic economic facts. In many climatic conditions, at least, to consume vegetable matter directly makes far less drain upon the natural ecosystem than to let animals use up most of the energy built up in the vegetation, yielding to us only their carcasses for food. For the bulk of the energy will have been dissipated by these animals simply by living for some months or years.

An important part of the needed birth control is of births to pet animals. To supply dog or cat food, whales are slaughtered, and in various ways the variety of nature is threatened. Feral cats and dogs are a danger to many wild species. Artificially transformed and maintained species need also to have artificial limits set to their reproduction if they are not to be an offence to the inclusive ecosystem. Since we are now too numerous simply to let nature take its way as it would do without us, we ought in all our important decisions to take thought as trustee for nature in this solar system. People have indeed no absolute control over nature, even on this planet, and negligible influence on the universe at large; nevertheless, on this planet we have at least, in military language, almost unlimited "nuisance value." We can wreak havoc upon the other creatures and upon our own kind. Were it not for the second of these two dangers, perhaps nothing much could be done to check the first. But the fact, now dawning upon all, that our greedy exploitation can boomerang upon ourselves or our children may in time lead our species to save some substantial portions of the wonderful tapestry of life on the surface of this globe. Yet the change of attitude must come rather rapidly if this is to occur. For the destruction of the nonhuman goes on rapidly indeed.

Think of the damage to wildlife of strip mining. Careful inquiry may perhaps show that areas so ruined can without impossible cost be redeemed. But has this been shown? There is the coal, perhaps the only safe (but how safe is it?) quickly available large new source of energy—until or unless atomic fusion is mastered, or sunpower feasibly exploited. Or could wind make a significant contribution? We need much better leadership than we are getting in this matter. If our somewhat alienated young could find a challenge here, we might begin to move fast enough in the right directions.

There are those who say that, since human beings are radically superior to nonhuman individuals, it is hypocritical or absurd to shed tears over ill-treatment of subhuman creatures, while ignoring that of many human beings, as of the poor or the ethnic minorities. This argument seems to be at its strongest if the ill-treatment in the subhuman case is indeed only of individuals. But the issue is not so clear if it is a question of species, threatened wholesale by slaughter or by habitat destruction. All animals, including people, die eventually, and although species also die in the end, the net result, until civilized humanity emerged, was not to decrease the beautiful variety of nature. Rather the contrary. But human actions now threaten a drastic decrease in the richness of nature's contrasting forms. Moreover, this impoverishment reduces the options open to human beings themselves. We are the only animals capable of being interested in, finding some use for, taking some delight in, *all* the forms of life. This is why zoos and botanical gardens exist. But wild nature has values that such artificial vessels can never more than very partially contain. Even from the selfish human point of view this is true. Thus one finds some of the famous species of songbirds in zoos, but so crowded together that what one hears from them is a discordant, confused, frustrating blend of noises, nothing remotely like the symphony of distinguishable voices one may hear at daybreak in a forest or wild savannah. And the whales and porpoises, those glorious large-brained creatures, are scarcely adapted at all to anything but life in the open ocean. How these creatures really live is very incompletely understood, and only recently has it become known that at least one species of whale has an elaborate song, in some respects far beyond any bird song. It seems a grim fact that the very century in which we began to understand the life of the other animals—think how the immense role of territory was generally known only after Eliot Howard published his book on the subject in 1920—is also the century in which the destruction of habitats began to threaten not dozens but hundreds or thousands of vertebrate species, not to mention similar or larger numbers of lower orders of creatures, animal and vegetable.[8]

My conclusion is that religion, philosophy, and science are best interpreted as supporting the idea that nonhuman forms of existence have intrinsic as well as instrumental values that we are ethically obligated to try to safeguard as best we can. There is also ground for thinking that indifference to the intrinsic values of the nonhuman is sure to lead to unnecessary losses in the extrinsic values of these other types of creatures for human beings. If we are merely selfish toward our fellow creatures, we shall probably, like all merely selfish persons, not do justice even to our own selfish interest.

The problem is one of education. Primitive man had a kind of culture that took the rest of creation into account. Our city populations, and our more and more commercially-minded farmers as well, are neither primitive nor scientific naturalists. They (many of them) are, to put it bluntly, uncultured persons, a somewhat new kind of barbarians. What is to be done? In Japan and in England, radio broadcasting takes bird song and similar topics into account. In the United States we seem in some respects on a lower level in such matters.

Yet we do have our Sierra Club and our Audubon Society. So those of us who love nature are not without allies and resources in the struggle to discharge our responsibilities as trustees for the life of this solar system.

## Notes

1 Immanuel Kant, *Lectures on Ethics*, trans. by Louis Infield (New York: Harper and Row, Publishers, Harper Torchbooks, 1963), pp. 239–40.

2 *The Collected Papers of Charles S. Peirce*, ed. by Charles Hartshorene and Paul Weiss (Cambridge: Harvard University press, 1935), par. 5.212; Plato *Timaeus* 30c–31b.

3 Alfred North Whitehead, *Adventures of Ideas* (New York: Macmillan Company, 1933), chaps. 11, 16; Martin Hidegger, *Sein und Zeit*, 7th ed., pp. 19f; Margin Heidegger, *Vom Wesen des Grundes*, 4th ed., p. 36, n. 55.

4 Karl R. Popper, *Objective Knowledge: An Evolutionary Approach* (Oxford: Clarendon Press, 1972), pp. 206–55; Peirce, *Collected Papers*, par. 6.59–.65.

5 Plato, *Phaedrus: The Laws* bk. 10.

6 See Charles Hartshorne. *Born to Sing: An Interpretation and World Survey of Bird Song* (Bloomington: Indiana University Press, 1973), pp. 6–9; see also Charles

Hartshorne, *Creative Synthesis and Philosophic Method* (La Salle: Open Court, 1970), p. 305.

7 See John B. Cobb, Jr., "Beyond Anthropocentrism in Ethics and Religion," in *On the Fifth Day: Animal Rights & Human Ethics*, ed. by Richard Knowles Morris and Michael W. Fox (Washington, D. C.: Acropolis Books, 1978), pp. 137–53.

8 Eliot howard, *Territory in Bird Life* (London: John Murray, 1920).

88

# IDENTITY, COMMUNITY AND THE NATURAL ENVIRONMENT

## Some perspectives from process thinking

*Clare Palmer*

Source: *Worldviews: Environment, Culture, Religion*, 2 (1998), 257–67.

### Abstract

Alfred North Whitehead's and Charles Hartshorne's process thinking presents a complex and sophisticated metaphysical underpinning for a theory of self and self-identity. Their construction of the self has significant implications for understanding of the (human) community and the natural environment. Process thinking, I argue, undercuts the idea of self unity; of self-continuity over time; and of self-differentiation from the world. When combined, these three elements mean that it is hard to separate the individual, personal self from the community and the natural world. I compare these implications from process self from the community what might seem similar implications from radical ecological philosophies. Although there are ethical and metaphysical differences between process thinkers and deep ecologists, both kinds of theory need to be treated with caution in application to our thinking about the environment.

### Introduction

As the general nature of the title suggests, this paper potentially covers a vast area. I will therefore focus here on only a few, limited aspects of process thinking on this topic, related almost entirely to the later work of the mathematician and philosopher A.N. Whitehead and to Charles Hartshorne (who in many ways is deeply influenced by Whitehead). Although process thinking can take a variety of forms, Whitehead and Hartshorne are, indisputable, pivotal members of this tradition. In particular, in the context of this paper, Whitehead and Hartshorne have developed some interesting ideas about human self-identity which have considerable significance for the ways in which (human) community and the natural environment might be conceived. These ideas, I will argue, bear a resemblance to those discussed by some deep ecologists – and may, thus, be vulnerable to the some of the same difficulties. In order to explore these questions, I will begin by briefly outlining some fundamental concepts of Whitehead's and Hartshorne's process thinking: first concerning the underlying nature and structures of the Universe; and secondly, the construction of value and, relatedly, what constitutes ethical behaviour.

## Underlying structures of the universe

Both Whitehead and Hartshorne reject the idea that the universe is fundamentally inanimate, and that life on earth (and other planets, if such exists) is life clinging to the edges of an insentient universe. Rather, both argue, the universe is fundamentally composed of tiny, fleeting moments of experience which Whitehead calls *actual occasions* and Hartshorne *organisms*. (I will use Whitehead's terminology in order to avoid confusion). The universe at all levels, from the atomic to the cosmic, is composed from these actual occasions; both they, and the universe which they form, are in a constant state of change or process. All actual occasions manifest individuality, process and subjectivity; and these key characteristics underpin Whitehead's and Hartshorne's conceptual systems. Whitehead maintains that actual occasions are 'the final real things of which the world is made up. There is no going beyond them to find anything more real'. (Whitehead 1978: 18.)

Whitehead sometimes characterises these actual occasions as 'drops' – a helpful image to use. A drop suggests *individuality* (in its discreteness) but allows for the sense that other, previously existing drops have emptied themselves into the new drop in order to create it as it is. The image of a 'drop' also suggests spatial extension, and the *process* of growth, development and change which characterises the actual occasion. 'Each actual thing is only to be understood in terms of its becoming and perishing. There is no halt in which the actuality is its static self, accidentally played upon by qualifications derived from the shift of circumstances. The converse is the truth'. (Whitehead 1948.)

Actual occasions are, Whitehead argues, like drops of experience of *subjectivity*. This is central to both his, and Hartshorne's, understanding of the universe. Everything that is coming to be in the world thus consists, most fundamentally, of subjectivity. Subjectivity is not confined to human beings, nor even to humans and nonhuman animals; the entire universe is composed of experiences (*panexperientialism*). Rather than arguing that humans (and perhaps nonhuman animals) exhibit radically different characteristics to the inanimate world, Whitehead maintains that humans display a particularly sophisticated version of what is universally present. Certainly, human consciousness is deeply significant for Whitehead, but: 'Consciousness is the crown of experience, only occasionally attained, not its necessary base'.

If this is how Whitehead and Harshorne interpret the universe on the 'micro-level' as it were, how do they interpret the 'macro-level'? How does this picture of fleeting, experiential actual occasions relate to the level of organisms in their usual sense and to everyday 'objects' for instance? Again, even at this macro-level, subjectivity, individuality and process are the key. The actual occasions are the drops which compose larger wholes; all the objects and persons around us are actually, most fundamentally, societies composed from actual occasions. But not all societies are the same: they have different degrees of cohesion and structure, and respond differently to the contexts in which they come to be. Primitive societies, in Whitehead's terms – for instance non-living objects, like tables and chairs – are fairly simple societies, little more than structured aggregations of actual occasions, which come to be and perish, come to be and perish, one after the other, with little change or evolution in relation to their circumstances. (Oddly, both Whitehead and Hartshorne include plants in this simple category.) Some societies, like animals and human beings, are, however, more complex. Whitehead argues that they have a degree of unifying control, a centre of governing experience, such that they may be called 'monarchical societies'. 'An animal body' he says 'in its highest example, is more analogous to a feudal society, with its overlord'. (Whitehead 1938b: 35.) Without such a unifying or controlling centre, animals, and humans could demonstrate no unified individual though or action. The governing centre, or 'controlling actual occasion' allows for the creation of novel, individual responses to situations encountered. But despite this sense of unity, it is important to note that the centre of control in a human being or an animal is no exception to the principle of universal change and process. In a fundamental sense individual human beings are no different from the rest of the Universe: they are in constant process; they are centres of experience; and they are fundamentally social, being composed from other, contributory centres of experience.

I want now to move on to look at the understanding of value and ethical behaviour for these process thinkers, before returning to consider how his understanding of the structure of human beings (in paricular) relates to interpretations of the self and self-identity in process thinking.

## Value in Whitehead and Hartshorne

It is impossible to explore Whitehead's and Hartshorne's understanding of value without first acknowledging the fundamentally theistic nature of their conceptual systems. For both thinkers, God has two aspects: what Whitehead calls a 'primordial nature' and what he calls a 'consequent nature'. The primordial nature, he maintains, is 'free, complete, primordial, eternal, actually deficient and unconscious' (Whitehead 1978: 343). The consequent nature, Whitehead describes as 'determined, incomplete, consequent 'everlasting', fully actual and conscious'. Fortunately, it is not necessary here to attempt to unravel in too much detail what Whitehead might mean by this. What is important in the context of value is that Whitehead understand the universe to be a part of God's consequent nature; that is, he adopts a form of *panentheism.* In God's consequent nature he (for Whitehead, God is 'he') experience the experiences of every actual occasion as it comes to be and perishes; more simply, we might say that God feels the feelings of an all-feeling universe. These feelings or experiences are woven together within God's consequent nature, which is always growing to receive them. Indeed, God's constant absorption of the experiencing actual occasions in the universe, Whitehead argues, contributes to the fulfilment of God's being; and it is the fulfilment of God's being which is of ultimate value.

However, not all experiencing actual occasions contribute equally to the fulfilment of God's being. Whitehead (1978: 105) maintains that 'God's purpose in the creative advance is the evocation of intensities'. The amount of fulfilment an actual occasion can provide for God – and hence the amount of value it ultimately produces – is dependent on how *intense* it is and whether its intensity is *ordered* (rather than chaotic). Not all actual occasions are capable of generating the same degree of intensity. To simplify here, Whitehead argues that the actual occasions which come to be in animals are more intense than those in (say) plants or rocks; and those in humans more intense than those in animals. They are thus capable of generating more fulfilment for God – and hence of creating more value in the universe (this hierarchical system of value generation is not, by far, all that can be said about the creation of value in process thinking; but it is nonetheless significant from an ecological viewpoint – although I will not be considering it further here).[1]

This description of value-generation in Whitehead's thinking hints at what ethical behaviour might mean in such a system. Hartshorne describes ethical behaviour as behaviour which concerns 'not the value of experiences themselves, but rather the instrumental value of acting so as to increase the intrinsic value of future experiences of those other than oneself'. (Hartshorne 1984: 10). Ethical behaviour (only possible for 'conscious beings', rather than for unconscious actual occasions) is behaviour which creates the greatest value possible for God. This may mean the sacrifice of some present ordered intensity in order to generate greater ordered intensity in the future; the renunciation of some value now in order to create more total value in the future (to contribute to God's fulfilment).

The ethical system outlined here is clearly consequentialist: to behave ethically is to act in a way which produces the best consequences – the production of ordered intensity of feeling for the fulfilment of God's being. It is indeed a particular form of consequentialism, where the aim is to produce greatest total value, a *value-maximising* approach (which resembles some forms of utilitarianism). Seen from this perspective, ethical actions are those which produce the greatest possible value for God's being.

Having briefly outlined the panexperiential, plural and temporally-changing nature of the process universe, and the consequentialist, maximising nature of process ethics, I want now to move on to bring these perspectives together in considering process understandings of the self and self-identity.

## Self and self-identity

As we have seen, for both Whitehead and Hartshorne, objects (such as tables and chairs) or persons which we normally think of as being fundamentally singular are, at a deeper level, more accurately viewed as societies of actual occasions, in a constant state of change. Of course, amongst humans, Whitehead concedes that a degree of unity is bestowed on human beings by the 'controlling actual occasion', allowing for unity of thought and individuality of action, whilst Hartshorne maintains that humans have some kind of individual essence. This allows Hartshorne to go so far as to talk about personal identity; but whilst he considers such language to be permissible he repeatedly points out that it is an abstraction:

> 'Personal identity is a partial, not a complete identity; it is an abstract of life, not life in its concreteness. Concretely, each one of us is a numerically new reality every fraction of a second'.
>
> (Hartshorne 1974: 201)

For Hartshorne, our language of individual identity and of consistent personality abstracts from the concreteness of actual change and from the plurality of actual occasions from which we are composed. If we change the focus of our thinking from the abstraction of personal self-identity to the concreteness of actual change and plurality, a transformation of our understanding of self (in particular self-interest) and therefore community and the natural environment will inevitably result. We will begin to see our own lives, and the lives of others, as changing processes; we would use his preferred language of 'personal sequence' rather than person. Though a common thread, which we may think of as self-identity, runs through a series of plural experiences (as the past empties itself into the present, and the present into the future) this thread is an abstraction from the concrete reality at any particular moment in time. To focus on the thread of self-identity, Hartshorne says, is 'an illusion of egoism' (Hartshorne 1974: 203). If we consider instead the concrete reality, rather than the abstract thread of personal identify, we find not a world of individual essences which, as it were, bump up as separate individuals against one another, but rather fluid experiences, not bound to any particular thread of identity, in a panexperiential universe.

This interpretation of the world, I think, weakens the idea of distinctive individual human self-identity in a variety of ways. It suggests the fragmentation of the *unity* of the self into a society of actual occasions; the erosion of the *temporal continuity* of the self by emphasising a constant process of change; and questions the *separation* of the self from the world, as it merges into a web of similar (if less intense) experiences in a panexperiential universe.

Hartshorne (and Whitehead, in whose footsteps he is closely following here) makes a number of observations about this way of viewing the world. Both philosophers recognise that it questions the idea of a physical self clearly distinct from the world. As Whitehead comments:

> The functions of the body shade off into those of other events . . . in many respects the demarcation is purely arbitrary, depending where on a sliding scale you choose to draw the line.'
>
> (Whitehead 1920: 107)

This leads Whitehead, in particular, to consider more closely the way we regard our bodies. He suggests that while we have a particular 'intimacy of association' with our own bodies, this is not *qualitatively* different, but rather only different in *intensity* from our relationship with the rest of the world.

Hartshorne develops this theme by exploring what the weakening of the idea of self means for our understanding of *self-interest*, and in particular our future-oriented actions. After all, if the self cannot be clearly captured and delineated, neither can self-interest. And if the self is in a constant process of change then a future self cannot be the same as the present self. This raises questions about what it might mean to be 'self-interested' since no future self is identical to the present self:

> Any future self, call it mine or not mine, which can benefit from my present act will be numerically a new and distinct unit of concrete reality. Hence self interest has no privileged metaphysical base whatever.
>
> (Hartshorne 1970: 191)

Whatever we do choose to do relating to the future cannot but be altruistic, because there is no self in the future which is identical to our present selves; our future actions will benefit another self, whether that is related by the (abstract) thread of personal identity to our present self or not.

So, what does this understanding of *self-identity* mean for Hartshorne and Whitehead's interpretation of *community* and the *natural environment*, and more particularly for ethical behaviour in relation to them?

As we have seen, this kind of process thinking seems to undercut the idea of *self unity*; of *self-continuity* over time; and of *self-differentiation* from the world. The combination of these three elements means that, most fundamentally, it is difficult to separate the individual, personal self from the community and the natural world. The self spills into the universe and the universe spills into the self. Whitehead argues that:

> In fact, the world is so intimately entwined in our own natures that unconsciously we identify our more vivid perspectives of it with ourselves. For example, our bodies lie beyond our individual existence. And yet they are a part of it. We think of ourselves as so intimately entwined in bodily life that man is a complex unity. But the body is part of the external world – continuous with it. In fact, it is just as much nature as anything else there – a river, a mountain or a cloud.
>
> (Whitehead 1938b: 29–30)

Whitehead's mention of what we 'think of ourselves' here seems to suggest, at least, a passing from a *physical* identification of self and world to a *psychological* identification of self and world. If humans begin to view their selves as more diffuse, part of a constant process of change, and as interdependent with rather than insulated from all that is around them, then what they identify with as their self can change, leading to psychological identification with the world around. This possibility is not explicitly developed by Whitehead, but has been proposed by more recent process writers, such as John Cobb: 'When we have existentially realised that we are continuous with the environment, that the environment is our body, then we find new styles of life appropriate to that realisation' (Cobb and Griffin 1977: 117). The (human) community and the natural environment are thus part of the physical self and (once this realisation has been made) can be psychologically incorporated into our understanding of self-identity. Thus process thinking leads to the idea of some kind of psychologically extended self.

This understanding of self-identity, community and the natural environment links closely with the interpretation of value and ethical behaviour in process thinking outlined earlier. As Whitehead says 'the importance of others melts into the importance of the self' (Whitehead 1938b: 161). Presently existing others (whether in the human community or the natural environment) cannot be clearly distinguished from the present self; whilst the future offers the possibility of different patterns of experience, of different intensity, and different selves. Ethical behaviour does not centre on the protection, or preservation, of particular threads of experience: as Hartshorne argues, it is the quality, degree and level of happiness that matters, not the thread of identity that does or does not connect it with a particular rational self. Sometimes, of course, greatest value may be achieved in the Universe by the protection of such threads of experience; but neither these, nor the human community, nor the natural environment, are the ethical focus for process thinking. The aim is at *cosmic* good, not at the good of any individual self or community; fulfilment of God's being, which incorporates the cosmos, should be the focus of ethical behaviour (one might argue at this point, although I know no process thinker who does, that if this notion of the psychologically extended self were to be pushed far enough in process thinking, one's self-interest would be explicitly identified with the *divine* or *cosmic* interest, and thus the aim of ethical behaviour would be to fulfil the cosmic self-resembling Warwick Fox's idea of 'cosmic identification' in his book *Towards a Transpersonal Ecology*).

## Deep Ecology and process thinking

As this mention of the work of Warwick Fox suggests, my account here of the nature of the self and self-identity in process thinking resembles those of the extended self in some forms of deep (or, I should say, transpersonal) ecology. Fox also argues for the 'dissolution of the boundaries' between the individual and her environment, and focuses on the importance of 'psychological identity beyond the confines of the person' (Fox 1990: 268). He divides this kind of identification into categories, which he calls 'ontological identification' (identification with what exists, that which stands out of nothingness) 'cosmological identification' (identification with the All, the underlying oneness of the cosmos) and 'personal identification' (identification with the closest and most personal beings). Although there are clearly similarities here to the kind of self envisaged in the work of Whitehead and Hartshorne, there are also some differences. These differences are interesting in themselves, but also raise the question whether process thinking is vulnerable to the same kinds of criticism as has been levelled at deep/transpersonal ecology, in particular in relation to the idea of the extended self.

Some of these differences, whilst fundamental and structural, probably have little practical effect (in terms of ethical behaviour for instance). For example, Naess, influenced by Spinoza, argues that existence has an underlying 'oneness'. Whitehead, who is also influenced by Spinoza, in contrast argues that Spinoza's thought should be inverted: 'Spinoza bases his philosophy upon the monistic substance, of which the actual occasions are inferior modes. The philosophy of organism inverts this point of view (Whitehead 1978: 81). For Whitehead, the foundational significance of the actual entities means that multiplicity is most fundamental to the Universe, rather than unity. This is clearly a difference of some metaphysical significance. But in practice, the outcome may be similar. Both Naess' and Whitehead's perspectives ultimately tend to dissolve the idea of the individual self – process thinking into the Many, Deep Ecology into the One.

However, other differences between process thinking and Deep Ecology are both metaphysically and ethically significant. The explicit theism of process thinking means that the aim of ethical behaviour is to contribute value to God's consequent nature. Thus process thinking has, as it were, a place where value is summed; and because of this, a way of deciding which action is ethically correct in every circumstance. Problems of ethical decision-making for process thinkers centre on ignorance about what God would find most fulfilling, rather than on fundamentally irresolvable value conflicts. In Deep Ecology – or at least many forms of it – God does not perform this role. A deep/transpersonal ecologist like Fox may argue for an extended self, and suggest that that extended self 'impartially identifies with all entities'. (Fox 1990: 268.) But how then should such a deep or transpersonal ecologist make behavioural decisions? At its simplest level, the aim would be at the realisation of the Many with whom one is identified (although this does not resolve the issue of how such realisation is achieved where the realisation of one organism is dependent on the demise of another). But also this analysis suggests, curiously, that despite a belief in the primordial multiplicity of the universe, the behaviour of 2 process thinker would be aimed at realising the One (i.e., fulfilling God's being) whilst despite its belief in underlying Oneness, the aim of a deep ecologist would be realisation of the Many.

In a world without 'tragic conflict' (to use Hartshorne's apt expression) this distinction would be of no practical significance. But where there is conflict and limitation, different practical problems result. Where one had (for instance) to make a choice which would harm some living beings and benefit others, a deep/transpersonal ecologist would face the problem of deciding which being to harm. If one identifies with all parties equally, as part of one's self, then how does one judge between them? (Deep ecologists have, of course, suggested various ways of handling this problem.) Whitehead's and Hartshorne's process thinking, in contrast, has a principle for making such decisions: that which produces the most overall value, in terms of valuable experiences, and is therefore most fulfilling to God's being. This, like other consequentialist approaches to ethics, raises problems of *justice* rather than adjudication. This problem is particularly acute in process thinking where living beings are more prominently characterised as flows of experiences rather than as individuals with strong self-identities. It is valuable experiences, not individual self-identities which are of ethical significance; and the maximising of such valuable experiences may result in harm to, or loss of, some (abstract) individual self-identities for the ultimate aim of fulfilling God's being.

This problem of justice is just one of the difficulties attaching to ideas of identity, community and the natural environment in process thinking. A further problem relates to the whole idea of the 'extended' self. The metaphysics of process thinking, as we have seen, undercuts the idea of a discrete, substantial human self and results in a diffuse self resembling, in some respects, the extended self suggested in some forms of Deep Ecology. But a variety of criticisms of such 'extended self' theories have been directed at deep ecologists. It has been suggested (e.g. by Sylvan 1985, 1990; Dobson 1990) that the idea is anthropocentric or even egocentric, and that extended self theory implies that the flourishing of the community and the natural environment is desirable only or primarily as instrumental to one's own flourishing. Luke (1988) argues that the idea of the extended self projects humanity's alienated self understanding into the natural environment and seeks to find its own permanence and meaning in it. Plumwood (1993: 174) maintains that the idea of the extended self means the loss of the natural environment as Other 'The essential tension between different and alike is lost; only one element, the Self is left'. Most generally, the criticism is that the extended self in some sense humanises the natural world, failing to recognise and to respect difference.

It is interesting to consider how far similar criticisms may be levelled at process thinking. First, inasmuch as more recent process thinkers such as Cobb have developed the idea of a psychologically extended self, the same critique may apply. But even where the specific idea of the extended self is largely undeveloped, as in Whitehead's work, the question of humanisation of the natural world (and indeed the Universe) is acute. This is not because Whitehead subsumes the world into an extended self, but rather because he assumes that the Universe is composed from experiencing actual occasions, occasions derived from his understanding of human experience. The universe is transformed into a place composed from rudimentary human beings. Human experience becomes – to use Geertz's terminology – both the *model of* and the *model for* the entire universe. As Eagleton comments about Hegel's system, he is 'modelling Nature itself after the freely self-generative subject, thus grounding that subject in a world whose structure it shares' (Eagleton 1990: 131). Even God, as the Christian theologian Colin Gunton points out, is modelled on human experience, which constitutes 'an example on which to found the generalised description of metaphysics' (Gunton 1972: 90). Furthermore, we might raise questions about what exactly Whitehead understands 'human experience' to be. Is there such a thing (or should I say, process) as *generic* human experience? Or just the particular experiences of individuals (in this case, Whitehead's own) which are regionally, temporally, racially, gender and species specific? If this is the case, the fundamental metaphysical premise of process thinking – that the world is composed from experiencing actual occasions – fails, like the extended self in Deep Ecology – to recognise difference. Where one might see the extended self in Deep Ecology as proposing a Great Me which encompasses the universe, process thinking proposes a multitude of tiny 'me's' which humanise and, by doing so, familiarise the universe.

## In conclusion

Whitehead's and Hartshorne's process thinking presents a complex and sophisticated metaphysical underpinning for a theory of self and self-identity (which in this paper I have been able to do little more than outline and summarise). This construction of the self has significant implications for understanding of the (human) community and the natural environment, and for what might, in these contexts, constitute ethical behaviour. However, a number of difficulties with such interpretations arise; some of which process thinkers share with deep ecologists who propose a form of 'extended self'; others of which relate to the generalisation of experience throughout the universe. These difficulties are, I would maintain, substantial enough to raise doubts about the suggestion that process thinking offers a metaphysical approach well suited to deal with environmental questions (see, for instance, Mar Gregorias, 1980). Indeed, given such difficulties, process thinking, both as a model *of* the world and a model *for* behaviour in it, should be treated with extreme caution.

## Note

1 But see extended discussion of this in Palmer 1998.

## References

Cobb, John and Griffin, David 1977. *Process Theology: An Introductory Exposition.* Belfast: Christian Journals Ltd.
Dobson, Andrew 1990. *Green Political Thought.* London: Unwin Hyman.
Eagleton, Terry 1990. *The Ideology of the Aesthetic.* Oxford: Basil Blackwell.
Fox, Warwick 1990. *Towards a Transpersonal Ecology.* Boston: Shambhala Press.
Gunton, Colin 1972. A Comparison of the Doctrine of God in Process Theology and in Karl Barth. D.Phil Thesis, Oxford.
Hartshorne, Charles 1970. *Creative Synthesis and Philosophic Method.* La Salle: Open Court.
Hartshorne, Charles 1974. 'Beyond Enlightened Self-Interest'. Etics **84**(3): 201–216.
Hartshorne, Charles 1984. *Omnipotence and Other Theological Mistakes.* New York: SUNY.
Luke, Stephen 1988. 'The Dreams of Deep Ecology'. *Telos* Summer 65–92.
Mar Gregorias, Paulos 1980. *The Human Presence: An Orthodox View of Nature.* Madras: Christian Literature Society.
Palmer, C. 1998. *Environmental Ethics and Process Thinking.* Oxford: Oxford University Press.
Plumwood, Val 1993. *Feminism and the Mastery of Nature.* London: Routledge.
Sylvan, Richard 1985. 'A Critique of Deep Ecology'. *Radical Philosophy* **40**: 2–12 and **41**: 10–22.
Sylvan, Richard 1990. 'A Critique of (Wild) Western Deep Ecology'. Discussion Paper in Environmental Philosophy no. 18 ANU.
Whitehead, Alfred North 1920. *The Concept of Nature.* Cambridge: CUP.
Whitehead, Alfred North 1938a. *Religion in the Making.* Cambridge: CUP.
Whitehead, Alfred North 1938b. *Science and the Modern World* [1925]. Harmondsworth: Penguin.
Whitehead, Alfred North 1978. *Process and Reality: The Corrected Edition.* New York: The Free Press, Macmillan.

# Part 4

# GAIA

89

# GEOPHYSIOLOGY—THE SCIENCE OF GAIA

*James E. Lovelock*

Source: Stephen Schneider and Penelope Boston (eds), *Scientists on Gaia*, Cambridge, MA: MIT Press, 1991, pp. 1–10.

There is growing recognition of the inadequacy of the separated disciplinary approach for the solution of planetary-scale problems. To understand even the atmosphere, which is the simplest of the planetary compartments, knowledge of geophysics is not enough; chemistry and biology are also needed. It might seem that research teams that included experts in each different discipline would resolve the problem, but anyone who has attended gatherings of experts knows that each expert speaks but does not or cannot listen. What might help would be a broader-based general science that provided an environment within which the separate disciplines could interact.

In many ways, contemporary concerns arising from the consequences of changes made by humans in the composition of the atmosphere and the nature of the land surface and biota echo similar concerns raised about the human body early in the development of medicine. In the late nineteenth century, the sciences of biochemistry and microbiology were well advanced but largely disconnected and not very helpful to those with disease. Advances in medicine were, however, vastly enabled by the existence of the general science of physiology, a science that was at once transdisciplinary and also recognized the essentially emergent properties of a living organism. If one were interested in how we maintain our core temperatures at 37°C, a biochemical approach to the problem would be fruitless. Temperature regulation is a systems control problem and within the remit of physiology. But by starting with physiology, the biochemical aspects involving, for example, oxidative metabolism, then fit naturally into place.

This chapter puts forward an analogous earth science geophysiology as the transdisciplinary environment for studying planetary-scale problems, particularly problems involving a wide range of disciplines and for which it is postulated although not proved that emergent properties exist. That is, for practical purposes it may be useful to consider the earth as if it were a living organism.

Before the nineteenth century, scientists were comfortable with the notion of a living earth. One of them was James Hutton, who has often been called the father of geology. Lecturing before the Royal Society of Edinburgh in 1785, Hutton said, "I consider the Earth to be a superorganism and that its proper study should be by physiology." He belonged to the Circulation Society, a scientific society that was inspired by the discoveries of physiology such as the circulation of the blood and the connection between oxygen and life. He applied these ideas to his view of the hydrological cycle and the movements of the nutritious elements of the earth.

James Hutton's wholesome view of the Earth was discarded early in the last century. I think this may have been a consequence of a growing interest in origins and in evolutionary theories for both earth and life scientists. For biologists, there was Darwin's great vision of the evolution of the species of organisms by natural selection. For geologists, there was the wholly independent theory that the evolution of the material environment was simply a matter of chemical and physical determinism.

The divorce of the earth and life sciences in the nineteenth century was inevitable. Not only was there a rapid increase in the supply of information about the earth as exploration and exploitation developed, but the techniques for studying organisms were very different from those for studying the ocean, the air, and the rocks. It must have been an exciting period of science. Few were inclined to stand back and take a broader view or try to keep alive Hutton's superorganism theory. What is remarkable is not the division of the sciences but that two distinct and very different theories of evolution could coexist even until today.

The reason the division has endured is, I think, a mutual acceptance by earth and life scientists of the anaesthetic notion of adaptation. Biologists have assumed that the physical and chemical world evolves according to rules laid down in the geology or the biogeochemistry department of their university and that the details of this material evolution, although interesting, need not concern them in their quest to understand the evolution of the organisms. Biologists are comfortable with the notion that whatever happens to the environment, organisms will *adapt.*

In a similar way, earth scientists were happy to accept without question their biological colleagues' idea of adaptation because it freed them of any need to constrain their earth models on account of the needs of organisms. After all, there were organisms living in hot springs at 100°C and others at the freezing point: a wide enough range for climatologists.

Adaptation is a dubious notion, for in the real world the environment, to which organisms are adapting, is determined by their neighbors' activities, rather than by the blind forces of chemistry and physics alone. In such a world, changing the environment could be part of the game, and it would be absurd to suppose that organisms would refrain from changing their material environment if by so doing they could leave more progeny. In his time, of course, Darwin did not know, as we do now, that the air we breathe, the oceans, and the rocks are all either the direct products of living organisms or else have been greatly modified by their presence. In no way do organisms simply "adapt" to a dead world determined by physics and chemistry alone. They live in a world that is the breath and bones of their ancestors and that they are now sustaining.

It was not until the present century that a minority opinion led by the Russian scientist, Vernadsky (1945), saw that the separation of the earth and life sciences had become too extreme. Vernadsky was the father of the modern science of biogeochemistry. He and his successors, like Hutchinson (1945) and Redfield (1958), recognized that life and the physical environment interact, that gases like oxygen and methane are biological products. Where they differ from proponents of the Gaia hypothesis is that they still accepted, without question, the dogma of mainstream biology, which is that organisms simply adapt to changes in their material environment, without then considering the consequences of such an adaptation in an environment modified by the organisms themselves. Vernadsky's worldview has been developed and expanded in coevolutionary theory and in biogeochemical models. Coevolution is rather like a platonic friendship. Biologists and geologists remain friends but never move on to an intimate, close coupled, relationship. Coevolution theory includes no active regulation of the chemical composition and climate of the earth by the system comprising the biota and their material environment; most importantly, it does not see the earth as alive in any sense, nor even as a physiological system.

## What is Gaia

Like coevolution, Gaia rejects the apartheid of Victorian biology and geology, but it goes much further. Gaia theory is about the evolution of a tightly coupled system whose constituents are the biota and their material environment, which comprises the atmosphere, the oceans, and the surface rocks. Self-regulation of important properties, such as climate and chemical composition, is seen as a consequence of this evolutionary process. Like living organisms and many closed-loop self-regulating systems, it would be expected to show emergent properties; that is, the whole will be more than the sum of the parts. This kind of system is notoriously difficult, if not impossible, to explain by cause-and-effect logic, as practicing inventors know to their cost. It is doubtful also if the fashionable and trendy Popperian falsification tests, so valuable for theories in physics, are really applicable to such systems. Consider, for example, the problem faced by someone unfamiliar with earth-based life of designing a test to show that a Lombardy poplar tree was alive. These trees are all males and, hence, can be propagated only by cuttings—90% or more of a fully grown tree is dead wood and dead bark, with just a thin skin of living tissue around the circumference of the wood. Then there is the question, "What does the word *alive* mean?" Biologists studiously avoid trying to answer it.

When biochemists examine a live animal, they know that many of its reactions and processes can be adequately described by simple deterministic physics and chemistry. But they also accept the legitimacy of physiology. They know that for an intact animal, homeostasis, the automatic regulation of temperature, and chemical composition, although it involves chemistry, are emergent properties. Such properties require physiology for their explanation and understanding. I think that the same can be said of the earth. If it is a superorganism, then explaining its reactions and processes requires physiology as well as chemistry and physics.

## Evidence for Gaia

The first mention of Gaia was a brief paper in *Atmospheric Environment* (Lovelock 1972). Earlier work in connection with the National Aeronautics and Space Administration planetary exploration program had suggested that atmospheric compositional evidence could provide prima facie evidence for the existence of life on a planet (Hitchcock and Lovelock 1967). Briefly, a dead planet would have an atmosphere characteristic of the abiological steady state and not far from chemical equilibrium. By contrast, a planet with life would be obliged to use its atmosphere as a transfer medium for waste products and raw materials. Such a use of the atmosphere would introduce disequilibrium among the chemical components, and this might reveal the presence of life. When the terrestrial planets are compared, Mars and Venus are found to have atmospheres dominated by carbon dioxide and close to the abiological steady state. Earth, by contrast, has an atmosphere in which profoundly incompatible gases such as methane and oxygen coexist. This disequilibrium reveals the presence of life. The persistence of the disequilibrium at a steady state for periods much longer than the residence times of the gases suggests the presence of an active control system regulating atmospheric composition. As we may soon discover, the unregulated injection of methane could be seriously destabilizing. Figure 1 illustrates the fluxes of gases through the Earth's present atmosphere in comparison with the fluxes expected of a dead Earth.

Until the San Diego meeting, Gaia theory had received little or no financial or other support from the scientific establishment. About five scientists worldwide worked on the topic part time. In such circumstances, it was not practical to strive hard to develop tests for the existence or otherwise of Gaian systems. It seemed better to go into the world and collect information, inspired by the predictions from the theory. Whether it was right or wrong seemed to matter less than that the quest was objective.

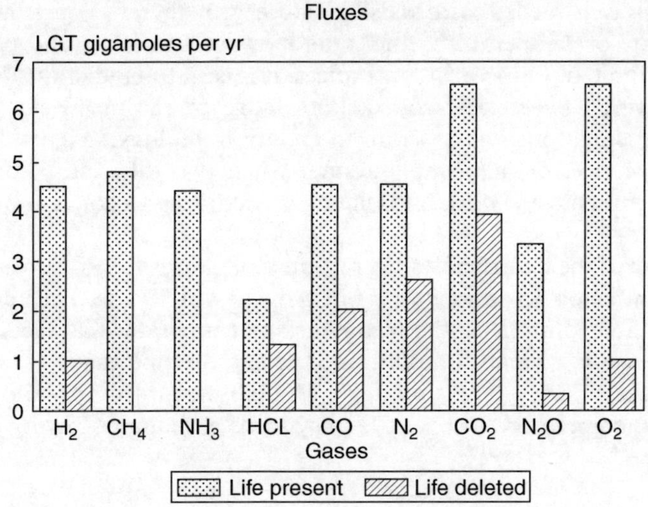


*Figure 1* The fluxes of gases through the present atmosphere compared with the fluxes of the same gases expected for a dead Earth. The vertical scale is in logarithmic decade units of gigamoles per year.

A good example of this was the expedition in 1972 aboard the research ship Shackleton from the United Kingdom to Antarctica and back to seek the presence of sulfur and iodine compounds in the ocean and determine whether it was possible to transfer these elements from the sea to the air and, hence, back to the land surfaces. This voyage found that the gases dimethylsulfide, methyl iodide, and carbon disulfide were ubiquitous throughout the ocean environment. It is relevant to note that before the expedition, peer review committees argued that the search for such compounds was pointless. This was a time when research funds were freely available, yet the expedition was not supported.

A more practical approach is to make models of Gaia and then see how well they can be mapped onto the observed systems. But the feedback loops linking life with its environment are so numerous and so intricate that there seems little chance of quantifying or of understanding them. It occurred to me late in 1981 to reduce the environment to a single variable, temperature, and the biota to a single species, daisies.

Imagine a planet like the Earth; it travels at the Earth's orbit around a star of the same mass and composition as our Sun. This planet spins like the Earth, but its atmosphere has few clouds and a constant low concentration of greenhouse gases. In these circumstances, the mean surface temperature is given by the Stephan-Boltzman expression of the balance between the radiation received from the star and the heat lost by radiation from the planet to space; the albedo of the planet determines its temperature. Assume that this planet is well seeded with daisies whose growth rate is a simple parabolic function of temperature, and that it is well watered and that nutrients are not limited. In these circumstances, it is easy to predict the area of the planet covered by daisies from a knowledge of the mean surface temperature and equations taken from population biology. Figure 2A illustrates the evolution of this simple system when two different-colored daisy species are present, one dark and one light according to conventional wisdom.

The upper panel of the figure shows the response of daisies to temperature; daisies do not grow below 5° or above 40° but grow best at 22.5°C. The lower panel illustrates the evolution of the mean surface temperature as the star increases in luminosity, a smooth monotonic temperature increase.

In Figure 2B, the same system is modeled as a close-coupled physiology. When the surface temperature reaches 5°, daisy seeds germinate. During the

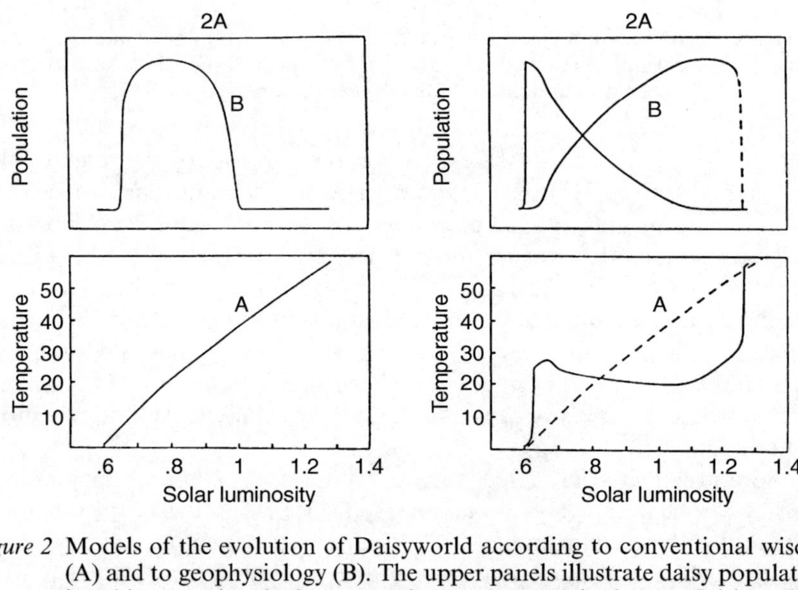


*Figure 2* Models of the evolution of Daisyworld according to conventional wisdom (A) and to geophysiology (B). The upper panels illustrate daisy populations in arbitrary units; the lower panels, temperatures in degrees Celsius. Going from left to right along the horizontal axis, the star's luminosity increases from 60% to 140% of that of our own Sun. *A* illustrates how the physicists and the biologists in complete isolation calculate their view of the evolution of the planet. According to this conventional wisdom, the daisies can only respond or adapt to changes in temperature. When it becomes too hot for comfort, they will die. But in the Gaian Daisyworld (*B*), the ecosystem can respond by the competitive growth of the dark and light daisies and regulates the temperature over a wide range of solar luminosity. The dashed line in the lower panel in *B* shows how the temperature would rise on a lifeless Daisyworld.

first season, dark-colored daisies will be at an advantage, because they will be warmer than the planetary surface. Light-colored daisies will be at a disadvantage, because by reflecting sunlight they will be cooler than the surface. At the end of the season, many more dark daisy seeds will remain in the soil. When the next season begins, dark daisies will flourish and soon warm, not just themselves, but their locality and, as they spread, the region and eventually the whole planet. The figure illustrates an explosive growth of both temperature and dark daisy population. The spread of dark daisies will eventually be limited by their decline in growth rate at temperatures above 22.5°C and by competition from light-colored daisies. As the star evolves, the dark and light daisy populations adjust according to the simple population biology equations of Carter and Prince (1981). The planetary temperature moves from just above the optimum for daisy growth at low solar luminosity to just below the optimum at high solar luminosity. Eventually the output of heat from the star is too great for regulation, and the plants die.

This simple model is a graphic illustration of a geophysiological process. The only criticism so far received has been the suggestion by biologists that in a real world there would be daisy species that "cheated," that is, took advantage by saving the energy needed to make the dark or light pigment and so took over the planet and returned the system to the model illustrated in Figure 1. Figure 3 is a model in which daisies having a neutral color, that

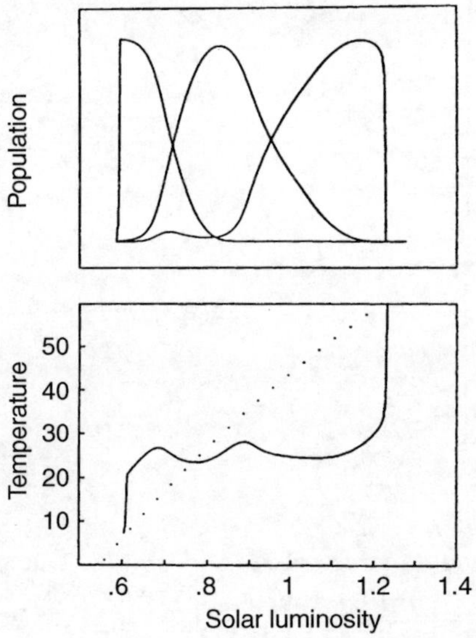


*Figure 3* The evolution of the climate on a three-species Daisyworld with dark, gray, and light daisies present. By comparison, the dashed line in the lower panel represents the temperature evolution in the absence of life.

of the bare planetary surface, are included. In this experiment, even when the neutral-colored daisies were given a 5% increase in growth rate, there was no indication of a planetary takeover or failure of regulation. At low temperatures, only dark daisies are fit to grow; at high temperatures, only light daisies are fit. Neutral daisies grow only when there is little need for regulation. An important point here illustrated is that Gaia theory and coevolution are not always mutually exclusive. Organisms do not strive ostentatiously to regulate their environment when regulation is not needed.

Figure 4 illustrates a model that included ten different-colored daisy species, their albedos ranging in evenly spaced steps from dark to light. The regulation of the mean surface temperature (lower panel) is more accurate than in the two- and three-species models. The middle panel shows the populations of the different-colored daisies and the upper panel the diversity index of the ecosystem as the model evolved.

The stable coexistence of three or more species in a population biology model is contrary to the experience of modelers in that field of science. Models of the competition of three or more species, like the three-body problem of astrophysics, tend to be unstable and chaotic. The stability of

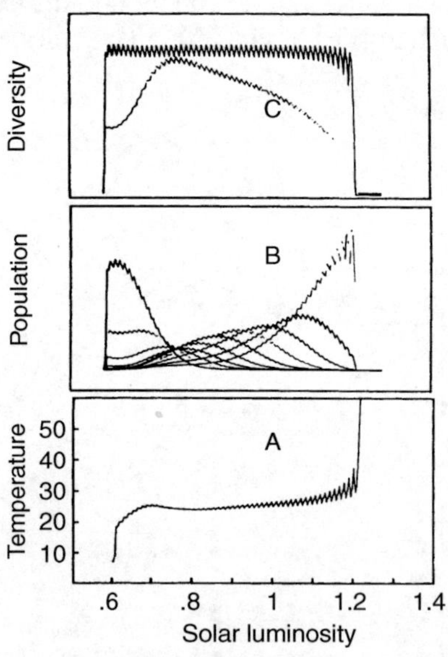


*Figure 4* The evolution of the climate on a ten-species Daisyworld. The lower panel illustrates planetary temperature, the dashed curve for no life present, and the solid line with daisies. The middle panel shows the populations of the twenty different-colored daisies, with the darkest appearing first (left) and the lightest last (right). The upper panel illustrates diversity, seen to be maximum when the system temperature is closest to optimum.

Daisyworld is even more remarkable given that no attempt was made to linearize the equations used in the model. Not only is the model naturally stable, but it will resist severe perturbations, such as the sudden death of half or more of all the daisies, and then recover homeostasis when the perturbation is removed. The models can include herbivores to graze the daisies and carnivores to cull the grazers without significant loss of stability.

Another scientist like James Hutton was Alfred Lotka, the father of theoretical ecology. Like Hutton, Lotka saw the science he founded develop in a way that he never expected or intended. The unwise isolation of biology from geology has led population biology into a mathematical cul-de-sac where the phenomena of complex dynamics are investigated rather than ecology. For over sixty years, theoretical ecology has ignored Alfred Lotka's wise advice. In his book, *Physical Biology*, written in 1925, he said that

> this fact deserves emphasis. It is customary to discuss the "evolution of a species of organisms." As we proceed we shall see many reasons why we should constantly take in view the evolution, as a whole, of the system (organism plus environment). It may appear at first sight as if it should prove a more complicated problem than the consideration of a part only of the system. But it will become apparent, as we proceed, that the physical laws governing evolution in all probability take on a simpler form when referred to the system as a whole than to any portion thereof.
>
> It is not so much the organism or the species that evolves, but the entire system, species plus environment. The two are inseparable.

Daisyworld, as I have described it, is just an invention, a demonstration model to illustrate how I thought Gaia worked and why foresight and planning need not be invoked to explain automatic regulation. But as we shall see when the details are fleshed out, it becomes a generality and a theoretical basis for Gaia. I would like to think of it as the kind of model Alfred Lotka had in mind but could not develop, because in his day there were no computers to carry out the immense task that the hand calculation of even a simple daisy model requires.

Andrew Watson and I (1984) described the mathematical basis of Daisyworld in *Tellus.* But at the time, neither of us realized its unusual properties nor the extent to which it is an expression of the general theory of Gaia. The paper explains the essential mechanism by which homeostasis is maintained. The Daisyworld thermostat has no set point; instead the system always moves to a stable state in which the relationship between daisy population and planetary temperature and that between temperature and daisy growth converge. The systems seek the most comfortable state rather like a cat as it turns and moves before settling.

Inventions often work well but are difficult to explain. Engineers and physiologists have long been aware of the subtleties of feedback. Homeostasis is possible only when feedback is applied at the right amplitude and phase and when the system time constants are appropriate. Both positive and negative feedback can lead to stability or instability according to the timing of their application.

Theoretical ecology models, notorious for their intractable mathematics, would not surprise an engineer who would see them in his words as "open-loop system," in which feedback was applied or happened by chance in an arbitrary manner. By contrast, geophysiological models, such as Daisyworld, include feedback, negative and positive, in a coherent manner. They are as a consequence robust and stable and happily accommodate any number of nonlinear equations and still prefer to relate with stable attractors.

Figure 5 compares the unstable and chaotic behavior (lower panel) of a model of an ecosystem of daisies, rabbits, and foxes according to population

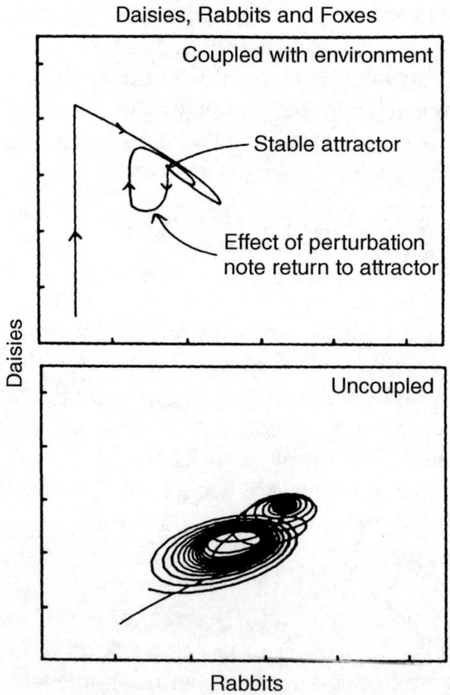


*Figure 5* Comparison of the stability of a Daisyworld that included rabbits to eat the daisies and foxes to cull the rabbits (upper panel) with the instability of a population biology model of daisies, rabbits, and foxes (lower panel). The model in the upper diagram included environmental feedback; that in the lower diagram did not.

biology with the calm stability of the same ecosystem (upper panel) when feedback from the environment is included, as in a Daisyworld.

But what of biogeochemical box models? Are these any more stable? Ann Henderson-Sellers drew my attention to a paper that suggested that biogeochemical models were also prone to chaotic behavior or to an unusual sensitivity to the choice of initial conditions. Geophysiology seems to be a way to avoid these distractions. The concluding model is of the period at the end of the Archean when oxygen first began to dominate the chemistry of the atmosphere. During the long period of the Archean, the biosphere was run by bacteria. The primary producers were cyanobacteria, and the oxygen they made was almost entirely used up to oxidize reducing compounds such as ferrous iron and sulfides present in and continuously released to the environment. The organic matter of the cyanobacteria was most probably digested by methanogens. In the Archean period, cyanobacteria would be like the white daisies of Daisyworld, tending to cool by removing carbon dioxide, and the methanogens would be like the dark daisies, tending to warm by adding methane to the atmosphere. A geophysiological model constructed this way settles down to a constant climate and bacterial population and sustains an atmosphere where methane is the dominant redox gas and traces only of oxygen are present. The continuous leak of carbon to the sediments and perhaps also hydrogen to space would have slowly driven the system oxidizing until quite suddenly oxygen became the dominant atmospheric gas.

In this model, as in Daisyworld, a key factor is the function that set the bounds of the environment for the biota. The same parabolic relationship between growth and temperature was used as in the daisy models but, in addition, a similar function was introduced for oxygen. Figure 6 shows how

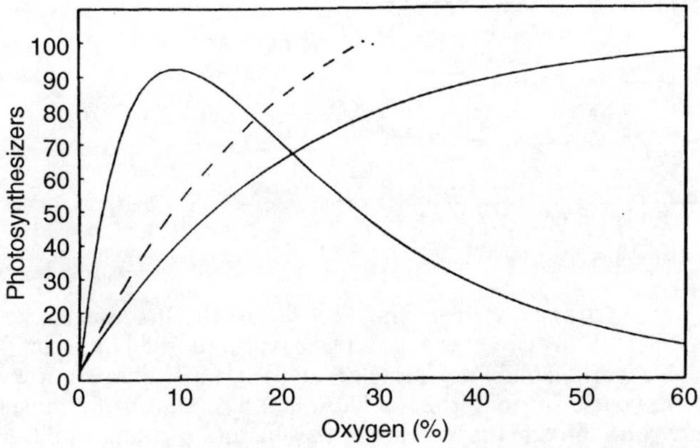


*Figure 6* The effect of oxygen on the growth of organisms (solid line) and the effect of the presense of organisms on the abundance of oxygen (dashed line). Where the two curves intersect is the level of oxygen at which the system regulates.

the growth of an ecosystem might increase as oxygen rises from zero. Oxygen increases the rate of rock weathering and, hence, the supply of nutrients, and also increases the rate of carbon cycling through oxidative metabolism. Too much oxygen is, however, toxic. The bounds for oxygen in the figure are set by two simple exponential relationships describing nutrition and toxicity.

Other bounds, such as those set by the limitations of pH, ionic strength, and the supply of nutrients, could have been included. Hutchinson saw the niche as a hypervolume negotiated among the species. In a similar way, I see the physical and chemical bounds to growth form a hypervolume whose surface intersects that of the hypervolume expressing the environmental effects of the species.

The model included geochemical data on the carbon dioxide cycle taken from Holland's book (1984). The rate of weathering was assumed to be a function of the biomass as well as of the abundance of oxygen and carbon dioxide. Figure 7 illustrates (in the lower window) the regulation of the

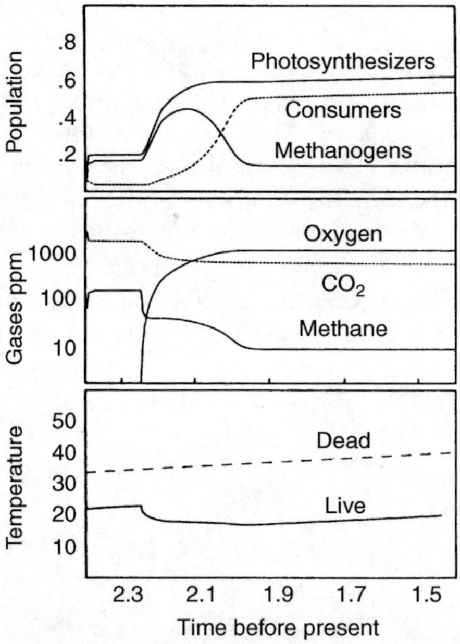


*Figure 7* Model of the transition from the Archean to the Proterozoic period. Lower panel shows climate with a lifeless world (dashed line) compared with a live world (solid line). Note the sudden fall of temperature when oxygen appears. The middle panel shows the abundance of atmospheric gases, carbon dioxide (dashed line), oxygen, and methane (solid lines). The upper panel illustrates the changes in population of the ecosystems as the transition is entered and passed. Note how both photosynthesizers and methanogens increase when oxygen first appears and how methanogens fall back to a steady level when the oxygen-breathing consumers (dashed line) become established.

temperature: constant during the Archean but falling to a lower steady state in the Proterozoic period after the appearance of oxygen as a dominant gas. The fall in temperature is due to the removal of most of the methane greenhouse effect. The middle window shows the abundances of the three gases—methane, carbon dioxide, and oxygen—during the evolution of the model. The upper panel shows the populations of the three main ecosystems: cyanobacteria, methanogens, and consumers.

Like Daisyworld, this model is just an invention and is not intended to describe the real world of those remote times. What it does illustrate is the remarkable mathematical stability of geophysiological models. Climate, three ecosystems, and three gases are regulated simultaneously and while the model is being continuously perturbed by an increasing solar luminosity. The stable homeostasis of the system is independent of a wide range of initial conditions and other perturbations. Perhaps most usefully, the values of the environmental quantities, temperature, and gas abundances it predicts are always realistic for the organisms. My purpose in making the model was to illustrate how I think Gaia works.

I do not disagree with those who propose that some, or even a large proportion, of the total regulation of any chosen earth property can be explained by deterministic chemistry and physics. Living systems use chemistry economically; they do not strive ostentatiously to do better than blind chemistry or physics—there is no need. The point about Gaia is that it offers a new way of looking at the earth and makes predictions that can be tested experimentally. Had it not been for the curiosity stimulated by thoughts on the mechanisms of Gaia, none of the important trace gases—dimethyl sulfide, carbon disulfide, methyl iodide, and chloride—would not have been sought and found when they were.

To conclude, Gaia theory provokes us to think about three things:

1 Life is a planetary-scale phenonmenon. There cannot be sparse life on a planet. It would be as unstable as half of an animal. Living organisms must regulate their planet, otherwise the ineluctable forces of physical and chemical evolution would render it uninhabitable.
2 Gaia theory adds to Darwin's great vision. There is no longer any need to consider the evolution of the species separetely from the evolution of their environment. The two processes are tightly coupled as a single indivisible process. It is not enough to say that the organism that leaves the most progeny succeeds. Success also depends on coherent coupling between the evolution of the organism and the evolution of its material environment.
3 Lastly, it may turn out that the gift of Gaia theory to geophysics is the reduction of Alfred Lotka's insight to practice. It may provide a new way to look at the earth mathematically that joyfully accepts the nonlinearity of nature without being overwhelmed by the limitations imposed by the chaos of complex dynamics.

## References

Carter, R.N., and Prince, S.D. 1981. Epidemic models used to explain biographical distribution limits. *Nature*, 213, 644–645.

Charlson, R.J., Lovelock, J.E., Andreae, M.O., and Warren, S.J. 1987. Ocean phytoplankton, atmospheric sulfur, cloud albedo and climate. *Nature*, 326, 655–661.

Hitchcock, D.R., and Lovelock, J.E. 1967. Life detection by atmospheric analysis. *Icarus*, 7, 149–159.

Holland, H.D. 1984. *The Chemical Evolution of the Atmosphere and the Oceans.* Princeton, NJ: Princeton University Press, 539.

Hutchinson, G.E. 1954. *Biochemistry of the terrestrial atmosphere. The Solar System.* Chapter 8. Kuiper, Ed., Chicago: The University of Chicago Press.

Hutton, J. 1788. Theory of the Earth; or an investigation of the laws observable in the composition, dissolution, and restoration of land upon the globe. *Royal Society of Edinburgh, Transactions*, 1, 209–304.

Lovelock, J.E. 1972. Gaia as seen through the atmosphere. *Atmospheric Environment,* 6, 579–580.

Lovelock, J.E., Maggs, R.J., and Rasmussen, R.A. 1972. Atmospheric dimethylsulphide and the natural sulphur cycle. *Nature*, 237, 452–453.

Margulis, L., and Lovelock, J.E. 1974. Biological modulation of the Earth's atmosphere. *Icarus*, 21, 471–489.

Redfield, A.C. 1958. The biological control of chemical factors in the environment. *American Scientist*, 46, 205–221.

Vernadsky, V. 1945. The biosphere and the noosphere. *American Scientist*, 33, 1–12.

Watson, A.J., and Lovelock, J.E. 1983. Biological homeostasis of the global environment: The parable of Daisyworld. *Tellus*, 35B, 284–289.

# 90

# FORMS OF GAIAN ETHICS

*Anthony Weston**

Source: *Environmental Ethics*, 9 (1987), 217–30.

James Lovelock's "Gaia hypothesis"—the suggestion that life on Earth functions in essential ways as one organism, as a single living entity—is extraordinarily suggestive for environmental philosophy. *What* exactly it suggests, however, is not yet so clear. Although many of Lovelock's own ethical conclusions are rather distressing for environmental ethics, there are other possible approaches to the Gaia Hypothesis. Ethical philosophers might take Gaia to be analogous to a "person" and thus to have the same sorts of values that more familiar sorts of persons have. Deep ecologists might find in the Gaia hypothesis a means by which to transform and reunderstand our concrete experience of the world. This essay canvasses some of the strengths, weaknesses, and possibilities of each approach.

## I The Gaia hypothesis

In the early 1960s NASA asked James Lovelock, an English chemist and inventor, to think about the sorts of experiments that might be included on the Viking spacecraft to test for life on Mars. Lovelock started thinking about the ways in which life on Earth reveals itself through its effects on atmospheric composition. In fact, he realized, the composition of the Earth's atmosphere is extraordinarily improbable. In the normal course of events without life, the best guess is that the Earth's atmosphere would consist almost entirely of carbon dioxide ($CO_2$), like the atmospheres of Earth's sister planets, Mars and Venus. In fact, however, our atmosphere contains 78% nitrogen, 21% oxygen, and next to no $CO_2$ at all. Even Earth's oceans depart dramatically from their expected composition in a lifeless world.[1] Lovelock argued that these drastic departures from chemical equilibrium must be attributed to the complex and interactive processes of living beings: life

* The author wishes to thank David Abram for much helpful criticism during the writing of this paper, and two anonymous reviewers for *Environmental Ethics* for spurring a very useful rewrite.

announces itself by radical local reversals of entropy. He therefore suggested that we could test for life on Mars without even going there: we could already tell, from telescopic tests of the Martian atmosphere, that Mars was lifeless. Naturally this conclusion was not popular at NASA.

Not only have the atmosphere and the oceans been transformed by life; life maintains the transformed environment in a steady state. Analyses of sedimentary rocks show that for the past three and a half billion years the Earth has maintained an almost entirely constant and hospitable temperature range—the oceans never boiled or froze, and even the Ice Ages affected only 30% of the Earth's surface. During those billions of years, however, the Sun's output of energy has increased at least 30%, and there have been regular perturbations in the Earth's orbit, volcanic eruptions, etc., which have also changed the energy inputs to the Earth.[2] As a result, powerful homeostatic processes must have been at work to keep temperature so constant: adjustments in the "greenhouse gasses" ($CO_2$ and ammonia), changes in the albedo (reflectivity) of the surface and/or atmosphere, and the like. The changes in levels of $CO_2$ and ammonia appear to have been regulated by biological processes. Likewise, the albedo of the surface seems to have been drastically altered, and quickly adjusted, by changes in the amount, type, and/or color of vegetation. Lovelock has developed a model in which a biota composed merely of two kinds of daisies—differently colored, and, therefore, with different light- and heat-absorbing properties, and therefore, capable of reproducing at different rates depending on ambient radiation—can homeostatically maintain temperature across a wide range of radiation inputs.[3] Earth's atmospheric composition has also been maintained at a remarkably optimum level—a little more oxygen and the forests would spontaneously ignite; a little less and all our energy would be spent on the next breath—and these oxygen levels too seem to have been regulated by complex life processes. Methane, for example, one crucial element in those processes, is produced from grass in the guts of ruminant herbivores.

These effects are collective. They are not a product of individual organisms, or even of species, but of life as a whole, on a planetary scale and evolving over eons. The biosphere as a whole, then, functions homeostatically: it creates and maintains the conditions for its own existence and growth. And, therefore, there is a sense in which the biosphere itself can be viewed as a *single organism*. This is the crucial step. Life on Earth is not merely a collection of individual species or organisms competing with each other and going their separate ways. Instead,

> the entire range of living matter on Earth, from whales to viruses, and from oaks to algae, could be regarded as constituting a single living entity, capable of manipulating the Earth's atmosphere to suit its overall needs and endowed with faculties and powers far beyond those of its constituent parts.[4]

All of its constituent processes, from farting herbivores to coral reefs (which Lovelock thinks help control the salinity of the seas) function in an integrated and self-adjusting way. The Earth itself is alive.

*Gaia* is the Greek Earth goddess, thus reborn, as it were, under the auspices of NASA. She is not, of course, wholly new. Whitehead and Spencer, and before them Spinoza and Hegel, conceived of nature in this organismic way. On the ecological side, Frederick Clements called the climax community an organism, W. M. Wheeler generalized his organismic views from ant colonies, and the ecology group at the University of Chicago in the 1930s and 1940s was frankly organismic.[5] Leopold wrote of the whole Earth as "an organism possessing a certain kind and degree of life," "a being that was old when the morning stars sang together and, when the last of us has been gathered unto his fathers, will still be young."[6] Gaia, then, is a new incarnation of an old intuition—but an incarnation that perfectly suits our age of general systems theory and interplanetary exploration. Lovelock himself waxes lyrical over the image of the Earth taken by the Apollo astronauts from the Moon:

> ... the evolution of *Homo sapiens*, with his technological inventiveness and his increasingly subtle communications network, has vastly increased Gaia's range of perception. She is now through us awake and aware of herself. She has seen the reflection of her fair face through the eyes of the astronauts and the TV cameras of orbiting spacecraft. Our sensations of wonder and pleasure, our capacity for conscious thought and speculation, our restless curiosity and drive are hers to share....[7]

Gaia is not only alive, but awake, and not set over against us, but, perhaps, constituted in her very awareness *by* us.

## II Gaian ethics according to Lovelock

The Gaia hypothesis might seem to lead directly to a strong environmental ethic. To invoke a goddess seems to invoke a religious ethical frame as well: surely we owe her obedience or at least respect, and deliberately harming her, even in small ways, becomes sacrilege. It may come as a surprise, then, that Lovelock does not draw these conclusions. His invocation of the goddess seems to have been meant to emphasize her powers, not our responsibilities, and he is, as cell biologist and co-worker Lynn Margulis puts it, "no admirer of most environmentalists," whom he even characterizes as "misanthropes" and "Luddites."[8] Let me begin with his conclusions, both because he is after all the originator of the Gaia hypothesis and as a kind of background and challenge to the more favorable readings which the next sections consider.

Since the Gaia hypothesis maintains that the biosphere is far more resilient and adaptable than environmentalists have often feared, Lovelock is unworried about most forms of pollution. Homeostatic adaptation can adjust for major changes in energy inputs, varying fluxes of atmospheric gasses and particulates. "They very concept of pollution," he writes, "is anthropocentric, and may even be irrelevant in a Gaian context. Many so-called pollutants are naturally present, and it becomes exceedingly difficult to know at what level the appellation 'pollutant' may be justified."[9] Natural oxidation of methane in the atmosphere, for example, produces a billion tons of carbon monoxide a year. Or again, ozone cannot be necessary to life, Lovelock argues, because for the first two billion years Earth had no ozone layer at all; indeed, he argues, too much ozone may be worse than too little, and ozone depletion may therefore be a natural process, since nitrous oxide and methyl chloride, both potent ozone depleters, are produced in nature in large amounts.[10] In its youth the Earth even included enough natural uranium to set off spontaneous fission reactions. There is much more in this vein. Even nuclear war, Lovelock argues, might not significantly disturb Gaia.[11]

Paradoxically, then, a Gaian perspective may undercut rather than reinforce many of the legal safeguards that environmentalists have established. It can be argued that nature is *not* fragile, on the whole, and therefore, that many of the protections we have enacted based on our fears of its fragility are probably unnecessary. It this way, the Gaia hypothesis threatens to knock the props out from under the utilitarian sorts of arguments that persuade most policy makers and the general public.

Gaia does, however, have "vital organs," like other organisms: the wetlands, the continental shelves, the rain forests. *Here*, according to Lovelock, Gaia *is* fragile. A large number of specialized algae live in the inshore waters of the continental shelf, for example; even so seemingly innocent a project as large-scale kelp farming might radically alter the populations of these algae and thus, for one thing, radically alter the production of essential trace elements in the sea and air.[12] Oil spills in the continental shelves, the destruction of the rain forests for short-run agricultural advantage, or the continued destruction of wetlands might similarly lead to disaster. In fact, then, there is a case for *some* environmental protection on the more or less utilitarian standard that Lovelock seems to be assuming. This is surely to the good, and we may become more sensitive to the special ecological roles of areas we often neglect. But the case is still very selective. The *rest* of Gaia—in particular, the lands and waters where most humans live, lands for which we are often especially concerned—is *not* vital, and apparently can be used pretty much as we please without threatening the Gaian balance of life. Thus, there seems to be no "land ethic" in Gaia, little protection for the deep ocean, no argument for relatively lifeless landscapes. Again, of course, such arguments may well be possible on other grounds, nonutilitarian or non-Gaian.

But the Gaia hypothesis itself offers such grounds, and indeed seems to preclude any strictly ecological ground for them.

Another worry has to do with the ethical status of individuals within Gaia. The problem here in a sense reverses the "species problem" familiar in environmental ethics. Critics have noticed that "animal liberation" arguments that rest on the disvalue of animal suffering and/or death apply only to individual animals, not to species as such. The death of a starling, from this point of view, is as bad as the death of an ivory-billed woodpecker, even though starlings are everywhere and ivory-bills are nearly extinct. It seems difficult to give an adequate and nonanthropocentric account of the value of species as such. The Gaia hypothesis, however, seems to have no way of valuing *individuals*. What matters is the operation of Gaia itself, the homeostasis of planetary life, and here individuals are clearly irrelevant. Actually even species are usually irrelevant: according to Margulis, 99.99% of the species which have existed are now extinct. At most the argument vindicates the value of phyla.

Maybe this only implies, again, that the Gaia hypothesis cannot ground a *complete* environmental ethic, which surely is no great surprise. But it also—so far, anyway—seems unable to ground much at all of what we expect of an environmental ethic. Even inessential and unattractive species like the snail darter ought to have a *little* value, and it is even more disturbing when the preponderance of species and natural formations turn out to be inessential. Even plastic redwoods do not appear to be a crime from a Gaian point of view. It would also be nice if something could be said about the value of humans which avoids exalting us as the "eyes" or "brains" of Gaia—surely a disturbing vision given the history of "reason" that these valuations of human beings invoke—but which, on the other hand, doesn't leave us as merely a rather inefficient way of making methane.

Like individual animals, individual natural formations also need a voice, and don't find it here. No case can be made for preserving a wild river or a virgin forest, since their Gaian significance is minimal. The sense of returning to roots or "sources," kinship with other animals, or the values of just plain silence, do not seem essential to a Gaian outlook as developed by Lovelock. Maybe, again, we are asking too much, but nonetheless it is striking how few environmental values seem to be invoked.

Finally, perhaps confirming many of the worries so far raised, Lovelock himself is fairly sanguine about "environmental engineering." The most striking example is his recent book, with Michael Allaby, called *The Greening of Mars*, which proposes literally to export Gaia's planetary-homeostatic mechanisms to reconstitute the atmosphere of Mars, making it habitable for humans.[13] Here the underlying utilitarianism is perhaps most explicit: we would remake Mars for our own purposes. This, of course, is ecological intervention on an absolutely massive scale: even to call it "intervention" is probably an understatement. More than a few eyebrows have been raised even outside the environmentalist community.[14] Certainly there are worries, as in any other engineering project, about unintended side effects. There is, in addition, a clear disrespect for Mars as it *is*, which surely ought to count for something. Although the ethics of the utilization of extraterrestrial objects is still mostly unexplored, surely we do not want to repeat the patterns of destruction and waste which humans have consistently brought to "unutilized" parts of our own planet.

Admittedly the idea has its attractions. Since the proposal is simply to "seed" Mars, it arguably urges a *sort* of respect for the planet: Mars is not conceived merely as an inert object to be acted upon mechanically: to "bring Mars to life" is to help *it* to come to life, not to force life upon it. This is not "engineering" in the usual sense (though it *is* rather like "genetic" engineering). It also acknowledges the degree to which we are creatures of this biosphere: the only way humans can live else where in the solar system is to take Gaia with us, or to let Gaia reproduce herself where, or before, we go. This is surely to the good when put against the usual fantasies of totally engineered space stations. But it is not *all* to the good. A disturbingly blithe interventionism remains.

## III Gaia and the ethic of persons

There are other possible readings of the Gaia hypothesis. Sometimes even Lovelock does not follow the line just sketched. I consider two alternative approaches to Gaia, encompassing between them a wide range of possibilities. They are not, of course, the only alternatives, but they may serve as stand-ins for two very general, and very different, types of environmental philosophy: contemporary philosophical ethics, which I take up in this section, and "Deep Ecology," which I take up in the next.

Utilitarianism has fallen sharply out of favor with ethical philosophers in the last decade. Favored instead is an approach built around the notions of rights, justice, and respect, notions that are in turn built upon some conception of a *person* or "moral subjects."[15] Initially this approach may seem to be even more anthropocentric than the usual interpretation of utilitarianism, and so to have little relevance to Gaia. In fact, however, there is a suggestive connection.

Gaia not only somewhat like an organism, but is also somewhat like a person herself. The biosphere can be viewed as an integrated being, according to the Gaia hypothesis, and it monitors its own states and adjusts accordingly. Personifying her, calling her by name, might therefore be entirely appropriate. If so, however, then we can invoke in her defense the already established, and already central, value of persons. We might change not our ethical conceptions but our conception of the person; we might challenge not the ethical centrality of persons, but the assumption that only humans (or only middle-sized organisms like ourselves) can *count* as persons.

This idea is implicit in some popular writing about Gaia, for instance, in Theodore Roszak's book *Person/Planet*.[16] It is explicit in some of the work of the ethical philosopher Kenneth Goodpaster:

> Indeed our moral universe might contain structures *inclusive* of persons respect for which is just as incumbent upon us morally [as is respect for persons]. Such, I would want to argue, is the biosystem as a whole: not as a mere collection of biotic particles, but as an integrated, self-sustaining unity which puts solar energy to work in the service of growth and maintenance. The history of evolution is the drama of the biosystem's successful self-protection.[17]

Although Goodpaster does not exactly claim that the biosystem *is* a person, the argument clearly depends on the very close *analogy* between persons and biosphere: thus Goodpaster's stress on its unity and growth, and indeed his very use of the active voice (the biosystem "puts" solar energy to use; it protects itself, etc.). Gaia, then, is supposed to be enough *like* a person to be entitled to the same sort of respect as individual human persons.

A further argument along these lines might take a cue from Lovelock's lyricism about humans as the "mind" of Gaia. Goodpaster stresses that Gaian includes human persons, but does not explain why this is relevant to *Gaia*'s value. Possibly, at the back of his mind, he is thinking what Lovelock argues explicitly, that humans play a specific and essential role within the Gaian economy, a role connected with our minds and so with the source of our individual personhood. Gaia then becomes quite literally a person; what little she otherwise lacks for personhood she gets from us.[18]

There is much to be said for this general approach. It is powerful both practically and conceptually because it appeals to a well-understood and widely accepted secular ethical paradigm. Ethical history has been a long struggle to morally acknowledge previously unacknowledged persons—barbarians, women etc.—and at each stage it seemed unthinkable to the then favored group that the hitherto unrecognized persons could actually *be* persons. But a slow expansion of the set of persons continues, with the inclusion of some of the (other!) higher animals, for example, now a real possibility. Acknowledging Gaian personhood may be a king of end point of this process. This line of argument also resonates with some pantheistic views, and it may suggest a distinctive place for humans within the larger system without necessarily devaluing other parts of the system.

On the other hand, there are questions and worries. One of the fundamental problems for environmental ethics is the extent which intrinsic values are monopolized by persons. This is why it has been so difficult to argue that nonpersons (individual animals, paintings, ecosystems on a non-Gaian view, etc.) have any value at all aside form "merely instrumental" value. Transitional persons too (fetuses, the newly dead) are thrown into moral limbo, thus creating much of the problematic of modern "applied ethics." Goodpaster avoids the problem by *accepting* the strong association of intrinsic value with personhood and then trying to invoke Gaian personhood to justify attributing her intrinsic value. It may, however, be that very association that needs to be challenged. Environmental ethicists have wanted to insist that other things besides persons—other things that are not even *like* persons—can also have intrinsic value. This is why environmental ethics was supposed to offer such a radical transformation in ethics. By pointing the way beyond anthropocentrism it might also point the way beyond what we might call *person*-centrism. But the present argument remains person-centric, and even if the intrinsic value of Gaia herself *is* secured by making her a kind of person, many other things are still left over—individual ecosystems, many animals, etc.—whose intrinsic value we will still have to defend in nonpersonhood terms.

There are also problems about whether Gaia really *is* person-like. Granted, homeostasis—self-monitoring and self-adjustment—is one important feature of persons. But it is also an important feature of certain nonpersons: amoebae, plants, even many machines. Lovelock himself uses a temperature-regulated oven as an example.[19] What distinguishes persons is more: some feature of mind, some kind of self-awareness, some capacity for autonomous choice. This is of course much-contested ground. The important point for us is that Gaia does not seem to qualify on *any* of these more restrictive criteria. Thus the case for Gaia's personhood is stretched, as is Goodpaster's case for her value. Possibly, too much weight is being put on *some* features of personhood, while others are conveniently forgotten.

A response to the last point might return to the idea that humans form an essential part of Gaia. Maybe we lend the self-awareness aspect of personhood *to* her. But this, frankly, seems to me to be not merely person-centric, but to retreat nearly all the way back to anthropocentrism. Once again *we* are the linchpins of the system, and once again it is out minds that make us so. Surely this sort of argument justly raises suspicion. Would Gaia truly lack what we most value in her if humans were not around? In any case, again, it leave the personhood monopoly unchallenged, and leaves an important set of natural objects and systems unspoken for.

*Must* we respect only persons? Why not respect for antiquity, for beauty, for complexity? Surely if we are to move beyond person-centrism, it is toward respect of these sorts that we must move, and Gaia is a superlative model for all of them. We may need an ethic of the awesome that recaptures some of the religious dimensions of nature, dimensions overlooked by the relentless humanism of contemporary philosophical ethics; or an ecosystemic ethic like Leopold's, which does not attempt to reinterpret ecological theories in the familiar ethical categories, but rather the reverse. Gaia is not a person, I would argue, but a new kind of locus of values, and perhaps a much older kind as well. The assimilation proposed by the contemporary ethics of the person is too simple and too easy.

## IV Deep Ecology and Gaian phenomenology

Arne Naess defines *Deep Ecology* as a "total field" conception, which, in Warwick Fox's words, "dissolves not only the notion of humans as separate from their environment but the very notion of the world as composed of discrete, compact, separate 'things'."[20] Deep Ecology, says Fox, "views humans as just one constituency among others in the biotic community, one particular strand in the web of life, just one kind of knot in the biospheric net." Deep Ecology thus sounds very close to some of Lovelock's formulations. Though I have stressed those points where Lovelock suggests that humans can or do play a special role in Gaian processes, the Gaia hypothesis itself essentially contextualizes humans within the vast movement and self-adjustment of life on Earth: it offers precisely Naess's "total field" conception. We can only understand ourselves as parts of much larger and older life processes, just as Deep Ecology insists.

Fox points out that Deep Ecology is not centrally an *ethic*. Instead its primary question is "How do we cultivate a deep ecological consciousness?"[21] Ethical questions, Fox says, look for conceptual answers; Deep Ecology, by contrast, looks to experiential answers. The crucial task is to change our "underlying perception of the way things are."[22] *Then* we can ask how our values and actions will change. How then might the Gaia hypothesis bear on those underlying perceptions?

A systematic answer has been advanced by my colleague David Abram in an *Ecologist* article called "The Perceptual Implications of Gaia."[23] Abram begins by insisting, with Lovelock, that we are quite literally *parts* of Gaia. Indeed, as Abram puts it, we live "in" this planet rather than "on" it: remember the centrality of the atmosphere in Lovelock's account. Thus, Abram writes,

> we must remember and reacquaint ourselves with the very medium within which we move. The air can no longer be confused with mere negative presence or with the absence of solid things: henceforth the air itself is a density . . . a thick and tactile presence. . . . We are immersed in its depths as surely as fish are immersed in the sea.[24]

What does it mean that we live in such intimacy with the Earth? It implies, according to Abram, that ordinary sensory perception, traditionally understood to be a passive and mechanical process, must be reconceptualized as a subtle form of communication between oneself and the living entity of which we are all parts. Indeed,

> the word "communication", so often associated with a purely linguistic interchange, has overtones of something rather more conscious and willful than what we are trying to describe. Here we are

> referring to an exchange far more primordial, and far more constant, than that verbal exchange which we carry on among ourselves. What is important is that we describe [perception] *as an exchange*, no longer a one-way transfer of random data from an inert world into a living organism, but a reciprocal interaction between two living presences—my own body and the vast body of the biosphere. Perhaps the term "communion" is more precise than "communication". For by "communion" we refer to a deeper mode of communication, more corporeal than intellectual, a sort of sensuous immersion. . . .[25]

We are constantly in communion, then, with the living world that encompasses us, part of an interactive webwork of communications that runs through us as it runs through all the animals and even, in Wordsworth's prescient phrase, "the round ocean and the living air." We participate in the planet's awareness, according to Abram, but we emphatically do not *constitute* it: there is no danger of anthropocentrism here.

Abram draws out parallels between this view of perception and the work of the perceptual psychologist J. J. Gibson, as well as the phenomenological investigations of Maurice Merleau-Ponty. But he also points it in a more ethical direction. If we exist in sensory communion with Gaia, then we can also know—directly—when and where she is stressed. When the Earth is torn open, I too feel the pain; when the sea is slowly poisoned, so am I; when unknown species are destroyed as the rain forest succumbs to the bulldozer, when unknown potentials and ways of seeing the world are lost forever, I too am made sick. My own body is entirely a part of the vaster physiology to which this is being done, and just as all the cells of a living body can be shocked and stressed by damage to some other part of the body, so I can be shocked and stressed by what is done to the Earth, even when it is done elsewhere and supposedly out of sight.[26] Abram thus takes Gaian organism very seriously indeed.

Like the assimilation of Gaia to the ethic of persons, Abram's account has important resonances, though very different ones. It honors native people's understandings of the Earth; it reappropriates a kind of animism, a way of seeing that Abram argues has been radically misunderstood by conventional anthropologists; it rescues those forms of perception that go beyond the narrow ambit of scientific empiricism from their present relegation to the "extrasensory"—a category which only reveals, as Abram points out, how narrow a view of the senses we have. It also restructures and enriches daily experience. We know that for too long we have divided human relations from our relations to other animals and to the natural world. But almost always this division is challenged *conceptually*. To challenge it on the *perceptual* level is, on the level of concrete experience, I think, absolutely imperative and long overdue. "Proprioception," for example, too long divided even from what we took to be "knowledge of our selves," now has Gaian equivalent.[27] The subtle and multiple ways in which we know our own bodies also open us to the wider living world. All of this makes for a powerful and unified *lived* environmental ethic.

Still, however, I have questions and worries. First, the argument is too metaphorical at crucial points. For us to *sense* the shock and stress of the Earth, for example, Gaia must literally make up one vast physiology. But intricate self-adjusting systems need not evolve specifically physiological means of intercommunication. In Lovelock's "daisy world," for example, the daisies don't communicate with each other at all; they just respond, individually, to changes in ambient radiation. More developed awarenesses may have more developed sensitivities to Gaian processes, but even so it is not clear that these sensitivities are analogous to physiological ones. Information may pass around the system without corporeal communion. Or again, Abram writes, quite correctly, that Gaia acts in ways which are "almost intelligent," that with the Gaia hypothesis "the locus of creativity shifts from the human intellect to the enveloping world itself."[28] But *almost* intelligent is not *actually* intelligent, and the creativity whose locus shifts out of the human intellect is not necessarily creative *awareness*. At least we need to hear more about alternative perceptual schemes before these metaphors will carry us as far as Abram wants us to go.

A second and still more fundamental reservation is that there is a sense in which Abram saves us from person-centrism only to leave with an admittedly broader but still restricted view. Having life, or being life-*like* in some extended sense or from a very general Gaian perspective, still seems to be a necessary condition for (intrinsic) value. But, just as I asked above why only persons should be worthy of respect, so we must also ask why only life should have intrinsic value. Environmental ethics in fact wants to defend the value of things that are not animate—at least in any usual sense—even if the Gaia hypothesis is true: rock formations, for instance, or the Moon, or Mars. Again, this is part of what makes environmental ethics philosophically provocative: we are trying to speak up for values that seem *utterly* different from the values the West is used to. And again it is not obvious that we should speak up for them by trying to assimilate them to values we *are* used to. We may need to resist the philosophical impulse to unify all values under a single genus; here we may need to stress difference, even radical difference, rather than similarity. Frank Lloyd Wright's work, for example, reveals what values lie in rocks as rocks, and the deserts of the southwest U.S. offer a natural rock beauty that is probably more Martian than Gaian. Persons are not the only things that have value—and neither is life itself.[29]

Certainly environmental ethics has at times been so preoccupied with arguments and concepts that we have lost sight of the experiences that we are in the end trying to speak for, experiences that may not even allow formulation into words, other than poetry: nights in the mountains sleepless for the silence, the awesomeness of a thunderhead on the prairie or of the boiling ocean at dusk, the quiet calm of a sunlit woods. On the other hand, we can speak for them without reconceiving perceptual experience itself; elsewhere I have laid out a framework for doing so.[30] Abram follows a far more strenuous and speculative path; the vistas are sweeping and startling, but the way through is not yet clear.

## V Cautionary conclusions

Philosophizing about the environment takes place in a wide variety of contexts and for a wide range of purposes. Why not read Gaia too in multiple ways? Use the suggestion of Gaian personhood, as far as it goes, to challenge the anthropocentrists; use Abram's Gaian reconception of the senses to challenge those who seem impervious to the natural unities they disrupt. There are many other uses as well. Understanding Gaian processes might, for instance, create a sense of perspective in us, a sense of humility much less encouraged by a world picture in which we seem to be the most admirable organism around. At other times and places the sheer scale of Gaia humbles: visiting the oceans we may think not merely of their size, but of the complexity of the oxygen modulation and nutrient cycling that the system owes to them. No doubt there are other lines as well. At the same time, of course, non-Gaian arguments must not be foreclosed. Individual species and ecosystems must often be spoken for in individual terms. If Lovelock is right that pollution is seldom a problem from a Gaian perspective, other arguments must be used against it. And if Mars should not be "greened," we presumably must be prepared to defend the value of what is *not* living, obviously in non-Gaian terms. Thus, Gaia is new support, a help, but not a new beginning, not a new "foundation" for environmental ethics.

Indeed, it is precisely the search for a "foundation" which seems to me to be the root of the problem. Partly the problem is its monism, the search for a foundation. Again, it is not at all clear that everything must have value in *one way*. The value of rocks may be just different in kind from the value of rabbits or of humans. The notion of "foundation" also suggests that environmental ethics requires unassailable starting points: the foundations must support everything else without themselves needing support. Environmental ethics has been preoccupied with the search for "intrinsic values" in particular: unassailable, "self-evident," and simple starting points. I have argued elsewhere,[31] however, that there are no such things as intrinsic values in this sense, nor should we wish for them. Everything in the concept of intrinsic value asserts independence, self-sufficiency: thus the Moorean test for intrinsic value consists precisely in thinking of the putative intrinsic value "by itself" and "in absolute isolation."[32] Surely this approach won't do when we are inquiring about the intrinsic value of specific species or natural formations: the whole point of ecology is that nothing in nature is so independent.

In this framework the Gaia hypothesis might seem to suggest that the biosphere as a *whole* has intrinsic value, since the Gaia hypothesis seems to suggest that the biosphere is the *smallest* entity we can find that is more or less independent (individual ecosystems are still *inter*dependent within Gaia). The immediate difficulty with this suggestion, of course, is that the biosphere is *not* in fact independent, but depends crucially upon the Sun (and Earth's own vulcanism) for its energies, and upon the Moon and the stars for its rhythms. Even apart from this difficulty, though, the Moorean conception of intrinsic value applies problematically to Gaia. The Moorean test is so preoccupied with self-sufficiency that self-sufficiency itself almost becomes the criterion for intrinsic value: we end up responding to Gaia not in all her complexity and awesomeness, but simply because she seems self-contained. Surely we need to question that insistence on independence itself. Our knowledge of ecological *inter*dependence could instead suggest that our basic values exist, like the elements of ecosystems themselves, in multiple relations, and gain their strength precisely from those relations rather than from their ability to stand alone.

Interdependent values are not necessarily intrinsic, and they are *not* foundational. Instead of being preoccupied with starting points, ethical argument should aim to articulate the connections and synergies between our values, and between our values and our beliefs, sources of inspiration, and so on. Moreover, many of these connections and synergies already exist. Bonds need strengthening, environmentally conscious areas of our cultural heritage need exploring, but it is not as if we simply have no environmental values at all until some intrinsic values in nature are finally established. And environmental ethics ought not be in the sort of shape that it must grasp at new ecological hypotheses for radical new starting points. Let us try to assimilate Gaia within our already existing rich and varied framework of environmental values—without substituting Gaia for them and without supposing that Gaia must have only *one* meaning or only *one* implication. And let us take our time.

## Notes

1 James Lovelock, *Gaia: A New Look at Life on Earth* (New York: Oxford University Press, 1979), chap. 3.

2 Ibid., pp. 19–25.

3 See Dorion Sagan and Lynn Margulis, "The Gaian Perspective of Ecology," *The Ecologist* 13, no. 5 (1983): 161–64. "Daisyworld" is also discussed, less technically, in James Lovelock and Michael Allaby, *The Greening of Mars* (New York: St Martin's, 1984), pp. 105–08.

4 Lovelock, *Gaia*, p. 9.

5 A very useful discussion is Donald Worster, *Nature's Economy: A History of Ecological Ideas* (Cambridge: Cambridge University Press, 1977), chap. 15.

6 Aldo Leopold, "Some Fundamentals of Conservation in the Southwest," *Environmental Ethics* 1 (1979): 138.
7 Lovelock, *Gaia*, p. 148.
8 Ibid., p. 123; Sagan and Margulis, "The Gaian Perspective," p. 166.
9 Lovelock, *Gaia*, p. 110.
10 Ibid., p. 116.
11 Ibid., p. 41. This was before the recent studies on "nuclear winter"; Lovelock may have changed his mind.
12 Ibid., p. 118f.
13 See, note 3.
14 See, e.g., Carl Sagan's review in *The New York Times Book Review*, 6 January 1985, p. 6.
15 See, e.g., B. A. O. Williams, *Moral Luck* (New York: Cambridge University Press, 1981), and John Rawls, A *Theory of Justice* (Cambridge: Harvard University Press, 1971), especially sec. 40.
16 Theodore Rozsak, *Person/Planet* (New York: Anchor/Doubleday, 1978), pp. 38–39.
17 Kenneth Goodpaster, "From Egoism to Environmentalism," in Goodpaster and Sayre, eds., *Ethics and Problems of the 21st Century* (Notre Dame and London: University of Notre Dame Press, 1979), p. 32. See also "On Being Morally Considerable," reprinted in Donald Scherer and Thomas Attig, eds., *Ethics and the Environment* (Englewood Cliffs, N.J.: Prentice-Hall, 1983), p. 39.
18 Peter Bunyard, "The Gaia Hypothesis and Man's Responsibility to the Earth," *The Ecologist* 13, no. 5 (1983): 159. Sagan and Margulis speak of humans as "a sort of nervous early warning system for Gaia" ("The Gaian Perspective," p. 166). In the "Epilogue" to *Gaia* Lovelock considers the possibility that whales have a consciousness similar to or exceeding humans: we may not exhaust Gaia's mind.
19 Sagan and Margulis write: "[S]ystems are said to be homeostatic if their variables . . . are regulated around a fixed set point. Examples of such set points might be 22 degrees C. for a room thermometer or 40% relative humidity for a room humidifier. If the set point itself is not constant but changes with time it is called an operating point. Systems with operating point rather than set points are said to be homeorrhetic rather than homeostatic. Gaian regulatory systems, like the embryological ones described by C. H. Waddington, are more properly described as homeorrhetic" ("The Gaian Perspective," p. 161). But personhood goes far beyond homeorrhesis, too.
20 Warwick Fox, "Deep Ecology," *The Ecologist* 14, no. 5–6 (1984): 194.
21 Ibid., p. 204.
22 Ibid.
23 David Abram, "The Perceptual Implications of Gaia," *The Ecologist* 15, no. 3 (1985): 96–103.
24 Ibid., p. 96.
25 Ibid., p. 98.
26 Indeed, in Abram's view, "the extreme mental anguish and psychological distress currently evidenced both inside and outside of our numerous psychiatric institutions may best be understood as expressions of stress in the enveloping ecosystem" (personal communication).
27 "Perception," writes Merleau-Ponty, "opens the world to me as the surgeon opens a body, catching sight, through the window he has contrived, of the organs in full functioning, taken in their activity, seen sideways" (*The Visible and the Invisible,* trans. Alphonso Lingis [Chicago: Northwestern University Press, 1968], p. 218). I am grateful to David Abram for pointing out this passage to me.
28 Abram, "The Perceptual Implications of Gaia," p. 98.

29 Abram's response is that the Gaia hypothesis requires us to reconceive what it is for something to be "alive" in the first place. Once we find ourselves in dynamic and reciprocal relation with any part of the world, we need to attribute to it some degree of "creative autonomy" too, and thus the contemporary limited notion of "life" is at least called into question. Mars, then, may be no less alive than Earth. Abram is surely correct that the concept of life is deeply problematic. But the distinction which I try to make here seems to me to remain. However we categorize the Badlands, or the Martian landscape, part of their value surely lies in what *distinguishes* them from rain forests and sea bird colonies and other such examples of "life" in a more usual sense. Do we really honor them by assimilating them all to one model?

30 See Anthony Weston, "Beyond Intrinsic Value: Pragmatism in Environmental Ethics," *Environmental Ethics* 7 (1985): 321–39.

31 Ibid.; on my general conception of the "space" for environmental values, see my article "Toward an Inclusive Ethics," *Bowling Green Studies in Applied Ethics,* vol. 8 (1986).

32 G. E. Moore, *Principia Ethica* (Cambridge: Cambridge University press, 1903), p. 187.

# 91

# THE MECHANICAL AND THE ORGANIC

## On the impact of metaphor in science

*David Abram*

Source: Stephen Schneider and Penelope Boston (eds), *Scientists on Gaia*, Cambridge, MA: MIT Press, 1991, pp. 66–74.

Many scientists and theorists claim that the Gaia hypothesis is merely a fancy name for a set of interactions, between organisms and their presumably inorganic environment, that have long been known to science. Every high school student is familiar with the fact that the oxygen content of our atmospheric environment is dependent on the photosynthetic activity of plants. The Gaia hypothesis, according to such researchers, offers nothing substantive. It is simply a new—and unnecessarily obfuscating—way of speaking of old facts. In the words of biologist Stephen Jay Gould: "The Gaia Hypothesis says nothing new—it offers no new mechanisms. It just changes the metaphor. But metaphor is not mechanism!"[1]

What Gould failed to state is that "mechanism," itself, is nothing more than a metaphor. It is an important one, to be sure. Indeed the whole process of modern science seems to get underway with this metaphor. In 1644 the brilliant philosopher Rene Descartes wrote, "I have described the earth, and all the visible world, as if it were a machine."[2] In his various writings, Descartes, developing a notion already suggested by other philosophers, effectively inaugurated that tradition of thought we call "mechanism," or, as it was known at that time, the "mechanical philosophy." And his metaphor is still with us today.[3]

But let us explore how this metaphor operates upon us. What are assumptions, explicit and implicit, that we wittingly or unwittingly buy into when we accept the premise that "the visible world" and, most specifically, the earth, is best understood as a very intricate and complex machine?

## The mind of a metaphor

First, the mechanical philosophy suggests that matter, itself, is ultimately inert, without any life or creativity of its own. The great worth of the machine metaphor is that it implies that the material world is, at least in principle, entirely predictable. According to this metaphor, the material world operates, like any machine, according to fixed and unvarying rules; laws that have been built into the machine from the start. It has no creativity, no spontaneity of its own. As a watch or a clock ticks away with complete uniformity until it runs down, so the material world cannot itself alter or vary the laws that are built into it. The laws of a mechanical world are preset and constant; if we can discover those laws, we will be able to predict with utter certainty the events of the world. Or so the mechanical philosophers thought in the seventeenth century.

The second assumption implicit in the mechanical metaphor is rather more hidden than the first. A machine always implies someone who invented the machine—a builder, a maker. A machine does not, in the manner of an embryo, generate itself. Clocks, carriages, or steam engines do not take form of themselves—if they did, they would be very wild and magical entities indeed, and we could not ascribe to them the fixity, uniformity, and predictability that we associate with any strictly mechanical object. If we view nature as a machine, then we tacitly view it as something that has been built, something that has been made *from outside*. This is still evident in much of the language that we use in our science today: we speak of behavior that has been "programmed" into an animal's genes, of information that is "hardwired" into the brain. But who wrote the program? Who wired the brain? As mechanists we borrow these metaphors from our own experience of build things—things that have been invented and constructed by humans—and then we pretend that the inventor or the builder (or the programmer) does not come along with the metaphor. But, of course, it does. If the material world is like a machine, then this world must have been constructed from outside.

This implication, I would claim, is precisely why the mechanical philosophy triumphed in the seventeenth and eighteenth centuries, to become part of the very fabric of conventional science. Mechanism gained ascendancy not because it was a necessary adjunct of scientific practice, but because it disarmed the objections of the Church, the dominant social and political institution of the time. The mechanical philosophy became a central facet of the scientific worldview precisely because it implied the existence of a maker (a divine interpreter) and thus made possible an alliance between science and the Church. But in order to make sense of this claim (and to better understand the power of the mechanical metaphor today) we must look briefly at the cultural forces and tensions that set the stage for the historical ascendency of the mechanical philosophy in seventeenth- and eighteenth-century Europe.

## A brief historical excursus

We moderns tend to assume that the adoption of the mechanical metaphor was a necessary precondition for the growth and flourishing of experimental science. Yet an attentive study of the various conflicts and debates that gave rise to the scientific revolution quickly calls such assumptions into question.[4] Until the latter half of the seventeenth century, the tradition of experimentation was *not* associated with the mechanical philosophy. On the contrary, the method of careful experimentation was associated with those who practiced it, those who developed and refined it to the level of an art, individuals who had a very different perspective from that of the mechanists. For these were the people who called themselves "natural magicians" and "alchemists." They viewed the material world, and indeed *matter itself*, as a locus of subtle powers and immanent forces, a dynamic network of invisible sympathies and antipathies. For the Renaissance natural magician Marsilio Ficino (1433–1499, founder of the Florentine Academy, and the first translator of Plato's works into Latin), for the Hermetic natural magicians Giordano Bruno (1548–1600) and Tommaso Campanella (1568–1639), for the brilliant physician and alchemist Paracelsus (1493–1541) and, indeed, for the entire alchemical tradition, material nature was perceived as alive, as a complex, living organism, with which the investigator—the natural magician, or scientist—was in relation. ("It is an error," wrote Campanella, "to think that the world does not feel just because it does not have legs, eyes, and hands.") The experimental method was developed and honed as the medium of this relation, as a practice of dialogue between oneself and animate nature. Experimentation was here a form of participation, a technique of communication or communion which, when successful, effected a transformation not just in the structure of the material experimented upon, but in the structure of the experimenter himself.[5]

Many of the great discoveries that we associate with the scientific revolution and, indeed, many of the scientists themselves, took their inspiration from this participatory tradition of natural magic—one need only mention Nicholas Copernicus, who wrote of the sun as the visible God, quoting the legendary Egyptian magician Hermes Trismegistus; Johannes Kepler, whose mother was imprisoned and nearly executed for practicing witchcraft—on the evidence of Kepler's own writings; William Gilbert, the great student of magnetism, which he termed "coition" as if it were a type of sexual intercourse that matter has with itself, and who, in his book *De Magnete*, published in 1600 (the year that Bruno was burned at the stake), wrote of the whole earth as a living body with its own impulse of self-preservation! And, of course, we must mention Francis Bacon, the "father" of experimental science, who saw his scientific method as a refinement of the tradition of natural magic and who wrote that through his work the term "magic," which "has long been used in a bad sense, will again be restored to its ancient and honorable meaning."[6]

How is it that we have forgotten this intimate link between experimental science and natural magic? How or why was this link with magic so obscured by the subsequent tradition of natural science? Why, for instance, did Isaac Newton, arguably one of the greatest of all natural magicians, find it necessary to hide and very publicly deny the vast alchemical researches that occupied him throughout his life?

Clearly, the Church in the sixteenth and seventeenth centuries felt itself threatened by this powerful tradition which held that the material world was a source of itself, this tradition—with roots in the high culture of Renaissance Neoplatonism as well as in the diverse folk knowledge of the present countryside—which spoke of the enveloping earth as a living being, a living matrix of spiritual powers and receptivities. Such a way of speaking threatened the theological doctrine that matter itself is passive and barren, and that the corporeal realm of nature was a fallen, sinful realm, necessarily separated from its divine source. (I do not refer here to Christian doctrine in general, but to the institutionalized Church of the sixteenth and seventeenth centuries—a period, let us remember, that saw hundreds of thousands of persons, most of them women, tortured and executed as "witches" by the ecclesiastical and lay authorities.)

The *true* source, according to the Church, was radically external to nature, outside of the earthly domain in which we are bodily immersed. The teachings of natural magic, however, with their constant reference to immanent powers, implied that the divine miracles reported in the Old and the New Testaments might be explained by subtle principles entirely *internal* to material nature. This was heresy—heresy of the first order!—because it enabled one to doubt the very agency and existence of the God outside nature. Clearly then, if natural experimentation was to become a respectable or even a permissible practice, it would have to find a new rhetoric for itself. It would have to shed its origins in the magical and participatory worldview and take on a new way of speaking more in line with church doctrine.

It was mechanism, or the mechanical philosophy, that provided this new and much safer way of speaking. For again, a metaphorical machine entails a metaphorical builder, a creator. Like the Church, the mechanical philosophy necessitated belief in a creative source entirely outside of the material or sensible world. And, like the church, the mechanical philosophy involved a denigration of corporeal matter, not exactly as fallen, sinful and demonic, but as barren, inert, and ultimately dead.

Here, then, was a perfect cosmology for the experimental scientists to adopt—one that would allow them to continue to investigate nature without fear or being persecuted, or even executed, for heresy. The mechanical metaphor made possible an alliance between seventeenth-century science and the Church. And thus mechanism became a central tenet of the scientific world view.[7]

## Mechanism and human privilege

We are now in a position to discern the third and most powerful assumption implicit in the mechanical metaphor. The only true machines of which we have direct experience are those that have been invented by humans. Hence, if the world really functions as a complex machine, then the one who built that machine must be very much like us. There is, in other words, an implied correspondence between humans and the one who built or programmed the complicated, vast machine of the world. We are, after all, made in his or her image. If the material Earth is a created machine, it falls to us—because we are not just created, but *creators* in our own right—to figure out how the machine works.

The mechanical metaphor, then, not only makes it rather simple for us to operationalize the world, by presenting nature as an assemblage of working parts that have no internal relation to each other—a set of parts, that is, that can be readily taken apart or put back together without undo damage; it also provides us with a neat justification for any and all such manipulations. The correspondence between the creative human mind and that which created the mechanical universe (between humans and God) ensures that the human researcher has a divine mandate to experiment upon, to operate upon, to manipulate earthly nature in any manner that he or she sees fit. The inertness of matter, the clear lack of sentience in all that is not human, absolves the researcher of any guilt regarding the apparent pain he or she may happen to inflict upon animals or ecosystems (such pain, as Descartes taught us, is entirely an illusion, for automatons cannot really *feel* anything).[8]

The mechanical worldview thus implicates us in a relation to the world which is that of an inventor, an operator, or an engineer to his or her machine (the very notion of "genetic *engineering*" can have sense only for a culture that maintains a mechanical view of nature). When the natural world is conceived as a machine, the human mind necessarily retains a godlike position *outside* of that world. It is this privileged position, the license it gives us for the possession, mastery, and control of nature, that makes us so reluctant to drop the mechanical metaphor today. If mechanism rose to prominence in the seventeenth century due to its compatibility with the belief in a divine creator, it remains in prominence today largely due to the deification of human powers that it promotes.

## The phenomenology of perception

But this deification, this human privilege, comes at the expense of our perceptual experience. If, at any moment, we suspend out theoretical awareness in order to attend to our sensory experience of the world around us (to our experience not as disembodied intellects but as intelligent, sensing animals), we find that we are not outside of the world, but entirely *within* it. We are thoroughly encompassed by the physical world, immersed in its depths. Hence our sensory relation to the world is hardly that of a spectator to an object. As sensing animals, we are never disinterested onlookers but participants in a dynamic, shifting, and ambiguous field.

Maurice Merleau-Ponty, the French phenomenologist and philosopher who has perhaps most carefully analyzed the experience of perception, underscored the participatory nature of this experience by calling attention to the obvious but easily overlooked fact that our hand, with which we touch the world, is itself a touchable thing, and thus is itself entirely a part of the tactile field that it explores. Likewise the eyes, with which we see the world, are themselves visible. They are themselves entirely included *within* the visible world that they see—they are one of the visible things, like the bark of a tree, or a stone, or the sky. For Merleau-Ponty, to *see* is also, at one and the same time, to feel oneself *seen*, and to touch the world is also to be touched by the world! Clearly a wholly immaterial mind could neither see nor touch things, could not experience anything at all. *We* can experience things, can touch, hear, and taste things, only because, as bodies, we are ourselves a part of the sensible field, and have *our own* textures, sounds, and tastes. We can perceive things at all only because we are entirely a part of the sensible world that we perceive. We might just as well say that we are organs of that world, and that the world is perceiving itself through us.[9]

But here the main point to get from Merleau-Ponty is that, from the perspective of our embodied, animal awareness, perception is always experienced as an interactive, reciprocal participation. The event of perception is never instantaneous—it has always a duration, and in that duration there is always movement, a questioning and responding, a subtle attuning of the eyes to that which they see, or of the ears to what is heard, and thus we enter into a relationship with the things we perceive.

When, for instance, a particular stone on the beach catches my eye, I may respond to this solicitation by bending to pick it up. I thereby discover that the stone is larger than I had a first thought—I now see that much of its bulk had been hidden beneath the sand. In order to heft it my body shifts its stance, legs and feet planting themselves a bit more solidly in the sand as I raise the stone to eye level. Now, as one hand moves over its surface, my fingers must adjust themselves to the particular texture of that surface; they must find the right rhythm, the right was to touch it if the stone is to disclose its subtle furrows and patterns. Likewise, only as my eyes find the right way to focus and question its surface will the stone begin to reveal to me the secrets of its mineral composition. As my body adjusts itself to the stone, the stone begins to speak its mute language, to subtly instruct and inform my senses. And the more I linger with this stone, the more I will learn. My experience, then, is of a reciprocal interaction, a mutual engagement of the stone by my body and of my body by the stone. And so it is with everything that we perceive, constantly, continually—the paved streets we walk upon, the trees that surround our home, the clouds that catch our gaze. Perception is always an active engagement with what one perceives, a reciprocal participation with things. As such, our direct perception always discloses things and the world as uncertain, animate presences with which we find ourselves in a sort of communication. That this is our native, human experience of things is attested by the discourse of virtually all indigenous, oral, tribal peoples, whose languages simply refuse any designation of things, or of the sensible world, as ultimately inanimate. If a thing has the power to "call my attention" or to "capture my gaze," it can hardly be thought of as inert. "If the moon was not in some sense alive, you would no longer see it," I was told by an old tribal shaman in the Mexican desert. By which he meant to say, I think, that simply to see or to perceive a phenomenon is already to be in an active relation with that phenomenon is already to be in an active relation with that phenomenon, and yet how could one be in a dynamic relationship with something if it was entirely inanimate, without any potency or spontaneity of its own? How indeed? By implying that matter is utterly passive and inert, mechanism denies our perceptual experience. It denies our sensory involvement in the world.

The scientist who holds a fundamentally mechanical view of the natural world must suspend his or her sensory participation with things. He strives to picture the world from the viewpoint of an external spectator. He conceives of the earth as a system of objective relations laid out before his gaze, but he does not include the gaze, his own seeing, within the system. Denying his sensory involvement in that which he seeks to understand, he is left with a purely mental relation to what is only an abstract image.

Likewise with any *particular* object or organism that the mechanist studies. There as well, she must assume the position of a disinterested onlooker. She must suppress all personal involvement in the object; any trace of subjectivity must be purged from her account. But this is an impossible ideal, for there is always some interest or circumstance that leads us to study one phenomenon rather than another, and this necessarily conditions what we look for, and what we discover. We are always in, and of, the world that we seek to describe from outside. We can deny, but we cannot escape being involved in whatever we perceive. Hence, we may claim that the sensible world is ultimately inert or inanimate, *but we can never wholly experience it as such*. The most that we can do is attempt to render the sensible world inanimate, either by killing that which we study, or by deadening our sensory experience. Thus our denial of participation ultimately manifests as a particular *from* of participation, but one that does violence to our bodies and to the earth.

Mechanism, then, is a way of speaking that denies the inherently reciprocal and participatory nature of perceptual experience. Thus it constricts and stifles the senses; they are no longer free to openly engage things like oak trees, bird song, and the movement of waves. We grow more and more obvious to the animate earth as our body becomes closed in upon itself; our direct intercourse with the sensible world is inhibited. Mechanism sublimates our carnal relationship with the earth into a strictly mental relation, not to the world, but to the abstract image of a finished blueprint, the abstract ideal of a finished truth.

This mentalistic epistemology, with its fear of direct relationship and its intolerance of ambiguity, is the mark, I suggest, of an immature or adolescent science, a science that has not yet come into its own. Although it sporadically fosters grandiose feelings or power and godlike mastery over nature, science as mechanism is inherently unstable, because it is founded upon a denial of the very conditions that make science possible at all. Such a science cannot last—it must either obliterate the world in a final apotheosis of denial, or else give way to another mode of science: one that can affirm, rather than deny, our living bond with the world that surrounds us.

## Toward an ecological epistemology

The Gaia hypothesis may well signal the emergence of just such a mature science—a science that seeks not to control the world but to participate with the world, not to *operate upon* nature, but to *co-operate with* nature. If the chemical composition of the air that we are breathing is, at this very moment, being actively monitored and maintained by all of the earth's organisms acting in concert, as a single, coherent, living metabolism, then the material world that surrounds us is not, in any sense, inert or inanimate. Nor are these insects, these trees, or even these boulders entirely passive and inert. For material nature can no longer be perceived as a collection of detachable working parts. It is not a created machine but rather a vast, self-generative, living physiology, open and responsive to changing circumstances. In short, it is an entity.

Of course, we may still attempt to speak of Gaia in purely mechanical terms, or try to conceive of Gaia as a strictly objective set of processes, straining thus to hold our science within the old mechanical paradigm to which we have become accustomed. We may be reluctant to give up the dream of a finished objectivity, and of the fixed reality to which it would correspond. Nevertheless, Gaia will never fit very neatly within the discourse of mechanism. A mechanism is entirely determined; it acts, as we have seen, according to a set of predictable and fixed rules and structures that it itself did not generate. Yet it is precisely such a formulation that Gaia, as an autopoietic or self-generating system, resists. Of course, we may say that Gaia is a machine, or a set of mechanisms, that is building itself. But then we will have given up, perhaps without realizing it, that part of the metaphor that makes mechanism so compelling. That is, a machine that generates itself could never be wholly predictable. For it must improvise itself as it goes, creatively. (We have no guarantee, for instance, that the so-called mechanisms that Gaia employs to regulate the salinity of the oceans, or to limit the influx of ultraviolet radiation into the atmosphere, are precisely the same that Gaia will be employing two centuries from now.) Gaia, as self-organizing entity, is no more and no less predictable than a living organism, and we might as well simply acknowledge the fact, and cease pretending that it is anything like a machine that we could build. The Gaia hypothesis suggests that the world we inhabit is rather more like a living physiology than it is like a watch, or a spaceship, or even a computer.

And we are entirely inside of, circumscribed by this organic entity. For the Gaia hypothesis indicates that the atmosphere in which we live and think is itself a dynamic extension of the planetary surface, a functioning organ of the animate earth. As I have written elsewhere:

> It may be that the new emphasis it places on the atmosphere of this world is the most radical aspect of the Gaia Hypothesis. For it carries the implication that before we as individuals can begin to recognize the Earth as a self-sustaining organic presence, we must remember and reacquaint ourselves with the very medium within which we move. The air can no longer be confused with mere negative presence or the absence of solid things; henceforth the air is itself a density—mysterious indeed for its invisibility—but a thick and tactile presence nonetheless. We are immersed in its depths as surely as fish are immersed in the sea. It is the medium, the silent interlocutor of all our musings and moods. We simply cannot exist without its support and nourishment, with its active participation in whatever we are up to at any moment.
>
> In concert with the other animals, with the plants, and with the microbes themselves, we are an active part of the Earth's atmosphere, constantly circulating the breath of this planet through out bodies and brains, exchanging certain vital gases for others, and thus monitoring and maintaining the delicate makeup of the medium.[10]

So simply by breathing we are participating in the life of the biosphere. But not just by breathing! When we consider the biosphere not as a machine, but as an animate, self-sustaining entity, then it becomes apparent that everything we see, everything we hear, every experience of smelling and tasting and touching is informing our bodies regarding the internal state of this other, vaster physiology—the biosphere itself. Sensory perception, then, discloses itself as a form of communication between an organism and the animate earth. (And this can be the case even when we are observing ourselves, noticing a headache that we feel or the commotion in our stomach caused by some contaminated water. For we ourselves are a part of Gaia. If the biosphere that encompasses us is itself a coherent entity, then introspection, listening to our own bodies, can become a way of listening and attuning to the earth.) Perception is a communication, or even a communion—a sensuous participation between ourselves and the living world that encompasses us. Yet we have seen that, phenomenologically, this is precisely the way that we commonly experienced perception—as an interaction, a participation or intertwining between ourselves and that which we perceive. Perception is never experienced as strictly objective, because to perceive anything at all is to be engaged by that thing, and to feel oneself influenced, however minimally, by the encounter. We have seen that mechanism denies this dialectic by assuming that the material world is ultimately a determinate object, incapable of open reciprocity and response. The Gaia hypothesis, on the other hand, ultimately affirms our perceptual experience, because it describes the sensible environment as open-ended and alive, which is precisely the way that our sensing bodies spontaneously experience the things around us. Thus the Gaia hypothesis enables, quite literally, *a return to our senses*. We become aware once again of our breathing bodies, and of the bodily world that surrounds us. We are drawn out of that ideal, Platonic domain of thoughts and theories back into this realm that we corporeally inhabit, this land that we share with the other animals, and the plants, and the microbial entities who vibrate and spin within our cells and the cells of the spider. Our senses loosen themselves from the mechanical constraints imposed by an outmoded language. They begin to participate, once again, in the ongoing life of the land around us.

## Conclusion

We are now in a position to contrast succinctly the epistemology of mechanism with the epistemological implications of Gaia. The mechanical model of the world entails a mentalistic epistemology, the assumption that the most precise knowledge of things is a detached, intellectual apprehension purged of all subjective, situated, or bodily involvement. It is an abstract, disembodied knowledge. Meanwhile, the Gaian understanding of the world—that which speaks of the encompassing earth not as a machine but as an autopoietic, living physiology—entails an embodied, participatory epistemology. As the earth is no longer viewed as a machine, so the human body is no longer a mechanical object housing an immaterial mind, but is rather a sensitive, expressive, thinking physiology, a microcosm of the autopoietic Earth. It is henceforth not as a detached mind, but as a thoughtful body that I can come to know the world, participating in its processes, feeling my life resonate with its life, becoming more a part of the world. Knowledge, ecologically considered, is always, in this sense, carnal knowledge—a wisdom born of the body's own attunement to that which it studies, and to the earth.

This view is entirely akin to that of Ludwig Fleck, the great epistemologist and sociologist of science, who wrote in 1929 that "Cognition is neither passive contemplation nor acquisition of the only possible insight into something given. It is an active, live interrelationship, a shaping and being reshaped, in short, an act of creation [or, we might add, cocreation]."[11]

Finally, we may wonder what science would come to look like if such an epistemology were to take hold and spread throughout the human community. It is likely, I believe, that scientists would soon lose interest in the pursuit of a finished blue-print of nature, in favour of discovering ways to better the relationship between humankind and the rest of the biosphere, and ways to rectify current problems caused by the neglect of that relationship. I have written of a science that seeks not to control nature but to communicate with nature. Experimentation might come to be recognized, once again, as a discipline or art of communication between the scientist and that which he or she studies.

Indeed, many scientists are already familiar with the experience of a deep communication or communion with that which they study, although current scientific rhetoric makes it rather difficult to admit, much less articulate, such experience. Moreover, the taboos against participation are much harsher in some scientific disciplines than in others. Physicists, from Heisenberg to Bohm, have generally been much freer to openly affirm such experiences than have biologists, and many have done so. Yet the freedom many physicists enjoy to speak of participatory or even mystical modes of awareness rests upon the fact that their objects remain transcendent to the world of our immediate experience. In other words, to mystically "participate" with subatomic quanta (in the manner of Heisenberg's recent interpreters), or to feel oneself fuse and participate with the ultimate origin of the universe (as do adherents of the strong version of the "anthropic principle") need not in any way move science, or society, to alter its assumptions regarding the determinate, mechanical character of the world accessible to our unaided senses (i.e., the surrounding landscape), and so does not directly threaten our assumed human right to control and to manipulate the natural world of our everyday experience. However, biologists and ecologists, geologists and climatologists, study this very world—the world that we can directly perceive—and they are for this reason in a much more precarious position politically. They cannot so readily acknowledge, much less discuss scientifically, their felt participation or rapport with the entities they study, whether insects or forests, for this would directly jeopardize our assumed human privilege and the many cultural practices currently justified by that assumption.

However, in a genuinely Gaian science, or in a genuinely ecological community of scientists, it would be manifestly evident that one is always already involved, or participant, in that which one studies. The effort then, would no longer be made to avoid or to repress this involvement, but rather to clarify and to refine it. Scientists, in other words, might begin openly to develop and cultivate their personal rapport with that which they study as a means of deepening their scientific insight.

The work of biologist Barbara McClintok, who was awarded the Nobel Prize for her discovery of genetic transposition, exemplifies the epistemology implied by a Gaian science. She insists that a genuine scientist must have "a feeling for the organism"—and not only for "living" organisms but "for any object that fully claims our attention."[12] McClintok describes a rather magical shift in her orientation that enabled her to identify chromosomes that she had previously been unable to distinguish. It is the shift to a participatory epistemology: "I found that the more I worked with them, the bigger and bigger the chromosomes got, and when I was really working with them, I wasn't outside, I was down there. I was part of the system. I was right down there with them and everything got big. I even was able to see the internal parts of the chromosomes—actually everything was there. It surprised me because I actually felt as if I was right down there and these were my friends. As you look at these things, they become a part of you. And you forget yourself."[13]

As Barbara McClintok came to perceive herself inside of the living system she was studying, so the Gaia hypothesis situates all of us *inside of* this world that we share with the plants and the animals and the stones. The things around us are no longer insert. They are our coparticipants in the evolution of a knowledge and a science that belongs to humankind no more, and no less, than it belongs to the earth.

## Notes

1 Gould's comments on Gaia were made during a lecture on evolution at the State University of New York at Stony Brook, in the spring of 1987.

2 Rene Descartes, *Principles of Philosophy*, part IV, principle CLXXXVIII; in *The Philosophical Works of Descartes*, translated by Haldane and Ross (Cambridge University Press, 1931).

3 Maintaining a position similar to Gould's, J.W. Kirchner in Chapter 6, this volume states that "If you try to treat a metaphor as a scientific principle, you will waste your time." Metaphors, he asserts, "are not testable." Kirchner provides, as an example, Shakespeare's phrase, "all the world's stage," and he suggests that such an assertion can never be tested. It would be testable, he writes, only if Shakespeare provided an independent definition of "stage," such as, for instance, "a raised wooden platform with footlights at the edge."

But let us resist taking Shakespeare's assertion in such an absurdly literal fashion. His metaphor for the world, *as* a metaphor, is eminently testable just as it is. Indeed, it has already stood "the test of time." The value or validity of a metaphor is readily tested according to the ability that it has to articulate our experience, to express succinctly a previously mute or unexpressed sensation (or complex of sensations). Kirchner might well reply that this is hardly the sort of test to which he is referring. Assessing the validity of a metaphor *as* a metaphor is clearly a somewhat subjective and imprecise affair; only by reducing the suggestive metaphor to a literal assertion can one attain a statement amenable to the precise criteria which, he assumes, is required by scientific analysis.

Yet Kirchner neglects to acknowledge that these very criteria by which he would judge all scientific propositions are themselves supported by a metaphor, and it is one which is no less suggestive and ambiguous than Shakespeare's. This is the claim that "the world is a machine"—the potent metaphor that, for three centuries, has provided the framework within which most scientific research has been conducted. This metaphor of a world-machine, certainly no more falsifiable than Shakespeare's theater of the world, has nevertheless proved itself by its immense heuristic value—its suggestion, for instance, that every aspect of reality should be susceptible of mathematical analysis, its tendency to render natural phenomena in a manner that opened them to technological intervention, and its concomitant ability to inspire and catalyze numerous research programs, many of them astonishingly fruitful. Still, this metaphor has also had its shadow side; it has deflected our attention away from many aspects of nature as well as of our own experience. Indeed, it may be that our mechanical view of the natural world has dangerously outlived its usefulness.

Kirchner makes the mistake of assessing Gaia according to the criteria of a science still structured by the metaphor of the machine. Condemning Gaia as "just a metaphor," he fails to recognize his own allegiance to a set of metaphors. Thus he is unable to discern the real level at which the hypothesis operates. He assumes that Gaia must be a hypothesis regarding some aspect of the determinate, mechanical nature he takes for granted, while in truth the Gaia hypothesis postulates an entirely alternative view of nature (and of our relationship to nature), and hence an alternative way of doing science.

The organismic metaphor of the world must be assessed, like any metaphor, according to the sense that it is able to make of our experience, its ability to articulate the previously inexpressable and in this case its power to catalyze new insights, and new research. Kirchner may maintain that this is a very fuzzy and unscientific way to "test" a theory. Yet such is the manner in which the metaphor of the machine was *itself* tested, by countless scientists, in the course of three centuries.

Originally illuminating for our experience and catalytic for our science, today the mechanical metaphor, taken alone, obfuscates our experience and, I believe, precludes an adequate response, by scientists and laypersons alike, to the ecological predicament in which we find ourselves. In this chapter I discuss some of the reasons for our reluctance to recognize, much less set aside, the mechanical metaphors that now limit our vision, and I propose some first thoughts regarding the value of Gaia as an alternative way of speaking, and hence of seeing.

4 For an excellent and finely documented historical overview of these controversies, see Brian Easlea's *Witch-hunting, Magic and the New Philosophy: An Introduction to Debates of the Scientific Revolution, 1450–1750* (Humanities Press, New Jersey 1980).

5 Campenella is quoted in Easlea, p. 105. A fine discussion of alchemy may be found in Frances Yates, *Giordano Bruno and the Hermetic Tradition* (Vintage Press, 1969). See also discussions of alchemy in relation to early modern science in Evelyn Fox Keller, *Reflections on Gender and Science* (New Haven: Yale University Press, 1985), and Carolyn Merchant, *The Death of Nature: Women, Ecology, and the Scientific Revolution.* (San Francisco: Harper and Row, 1983).

6 From *Of the Dignity and Advancement of Learning*, cited in Easlea, p. 128. See also P. Rossi, Francis Bacon: *From Magic to Science*, translated by S. Rabinovitch (Routledge, 1968).

7 On this reading mechanistic science went hand in hand with a Christian metaphysics. The schism that we have come to assume today between the scientists and the theologians, or between science and religion, only really got

underway with the publication and dissemination of the *Origin of Species*. For Darwin was beginning to speak of a sort of creative power inherent in nature itself; he wrote of a "natural" selection—a selective power not outside of nature but internal to nature. Of course, by using the metaphor of selection he was still propagating a metaphysics somewhat similar to that of the Church (in which he had been steeped as a young man): "Selecting" is the kind of thing that an anthropomorphic divinity does; and we can see from newspaper articles of that time that many readers interpreted Darwin's use of the term "selection" as a sort of indirect argument for the existence of God. His correspondence indicates that Darwin himself remained somewhat attached to the idea of a transcendental divinity; it may well be that his use of the term "selection," with all its associations of humanlike *will or choice*, helped him to reconcile his revolutionary theory with his religious beliefs. (See Robert M. Young, *Darwin's Metaphor*. 1985, Cambridge University Press, pp. 79–125.) Nevertheless, Darwin's work was the first in the modern era to imply a creativity inherent in nature itself, and this was a blow to the church . . . We now are beginning to discern that if the so-called environment selects the organisms that inhabit it, so those organisms also selectively influence that environment; perhaps, then, given this more open, circular causality, "selection" is not such a useful term. The interaction is a much more reciprocal phenomenon than that suggested by the metaphor of selection—it is more of a sort of dialogue wherein the environment puts questions to the organism and the organism, in answering those questions, poses new questions to the environment—which that environment, in turn, answers with further questions. It is precisely this sort of open dialectic, this mutual participation between the organism and the earth, that the Gaia hypothesis is beginning to the matize and articulate.

8 Descartes' major follower, Nicholas Malebranche, wrote succinctly that (nonhuman) animals "eat without pleasure, they cry without pain, they grow without knowing it; they desire nothing, they fear nothing, they know nothing" (cited in Easlea, p. 128). The mechanical philosophy was the principle and oft-cited justification for the vivisection experiments that began to proliferate in the seventeenth century (and that continue, in one form or another, in numerous laboratories today).

9 Maurice Merleau-Ponty, *The Visible and the Invisible*, edited by Claude Lefort, translated by Alphonso Lingis (Evanston: Northwestern University Press, 1968). See also Merleau-Ponty's seminal text, *The Phenomenology of Perception*, translated by Colin Smith (Routledge and Kegan Paul, 1962).

10 David Abram. "The Perceptual Implications of Gaia." *The Ecologist*, Vol. 15, No. 3, 1985.

11 Ludwik Fleck. "On the Crisis of 'Reality.'" In *Cognition and Fact: Materials on Ludwik Fleck*, edited by Robert Cohen and Thomas Schnelle (D. Reidel Publishing Co., 1986), pp. 47–57. Fleck's brilliant writings on the genesis of scientific "facts" were a major (and at first, unacknowledged) source for Thomas Kuhn's later work on the structure of scientific revolutions.

12 Evelyn Fox Keller. *Reflections on Gender and Science* (New Haven: Yale University Press 1985), p. 166.

13 Cited in Keller, p. 165.

# Part 5

# COSMIC AWARENESS

92

# LETTING THE WORLD GROW OLD

## An ethos of countermodernity[1]

*Freya Mathews*

Source: *Worldviews*: *Environment, Culture, Religion*, 3 (1999), 119–37.

**Abstract**

'Nature' may be understood, for environmental purposes, as whatever happens when we, or other agents under the direction of abstract thought, let things be. From this point of view it is accordingly never too late to 'return to nature'. To do so is not to restore a lost set of things or attributes, but rather to allow a certain process to begin anew. This process recommences whenever that which already exists – whether it be of nonhuman or human provenance – is permitted to endure. 'Environmentalism' is thus conceived in broadly Taoist rather than ecological terms, as involving the affirmation of the given.

### 1 What is nature?

What is nature, and how are we to live *with* it rather than *against* it, as ecophilosophers enjoin? I would like to outline my current answers to these two questions, and then explain, briefly, how I think these answers have been shaped by my experience of living in this Great South Land. My understanding of nature and of the appropriate form for our relation to it is ultimately traceable to a metaphysics that could be broadly described as panpsychist, in that it attributes an internal principle, or subjectival dimension, to matter generally. I have explored, and am still elaborating, such a metaphysic elsewhere, and do not propose to detail these studies here.[2] My purpose is rather to consider the implications of such a broadly panpsychist premise for specifically environmental approaches to 'nature'.

There have of course been multifarious answers to the questions, 'what is nature?' (Rolston 1988; Brennan 1998; Seddon 1997). One of the most common has been that 'nature' denotes the domain of physics – either the physical universe in its entirely or the laws that undergird it. In this widest sense, nature will not be of particular interest to environmentalists, since such a nature cannot be threatened, and does not, on the face of it at any rate, stand in need of human protection or conservation. In what sense then is nature understood by environmentalists as in need of defence? It is often assumed that, in speaking about nature, environmentalists are referring to the domain of biology, the realm of living things – forests and swamps, for instance, or plants and animals generally, or wildlife in particular – since it is this biological realm which is clearly under threat from the engines of industry and the appetites of global markets today. However, where the cosmological definition is too wide, this strictly biological definition of nature will usually turn out to be too narrow for environmental purposes, since many environmental battles are fought over inorganic, or only partly organic, features of the environment, such as rivers, dunefields, caves, mountains and rock formations, as well as over organic systems.

What environmentalists in fact usually seem to have in mind, in their references to nature, is parts or aspects of the world which have not been created or unduly modified by human agency. Is this a tenable environmental definition of nature? Implied in the definition is a distinction between the artefactual and the natural. I however, like many others, would immediately question the validity of defining the natural in contrast to the artefactual in this way, on the grounds that, since human beings, as biological organisms, surely belong to nature, and since making things comes to us as naturally as eating and drinking do, our handiwork itself has as much claim to be considered part of nature as the handiwork of spiders, insects and marine life does (Mathews 1996). Therefore to regard trees and rocks and animals, not to mention webs, hives, termite mounds and coral reefs, as falling within nature, and cars and ships and fax machines as falling outside it, is to over-simplify the issue.

However, there *is* a distinction buried within this over-simplification which does, I think, call for further investigation. This is the distinction between what happens when things are allowed to unfold in their own way, or run their own course, and what happens when, under the direction of abstract thought, agents intentionally intervene to change that course of events for the sake of abstractly conceived ends of their own. From the point of view of this distinction, nature might be understood as whatever happens when we, or other agents under the direction of abstract thought, let things be, while artifice is what happens when such agents redirect events towards their own ends.

It should be pointed out that the distinction between processes under the direction of abstract thought and processes which unfold of themselves does not coincide with the distinction between conscious and unconscious processes, or even living and non-living ones. That is, it is not a distinction between processes which occur in, or as a function of, consciousness, and processes which occur in, or as a function of, unconscious matter. All things, whether conscious or unconscious, living or non-living, have laws of their own unfolding. In the case of the non-living, these laws may be embodied in the physical and chemical dispositions of the things in question. In the case of living beings, they may reside in biological or instinctual dispositions, which may be described as their immanent telos or conatus.[3] Even conscious beings such as humans are imbued with an underlying conatus, an inherent 'law' of self-realisation which is not a product of our own reflection or ratiocination, but which already informs our instincts, reflexes, and other spontaneous responses.[4]

All these processes however, whether in ourselves or in the world, are capable of being re-routed by the interventions of intentional agents, where by describing the agents as 'intentional' here, I mean that they are acting in order to realise abstractly conceived end-states rather than unthinkingly following their own immediate impulses. Thus inorganic physical processes can be redirected by intentional intervention; the behaviour of organisms can likewise be modified; and spontaneous impulses in human beings can be suppressed or sublimated by the intentional interventions of the individuals themselves or by societies. At each level a contrast can be drawn between an undisturbed state of affairs, in which everything unfolds or behaves in accordance with its intrinsic dispositions or an innate conatus, and a state of affairs in which the dispositions of things are diverted by agents to produce end-states which match those agents' abstract ideals or imagined scenarios. The undisturbed unfolding of events may be described as 'natural', according to the present definition, and the intentional re-direction of such a course of events as 'artificial'.

The contrast between the natural and the artificial, drawn this way, though not absolute,[5] is I think, a significant and defensible one for environmentalists. For the primary aim of environmentalists is to protect the biosphere from the depredations of modern civilisations, and it is arguable that when all things are left to realise themselves in their own way, the life process will be assured, at least in the long term. This conclusion is supported by commonsense and evolutionary theory: 'nature knows best' because life has had countless millennia to adjust itself to terrestrial conditions. But in many ecophilosophies, this faith in the capacity of the life process to take care of itself rests on deeper metaphysical foundations, involving an attribution of purpose or intelligence to reality. My own argument for this position rests on a view of the world as constituting on the one hand a subjectival whole (the One) imbued with its own conatus, or impulse towards self-realisation, and on the other hand a manifold of relatively self-realising individuals (the Many), whose conative impulses generally intermesh to further the self-realisation of the whole (Mathews 1998).

To affirm the view that nature knows best is not to deny that when things are left to their own devices, the interests of individual organisms and smaller-scale ecosystems might not be subordinated to those of greater ecosystems or the biosphere as a whole: competition, predation and disease may further the ends of the greater systems at the expense of the smaller ones. But by and large, when the Many are following their own innate individual patterns or principles of self-realisation, the self-realisation of the One is assured. If this is accepted, then to allow the Many to unfold in their own distinctive ways is presumably the surest approach to the protection of life on earth.[6]

How different is the outcome when agents under the direction of abstract thought take control of events however. In accordance with their abstract ideas, such agents may make arbitrary interventions: they may kill off a particular species for the sake of its coloured feathers, for instance, or pollute a lake with manufacturing dye for the sake of people who have decided they want purple hair. With our powers of abstract thought we can conjure up inexhaustible numbers of ways the world could be, ways ranging from the demonic to the absurd and whimsical through to the humanitarian or utopian; and with the same powers of abstract thought we can, increasingly, devise means for making the actual world conform to our fantasies, at least in the short term.

Historically speaking, abstract thought has developed hand in hand with civilisations dedicated to the remaking of the world in accordance with variable human interests, values, dreams and conceits. By 'civilisation' here I mean that form of social organisation that was initially characterised by a settled way of life involving both the building of cities and the practice of agriculture and animal husbandry. The salient feature of civilisation as a form of life is, in this context, its refusal to live within the terms of the given, as earlier hunter-gatherer societies had by and large done. Instead, civilised societies have sought to change those terms – they have striven to re-shape the world to suit their own purposes. They imagine categories of possibility beyond the already manifest – new kinds and classes of things in addition to those already in existence: houses to provide shelter, for instance, and wheels for transport, and animals subdued into pulling the wheeled vehicles, and succulent leafy vegetables growing in neat rows right at the back door of the houses. In order to conceive of these new kinds and classes of things, it was necessary to abstract quite radically from the world as it was, to look beneath the appearances at the underlying structures, and then to imagine ways those structures could be reassembled. This kind of thinking involves a different level of abstraction from that of the hunter imagining a kangaroo roasting on the fire at the day's end. Hunters and gatherers seek to satisfy their material and spiritual needs and desires within the categories of the given. They carefully study all the subtle particularities of life's tides, and learn to swim intelligently with those tides in order to achieve their ends. Civilised societies, in contrast, seek to transcend those particularities – the various vicissitudes of fortune, the reversals of the seasons, the manifold burdens of the flesh – by remaking the world to their own design. Rather than swimming with the tides, they try to uncover the laws of hydrodynamics and gravitation, in the hope of re-orchestrating and maybe even abolishing tides.

At the level of self, latterday civilisations have involved the re-deployment of our instinctual self-realising energies in the service of socially prescribed ideals of individual selfhood. Our intelligence has been harnessed to abstract self-images and self-stories rather than to the deeper imperatives emanating from our conatus. Instead of letting ourselves be, we compulsively make ourselves over to match external ideals, where this can result in all kinds of aberrations, from neurotic individualism to fanatical collectivism or patriotism.

At the level of both world and self then, civilisation tends to induce a preference for the abstractly imagined over the given, and seeks to substitute the possible for the actual. This tendency of civilisation has become progressively more pronounced in the modern period, with its repudiation of tradition and its exponentially increasing power to manipulate matter, until it has resulted, today, in the dizzying regime of international capitalism, involving almost incessant, runaway erasure of the actual in favour of the possible, regime euphemistically going under the name of 'development'.

When human beings are freed from excessively abstract thought however, and restored to their conative nature, then the world is relatively safe from such 'development'. For the conative natures of all beings are, from an environmental point of view, relatively trustworthy, simply because, as I have already noted, all have evolved in concert with one another, with the consequence that their natures are mutually cross-referencing. Self-realisation is achieved, at the conative level, within the parameters of the already existent rather than the merely possible, and thus, at a conative level, all beings have a broad interest in maintaining the world as it is.

To recapitulate then, when 'nature' is defined, as it has been here, in contrast with artifice, it is arguable that generally, on the terrestrial scale, though not perhaps in every conceivable circumstance, nature knows best.[7] By this I mean that when things are left to themselves, the biosphere is likely to fare better than it is when things are subject to out arbitrary interventions. There will of course be casualties at the level of individuals and smaller-scale systems when things are left to unfold in their own way, and as beings with a constitutive interest in our own self-preservation and the preservation of our nearest and dearest, we are on occasion called upon to intervene in these unfoldings, just as hunter-gatherers do in the course of their daily lives.[8] Such interventions will not count as arbitrary however, and will be compatible with a general attitude of letting things be, an attitude which renders the overall project of life on earth relatively assured.[9]

## 2 Living with nature

The trap for environmentalists, in thinking about nature, is to reify it, to conceive of it in terms of things rather than processes. When we think of it in this way, we understand it as consisting of all those things which are not the product of abstract human design: forests, swamps, mountains, oceans, etc. We then contrast nature with the human-made environment, consisting of cities, artefacts, technologies, etc. We make the same mistake in thinking about nature at the level of the self: the *natural* self is equated with the body, the instincts, intuitions, emotions etc., and this is contrasted with the civilised self, consisting of the controlled rational ego. The environmentalist's defence of nature is accordingly read as a project not only to save existing swamps, forests, etc, but to restore lost ones. Introspectively it is taken to imply a counter-cultural ethos of spontaneity, eros, intuitiveness and instinctuality. From the present point of view, this is a mistaken reading. To 'return to nature' is not to restore a set of lost things or attributes, but rather to allow a certain process to begin anew. This is the process that takes over when we step back, when we cease intervening and making things over in accordance with our own abstract designs. Such a process can recommence anywhere, any time. It is not logically tied to those aspects of the world that we mistakenly reify as nature – the forests, swamps, instincts, bodily functions, etc – but can start to unfold again in the midst of the most intensively urbanised and industrialised environments on earth and in the most controlled and civilised of persons.

In a world already urbanised, 'returning to nature' means not tearing down the cities and factories, and planting woods and gardens in their stead. Such action would merely perpetuate the cycle of making the world over in accordance with abstract designs – albeit in this case ecological designs – and would reinforce the mind-set involved in living against nature. Rather, 'returning to nature' in an urbanised world means allowing this world to go its won way. It means letting the apartment blocks and warehouses and roads grow old. Yes, we shall have to maintain them, since we shall need to continue to use and inhabit them. Inhabitation will also call for adaptation and aesthetic enhancement. But this is compatible with a fundamental attitude of letting be, of acquiescence in the given, and of working within its terms of reference, rather than insisting upon further cycles of demolition and 'redevelopment'. Gradually such a world, left to grow old, rather than erased for the sake of something entirely new, will be absorbed into the larger process of life on earth. Concrete and bricks will become weathered and worn. Moss and ivy may take over the walls. Birds and insects may colonise overhangs and cavities within buildings. Green fingers will open up cracks in pavements. Bright surfaces will fade, acquiring natural patinas. Under the influence of gravity, the hard edges of modern architecture will soften, and imitate the moulded contours of landforms. Given time, everything is touched by the processes of life, and eventually taken over by them, to be fed into the cycle of decay and rebirth. Left to itself, the living world reclaims its own. Things which initially seemed discordant and out of place gradually fall into step with the rest of Creation. Old cars take their place beside old dogs and old trees; antiquity naturalises even the most jarring of trash.[10]

When the world is allowed to grow old, when things are retained, or left to unfold in their own way, then it is possible truly to *inhabit* places, to come to belong to them, in ways that are undreamt of in change-based societies. As years pass, and places retain their identities, they can, if we let them, come to be inscribed with our histories and the histories of our families and communities. They acquire meaning for us as our life experiences are woven into them. Here, on this road going down to the creek, where I walk my dog every week, is the house my great-grandfather built. I have a faded photograph of it on lock-up day, sometime late in the 19th century. Around the corner is the store my grandfather ran, and over there is the park in which my parents walked, each evening, holding hands, for sixteen years. Here, alongside it, is the cemetery where I roamed in my gothic youth, looking for the grave of that same great-grandfather, keeping trysts in the peppercorn groves, composing poems about roses. And it was along the tree-lined avenue at the edge of this cemetery that I pushed my baby son to crèche. Layers and layers of significance accrete as our lives unfold amidst familiar spaces, significance that can for us never be reproduced in any other setting. The setting itself infiltrates our identity. This irreplaceable significance of our own place or places for us binds us to them. We become their natives.

This belonging is reinforced in another way when we let our world – whether urban or rural – grow old. For when the lay-out and structures and constellation of physical features that define a particular place are allowed to endure for a long time, then not only can it become interwoven with our individual and collective identity in a way that binds us to it, but, from a panpsychist point of view, *it* can come to know *us*. In time, and only in time, a place can, if we commit to it, come to accept us, open its arms to us, receive us – it agrees to be our place, attentive to us, attuned to us. We become its people. The land, or place, claims its own. It can never receive the casual or expedient sojourner or stranger in such familiar fashion. In this way too then, through time, and the reinhabitation of places that are allowed to be, we become native to our world.[11]

The self-realisation of the biosphere – which is to say, the unfolding of nature on earth – involves a pattern of gradual but continuous change – a pattern of ageing and decomposition followed by spontaneous reconstitution into new forms. This is what happens to things when we let them be. Artifice, as here understood, correspondingly consists in any regime of abrupt, wholesale change, change that involves the erasure of one environment, or order of things, and its replacement with an entirely new one. Such regimes generally come about only at the instigation of agents in the grip of abstract ideas or images which they are intent on actualising irrespective of context – irrespective of what existed before and what surrounds the new 'development'. In this sense an old factory site, overlaid with grime and saturated with heavy metals, but in its cracks and neglected crannies also burgeoning with hardy and creative biological and social forms of life, is more natural than a town planner's lush park stocked with store-bought indigenous plants and subject to wholesale redesign or 'redevelopment' at any time.

I should note here that by making the point that 'nature' in its deeper sense connotes not trees and grass and wildlife, but the processes which occur to any and all things when they are no longer subject to intentional control, I am by no means wanting to say that the conservation of trees and grass and wildlife – which is to say, environmentalism in its traditional form – becomes superfluous. Existing ecosystems should, like cities and selves, be allowed to unfold in their own way, free from undue human disturbance. Where such ecosystems have already been modified by the introduction of exotic species however, the present view entails that the new and old species should in principle – though there may be many countervailing considerations in practice – be left to sort it out. To respect nature in this connection does not imply that we should eradicate the exotics and restore the indigenous. It means that we should forego interventionist 'management' and allow natural processes to reassert themselves. It may be deeply distressing to watch native plants and animals disappearing under the onslaught of aggressive invaders, and there may be many compassionate and practical reasons for attempting to temper this onslaught, but, to the extent that we opted out of the whole affair, we would at any rate have the satisfaction of knowing that what we were witnessing was in fact a return to nature. In this scenario it is likely that some of the original species would decline, and new ones would steal their niches, but as soon as competition had stabilised, speciation would begin again in situ, because we would no longer be intervening to reverse this trend.

A certain stepping back, then, is what is involved in 'returning to nature' in the outer world. At the level of self, making the transition from civilisation to a more natural state is no more a matter of trying to reinstate an instinctual, free-and-easy, impulsive regime than returning the world to nature is a matter of restoring lost forests and swamps. To try to transform to tensed, guarded, rationally-minded self of civilised society in this way would only be to perpetuate the process of control – the process of making the self over to match a socially approved ideal. At deeper psychic, and perhaps somatic, levels, such an attempt, with the self-rejection it implies, would presumably only exacerbate the tension to which the self in question is subject. The way for a self-censoring self to 'return to nature' is simply for it to stop altogether the business of attempting to make itself over in accordance with abstract ideals, and surrender instead to what it already is. When we give up being dissatisfied with ourselves, and reconcile ourselves to our 'unnaturalness' – our tedious up-tightness, for instance – then, ironically, we start to relax anyway; as we stop forcing ourselves to follow the latest social prescriptions, our own instinctual conatus has a chance to make itself felt again. Gradually we become re-animated with our native will to self-realisation.

The point I am making here, at the level of both self and world, if that it is never too late to return to nature. No matter how artificial our self or world has become, they can always, at any given moment, become subject again to natural processes, simply by our decision to call a halt to 'development' and 'progress' and 'self improvement', and to allow things to remain as they are, to be retained rather than replaced. In saying this I am not of course intending to ban change altogether, but to insist that change should not disrupt the general unfolding of things. It should not raze the old and superimpose on the space that is left something unrelated to what preceded it. Change should carry us gently and smoothly into the future, respecting the cycles of creation, decay and regeneration. It should grow from within the shell of the given.

It might be objected at this point that the attitude of letting things be that I am recommending here is too passive to be of use to the environment movement, that in the end it amounts to little more than a laissez-faire acquiescence in the political status quo. To dispel this fear, let me explain in a little more detail how such an attitude would, if adopted by a significant proportion of the populace, in its quiet way thoroughly disable the present world-destroying order of capitalism, by systematically negating the following values on which that order rests.

### *Consumerism*

When we embrace those things that are already at hand, we do not seek to replace them with new ones. Such embracing of the given is thus an antidote to the culture of disposability and conspicuous consumption fostered by capitalism. From the viewpoint of letting things be, we would be most pleased, not with our brightest and newest things, but with those that were our oldest and most well-worn, things which had long figured in our lives, and mingled their identity and destiny with ours. 'Keeping up with the Joneses', if it applied at all in the letting-be scenario, would entail having fewer and older things than the Joneses. (Of course it would not apply, since in the new scenario we would not be measuring ourselves against the kind of social expectations personified by the Joneses.) Acquisitiveness, and hence consumerism, melt away in the face of an attitude of letting be.

### *Commodification*

When we value things and places for the meaning that our own lives have invested in them, via our relationship with them, this removes them from the market place. They cannot be replaced by other things and places, even things and places of the same type, since the substitutes will not share our history nor hence be imbued with the same meaning for us. From this point of view, I could no more buy or sell things or places which had become part of the landscape of my life, part of my very identity, than I could buy or sell members of my family. Thus the pool of commodities is continually diminished.

### *Productivity*

When we embrace the world as it is, and are no longer forever seeking to make it better, according to abstract (generally egocentric or anthropocentric) conceptions of the good, then greed is effectively abolished. We no longer crave bigger and better houses, cars, roads, cities, whatever. We are instead attached to what is already given. There is thus no call for ever-increasing productivity.

### *Progress*

When people no longer believe that the world can always be improved, the slate wiped clean and a better world, a better society, inscribed on it, then the ideological rationalisation for capitalism, viz: that it can continue to improve peoples' 'standard of living' indefinitely, collapses.

### *Efficiency*

In late capitalism, efficiency – patently a nation pertaining to means – has acquired an almost fetishistic status. Tools (where this includes all kinds of techniques and procedures as well as implements and technologies) are valued not so much for what they do as for their efficiency, and they are retained only so long as their efficiency is perceived as maximal. When the attitude of letting be is assumed however, tools are valued not merely for their efficiency, but for their meaning. I may continue to use an old plough, or a leaky fountain pen, or a certain laborious method for making dough, simply because this is the plough, or pen, or method, that my mother or grandfather used. Efficiency may still be a consideration, but it will be only one factor determining the means I choose to achieve my ends.

### *Industry/business*

These are the two definitive modalities of capitalism – industriousness and busyness – both connoting a certain kind of externally driven, externally focussed, hectic state of doing or acting. Those who are busy and industrious act on the world, they take initiatives and make things happen. When we assume the attitude of letting be however, we let the world do the doing, and we fit in with it. We favour 'inaction'[12] which is not passivity, but action which is effortless because emanating from our own conatus and meshed with the conatus of other beings rather than driven by external social expectations or ideologies.

### *Development*

When we understand 'development' in terms of the transformation and regeneration that eventually transpires when things are left to grow old, to unfold in their own way, then we will not tolerate the erasure of the given which is the precondition and prelude to 'development' in the capitalist sense, i.e. the replacement of the given by the decontextualised abstractly imagined new.

### *Profit*

If we do elect to let things be, it is on the assumption, as I explained earlier, that nature knows best – that nature, left to itself, conserves itself, does not exhaust itself, but rather replenishes itself, accordance with the law of birth, decay and rebirth. To sustain itself in this way, nature returns *everything* to the life cycle, it recycles everything. There is no 'surplus' in this system, and hence no accumulation. The law of return makes nonsense of the notion of 'profit'. 'Profit' in one part of the system merely signals loss and depletion in another part.

### *Automation*

For the capitalist, labour is merely a means to production; if automation provides a cheaper, more efficient means, it will be preferred. From the viewpoint of letting things be, human labour is, or can be, a vehicle for meaning. Things become significant to us partly as a result of our building, making, repairing or decorating them ourselves. How much more of a presence in our lives is a church – like the Russian Orthodox Church in my neighbourhood – which is built by the hands of the parishioners themselves, over a period of many years, than one which is contracted out to professional builders and erected 'efficiently' in eight weeks. To 'mix our labour' with things is, as Locke said, though with entirely different intent, to make them ours, in a sense analogous to that it which our family is ours. To make things ourselves, or to have them made by the hands of others, then, is in certain respects preferable, from the present point of view, to mechanisation of the processes of production.

### *Property*

When people honour the world as it is – honour its immanent telos, its capacity to unfold in its own way – they no longer seek to *own* the world, but rather to *belong* to it. They belong to their world by being faithful to the things it contains, keeping and tending them and letting the world manifest through them as they endure. The world expresses itself, reveals itself, through the changes it induces in these things, through the lichen on the walls, the cracks in the glaze, the slow, stooping, inevitable return to earth. By continually replacing things we never witness the way the world reclaims its own, so we miss out on knowing it, encountering it. Strangers to the world, we do not belong, we are anything but natives. We comport ourselves as invaders, conquerors, buying up the matter which means nothing to us, and trashing it when we are tired of it. We treat ourselves, our own bodies, in the same way, truculantly professing to own them and reluctant to allow them to be reclaimed by the world, reluctant to see the world tenderly revealing itself to us through them, through the fading and crazywork and mute surrender of flesh to gravity. But of course, at the final call, the world claims us anyway, and we go, back into the earth, but no wiser, and a lot lonelier, strangers to the end.

To assume a panpsychist worldview, and to express this worldview via an attitude of letting be, is thus inevitably to extricate oneself, to a significant degree, from the ideological grid of capitalism. It is to begin to shift towards an entirely different form of praxis and of management of the material dimension of life – in other words, to an entirely different ‘economics’, or way of ensuring the satisfaction of our material wants and needs. Since economics involves a certain engagement, on our part, with the world of matter, an economics with panpsychist presuppositions will obviously differ from an economics based on a mechanistic worldview. For when the material world is viewed as a subject, or as a manifold of subjects, then economics constitutes an opportunity for encounter: the ways in which we utilise matter must not conflict with, but enhance, our intersubjective engagement with it. An economics in accordance with the principle of letting be is an economics of the given, an economics which respects the world as it is, and finds metaphysical sufficiency in it.

## 3 The politics of reinhabitation

I have argued that when we assume the attitude of letting things be, we cease to be interested in consumerism and the pursuit of wealth and become defectors from capitalism. We become children of nature again, true natives, of our cities, of our world. When a significant proportation of the population adopts this attitude, the capitalist system, robbed of its consumers, would presumably wither away. But is such ‘inaction’ enough? In a society in which this is not the majority position, a society dedicated to relentless progress and development, with the insatiable bulldozers and chainsaws and toxic spills that this entails, can we stand by and ‘do nothing’? Wouldn’t our attitude of letting things be merely abet the destruction in this scenario? Isn’t *resistance* rather than acceptance the appropriate political response to a system which will not let things be?

I think there is indeed a place for resistance in the politics of letting be. The impulse to let things be, after all, springs from cherishing the given, embracing the world as it is. This commitment to the given is primarily a commitment to *things*, to concrete particulars and places. It implies a commitment to protect those things and places against arbitrary erasure undertaken for the sake of abstract ideals. We will thus defend the things and places to which we have pledged our loyalty, and we will do so, not in the name of an abstract ideology, but simply because a world which is allowed to grow old claims us as its own, and we shall spring to its defence as surely as natives have perennially sprung to the defence of their lands.

But how can we neo-natives, a tiny minority, conduct such defence against the forces of capitalism? Our defensive resources are no greater than those of a small country facing a large-scale foreign invasion. But just as civilian defence seems to be the principal strategy for small populations in this situation, so such a form of defence might provide a key to imagining resistance from the perspective of letting things be. Civilian defence, of course, involves bands of individuals from the occupied territories withdrawing to secret, often inaccessible, locations in the countryside or city, and, through sniping and ambush, making it impossible for the invaders to move freely through the occupied regions. By analogy, followers of the politics of letting be could block the expansionism of capitalist interests by *reinhabiting* places earmarked for the inevitable 'development'.

What does reinhabitation mean in this context? It signifies people acknowledging particular places as the inalienable landscape of their lives. Just as civilian defense requires of citizens that they get to know the lie of their own land intimately, so that they can live off it, and disappear into it when the heat is on, so the politics of reinhabitation requires of its followers that they explore and research their own homeplace or region thoroughly, discovering its secrets and substrata, checking titles offices and municipal and historical records, for instance, so that the offensives of planners and developers can be anticipated. It involves finding means of sustenance, on every level, within their own neighbourhood. To reinhabit place in this way is to enter into an indissoluble metaphysical relationship with it. And to acknowledge the metaphysical relation between people and places is to begin to resacralise those places, a resacralisation which can be consummated and proclaimed in art and poetry, in festivals and ritual performances. Every place could eventually, in principle, be thus 'sung up', and assume the status of a sacred site for some band of people, a site sacralised and immortalised in song, verse and dance.

To reinhabit the world in this way, ensuring that every place was unnegotiably, metaphysically home to someone, would provide a formidible obstacle to a system which has relied, for its justification, on its rendition of the entire world as, at a metaphysical level, a vast terra nullius, spiritually belonging to no-one, and hence up for grabs. If each place had its own true inhabitants, who could demonstrate its unique metaphysical significance for them, and who could not be bought out or 'compensated', where would the developers go? The rules of the capitalist game no longer allow, at least explicitly, for bona fide natives simply to be shot, or forcibly removed. Their relationship to place has to be accommodated. So a politics of reinhabitation calls for us to find creative and forceful ways of both re-establishing and proclaiming ourselves as natives in the the midst of the industrial and urban devastation of the world today. It also calls on us to recognise that global capitalism represents nothing short of an *invasion* of our world, and requires that we organise and comport ourselves with corresponding tenacity in our world's defence.

## 4 From modernity to countermodernity

From a panpsychist perspective, respect for nature, as we have seen, is not a matter of protecting only ecosystems, but *all* material things, from undue human disturbance, including things that do not usually arouse the concern of environmentalists, even non-anthropocentric environmentalists, such as deep ecologists. This view of nature, and what it is to live with rather than against it, implies an ethos that is far more encompassing than that of the traditional environment movement. It is an ethos as encompassing in fact as the ethos of modernity that it seeks to reverse. For the hallmark of modernity is radical change – in the form of development, control, management, design, intervention, progress, improvement, even salvation. (This is reflected in the very etymology of the word, 'modern', which is derived from 'mode', meaning 'of the present', as in 'à la mode', keeping up with the latest. Modernity is that period which can be characterised in terms of its commitment to the ever-emerging new, its dissatisfaction with the given, its radical discontinuity with the past and its dissociation from tradition.) The ethos of letting be challenges modernity head on, trusting as it does the innate wisdom of things, and eschewing as it does the definitive ambition of modernity, to remake the world in accordance with abstract ideas. From the present point of view, not only is environmentalism, even in its deep-ecological forms, missing the larger metaphysical picture in its approach to modernity; it is also itself deeply entangled or imbued with the modernist ethos in its understanding of its own mission; it needs to extricate its legitimate concern for nature from heroic modernist assumptions about its own world-changing, world-saving role.

I am suggesting that instead of perpetuating this profoundly modernist ethos of changing or saving the world, the environment movement could assume what I would venture to describe as a *counter*modern attitude of letting things be. We could step right outside the presuppositions of modernity, and dare *not* to try to make things better, at any rate if 'making things better' is a rationalisation for continually replacing one regime with another. When we say, 'let's fix the world up – let's pull down these slummy old tenement blocks and build a brand new eco-permacultural-urban-village in their stead' – we are just as much in the grip of the old ethos of domination and control as the city fathers were. We are rejecting the given in favour of an abstract or imagined alternative of our own – we are refusing to 'let things be' – and it is this hubristic mentality which is the motor of modern civilisation and the source of the environmental crisis. In remaining in the grip of the old ethos, in nursing the desire to make things better, we are simply continuing to water the deeper modernist roots of the present predatory economic system.

An ethos which tries to avoid the pitfalls of this mentality will, of course, be an ethos of conservatism rather than radicalism. This conservatism has always been implicit in the environment movement, as plainly betokened by the fact that the term 'conservation' is often regarded as synonymous with 'environmentalism'. However, such conservatism does not imply that the attitude of letting be is aligned with the political right.The political right has historically, of course, been conservative, that is, committed to the given, while the left has been opposed to the given, and committed to the abstract possible or ideal. The historical right however, though conservative, differed from the present position inasmuch as it was socially rather than ontologically motivated – it was at heart a defence of social, political and economic privilege, rather than a defence of the world for its own sake. In other words, the right insisted on the preservation of traditions and institutions because it was through these traditions and institutions that the upper classes retained their privileges. A degree of ontological conservatism – the conservation of architecture and landscapes, for instance – was implicit in this position, but this ontological conservatism was in reality a mere spin-off from a self-interested politics of oppression. The historical left rejected this politics of oppression, and demanded the overthrow – and ongoing readjustment – of the existing social order, in order to end the systematic privileging of the powerful few at the expense of the many. This revolutionary or radical politics however sustained an unremitting antagonism not only to traditions and institutions, but to the world that these traditions and institutions had built, and this legacy has served, in the long run, to legitimate a rapacious contempt for the given in all its social and ontological forms, where this contempt is the hallmark of late capitalist modernity.

The attitude of letting things be, in contrast, is conservative out of genuine respect for the world, for the capacity of things to unfold in their own way. Its conservatism is ontologically rather than socially motivated, and it extends primarily to material things rather than to cultures, traditions and social institutions.[13] This attitude certainly does not spring from a desire to preserve the privileges of the few, as that of the historical right has done. Indeed, it tends, almost incidentally, towards a de facto form of non-hierarchism, inasmuch as it is antiethical to the accumulation of wealth, as I have explained above. It achieves this 'equalising' effect, however, without recourse to the radicalism, to the ethos of intervention and overthrow, of the historical left. So, without itself resorting to the ideality of morality, the ethos of letting be reconciles something of the custodial role of the right with something of the moral intent of the left. In this respect it is in fact the reverse image of the market-driven politics which is achieving hegemony in the industrialised world today, and which is also neither of the left nor the right. This latter politics – the new 'economism' – combines the old rightwing investment in the perpetuation of minority privilege, and moral indifference to the suffering of the majority, with the old radicalism of the left, its dissatisfaction with the given, and its dedication to building ever new, ever more extravagant, worlds. In other words, 'economism' has managed to combine the downsides of both the right and the left. In this unholy new regime, nothing is sacred, everything – every being, every object, every place, every institution, every element of culture and society, every relationship – is subject to obliteration or co-optation by a commerce that sustains fewer and fewer. The attitude of letting things be effectively inverts this nightmare and, by stepping outside the game of left and right altogether, inadvertently combines the essence of the upside of each of the old right and left.

The aim of such a conservative ethos then is not a brave new world, but an old world, a world unfolding naturally, redolent with meaning, beauty, and its own life and terms. The only way of achieving such a world, without engaging in further interventions, is, as I have explained, to let the present world grow old – to let the cities weather and fade, and the ivy creep up the walls.

That the ethos of letting be is *counter*- rather than *post*-modern is clear from the fact that its opposition to modernity is based not on the deconstruction of metaphysics, as postmodernism is, but on the substitution of some version of panpsychist metaphysics, broadly construed, for the materialist metaphysics of modernity. The project of modernity – of remaking the world according to our own abstract specifications – rests on the materialist assumption that matter is sheer externality, devoid of its own informing telos – putty for us to shape to our own designs. The relativist outlook of postmodernity presents no challenge to this view: by insisting on the entitlement of different cultures to live by their own metaphysical lights, postmodernists imply that there is no way the world is in itself, and to that extent they indirectly underwrite the message of modernity: the world has no informing meaning or purpose of its own. It follows from the meta-level assumptions of postmodernism then, just as it follows from the object-level assumptions of mechanism, that we can do with the world as we will. The present position overturns both these sets of assumptions, and thus puts an end to the project of modernity, by reintegrating human subjectivity and agency into the subjectivity and agency of re-awakened world.

## 5 Discovering the Tao in Australia

The position that I have outlined in this paper is basically a Taoist one. Nature, according to my definition, is more or less equivalent to the Tao: it is the wise way the world unfolds when left to its own devices (Mathews 1996). The Taoism of Lao Tzu does not announce itself as panpsychist, but clearly a world animated by the Tao is one which is possessed of some intelligent inner principle, a principle that can be trusted to guide us into the deepest channels of life.

It is not my place to comment in detail here on the affinities between the broadly Taoist attitude of letting be and certain fundamental characteristics of Aboriginal thought. Suffice it to say that Aboriginal cultures evince a powerful engagement with the given that ensures their continuity with their own past but also their flexibility in the face of an almost unimaginbale scale of externally imposed change. For while not *craving* the possible and the ideal, they exhibit a genius for accommodating the new once it has become actual. One of the flashpoints in the evolution of my own thinking occurred when a non-indigenous friend who had married into an Aboriginal family in the far north-west of Australia told me about an elder there who included *motorboats* in his Dreaming stories. Years later I happened to find myself living for a while in the very community to which the old man – by now deceased – had belonged, and I was enchanted by the way in which the people in this community refused nothing. They accepted – though they never craved – anything and everything that drifted their way, all the trappings and junk of modern civilisation. But in the process of accepting this tawdry stuff, they also uncannily Aboriginalised it, so that it assumed an entirely different significance in the context of their community from its intended significance within the framework of a capitalist culture. Somehow, through this affectionate trust in the given, the everyday was rendered numinously spiritual, and the spiritual unpretentiously everyday.

Mary Graham has declared that one of the most taken-for-granted assumptions of Aboriginal thought is that spirit is real; another is that land is all there is (Graham 1992). That is to say, spirit has a status, in Aboriginal thought, as incontestable as that of energy and matter; and since there is no heaven and hell, and since theories and ideas, however dazzling, are not real, land is ultimately the only thing that exists. If 'land' is expanded to encompass the concrete given – all that is actual in a physical sense – then I think that the attitude of letting be follows from these twin premises: spirit animates the given rather than existing in the realm of the abstract, so we connect with spirit by engaging – and not unnecessarily interfering – with the given. By embracing the given even in its most adulterated forms, we reinhabit our own contemporary, mundane reality in the same kind of profound way that traditional Aboriginal peoples inhabited their reality, the still edenic land.

Aboriginal peoples learned to inhabit the real rather than escaping into the ideal by remaining attuned to the wisdom that this peaceful old land imparts to those who pay attention to it. I believe that my own early glimpses of this way of living came to me not in the first instance through explicit Aboriginal influences – though these influences later helped to bring it to consciousness – but through the opportunities for such attentiveness that were vouchsafed me as a child. Born into this intimately companionable land that has for so long been singing along, humming along, with its human inhabitants, non-Aboriginal Australians might also, if we collectively pause to feel the resonance of the endlessly poetic communiques that surround us, rediscover, in a contemporary context, some of the fundamental aspects of the Aboriginal relation to the world.

## Notes

1 An outline of this paper was presented at the Environmental Justice Conference at University of Melbourne, 1997.

2 My argument for a broadly panpsychist view of matter was begun in my book, *The Ecological Self* (Mathews 1991), and developed further in Mathews 1998. I am currently completing a book entitled *The Tower Spoke: a Contemporary Panpsychism*, in which the ideas of the latter article are greatly expanded. The present article is extracted from a chapter of a second book in preparation; in this book the implications of panpsychism for environmental praxis are drawn out.

3 I have expanded elsewhere on the thesis that living systems, including individual organisms, may be characterised as self-realising systems imbued with their own conatus. (Mathews 1991, Ch 3).

4 To defend this essentialist claim against the social constructionism of contemporary postmodernism and poststructuralism would take me too far afield here. Suffice it to say that at a minimal level, this claim is self-evidently true: people's hunger, thirst, sociability, disposition to learn and use language, to build cultures, to live collectively, all operate independently of ideology – which is not to say that they may not become disturbed by ideology in dysfunctional circumstances.

5 It could be said that it is part of the 'disposition' of things to be susceptible of manipulation by intentional agents, and to respond to such manipulation in determinate ways. While this is no doubt a fair semantical point, it overlooks the significance of the patterning that occurs when things are not subjected to manipulation compared with the failure of such patterning when they are subject to arbitrary human interventions.

6 This is not to say that nature necessarily has an answer for everything. The scenario that immediately springs to mind is that of the life-shattering asteroid. It is not obvious that the biosphere has a 'contingency plan' for this circumstance. On the other hand, the likelihood of this catastrophe occurring is very small.

7 This faith in nature is also justifiable to some extent even on the scale of the individual, since, as I have explained, we and other individual organisms are perhaps, although not invariably, more likely to realise ourselves, in the sense of satisfying our fundamental needs and impulses, when we are not required to act out abstractly conceived roles and identities.

8 In Mathews 1994 I have explored at much greater length the tension between the position implicit in deep ecology, namely that nature knows best, in the sense that it can best look after its own larger, long-term interests, and the imperative that we as

individuals confront, to look after our own interests, and those of our loved ones, in the face of sometimes hostile natural processes.

9 It has been put to me (by my colleague, Robert Young) that most people would happily employ abstract thought to avert a natural disaster that was going to kill a lot of people, even though this led them to move against nature in the present sense – to redirect the natural course of events, by large-scale engineering measures, for instance. I agree that most of us would react in this way if we were faced with the choice in these terms. But to ascertain our true values in this connection it would be more telling to confront us with a choice between two types of society: one which possessed such measures, and was consequently destined eventually to despoil the planet's biosystem, and one which did not, and was consequently set to live within the ecological terms of that system indefinitely, even though it was unable either to avert catastrophes, or to save as many individual lives in the short term, as the other society. Couched in these terms, I do not think it is nearly so obvious which choice most people would make.

10 I have observed a similar attitude to junk and rubbish in Aboriginal communities in Australia. Lots of litter and old machinery and cars and fridges and so on are left lying around in some of the remoter communities, apparently on the assumption that these things, like all others, have their place in the world, so there is no point in pretending that they do not exist by putting them out of sight. They will in time, again like everything else, be received back into the land, back into the cycles of life.

11 For a comparable account of the way the land claims its native sons and daughters in the Aboriginal world, see Deborah Bird Rose 1992, and this issue.

12 I intend 'inaction' here in the Taoist sense of the term.

13 There is a great deal more to be said than I can say here concerning the ultimate scope of the principle of letting be. On the face of it, the principle should apply only to concrete things, because its justification is that concrete things have their own informing principles of unfolding, principles that are clearly not possessed by ideas or abstract things. However, in social and cultural contexts, no clearcut division can be made between ideality and concreteness, since ideas inform the corporeality of human beings and their built environments. A more extended discussion of the scope of the principle of letting be is included in the chapter from which this paper is extracted. For present purposes it will suffice to emphasise the ontological rather than the sociocultural scope of the principle.

## References

Brennan, Andrew 1998. *Thinking about Nature*. London: Routledge.

Graham, Mary 1992. Interview on *Aboriginal Perspectives*, Caroline Jones and Stephen Godley, ABC Religious Program.

Mathews, Freya 1991. *The Ecological Self*. London: Routledge.

Mathews, Freya 1994. 'Relating to Nature', *The Trumpeter* **11**(4): 159–166.

Mathews, Freya 1996. 'The Soul of Things', *Terra Nova* **1**(4): 55–64.

Mathews, Freya 1998. 'The Real, the One and the Many in Ecological Thought', in David Cooper and Joy Palmer (eds), *The Spirit of the Environment*, pp. 57–72. London: Routledge.

Rolston, Holmes, III 1988. 'Following Nature', in *Environmental Ethics*, pp. 32–44. Philadelphia: Temple University Press.
Rose, Deborah Bird 1992. *Dingo Makes Us Human.* Cambridge: Cambridge University Press.
Seddon, George 1997. 'The Nature of Nature', in *Landprints*, pp. 7–14. Melbourne: Cambridge University Press.

# 93

# THE NEO-STOICISM OF RADICAL ENVIRONMENTALISM

*Jim Cheney**

Source: *Environmental Ethics*, 11 (1989), 293–325.

Feminist analysis has convinced me that certain tendencies within that form of radical environmentalism known as Deep Ecology—with its supposed rejection of the Western ethical tradition and its adoption of what looks to be feminist attitude toward the environment and our relationship to nature—constitute one more chapter in the story of Western alienation from nature. In this paper I deepen my critique of these tendencies toward alienation within Deep Ecology by historicizing my critique in the light of a development in the ancient world that is disquietingly similar to the rise of Deep Ecology in recent times—namely, the rise of Stoicism in the wake of the breakup of the ancient polis.

## Introduction

(1) *Overview*. Much of the male-authored radical environmentalism known as Deep Ecology consitutes a resurgence of Stoic sensibility. Those who are familiar with the literature of Deep Ecology will note the peculiarly contemporary ring to the following account of ancient Stoicism:

> The veneration of the cosmos is the veneration of the whole of which man himself is a part. The recognition of and compliance with his position as a part is one aspect of man's proper relation to the universe in the conduct of his life. It is based on the interpretation of his existence in terms of the larger whole, whose very perfection consists in the integration of all its parts. In this sense man's cosmic piety *submits* his being to the requirements of what is better than himself and the source of all that is good. But at the same time man

* The author would like to thank Tom Birch, in particular, for extensive comments on an earlier draft, from which this paper benefited greatly. The paper also benefited from comments by the following people: Bill Devall, Warwick Fox, Eric Katz, Marti Kheel, Carolyn Merchant, Andrea Nye, George Sessions, Karen Warren, Michael Zimmerman, and an anonymous referee.

> is not just a part like other parts making up the universe, but through the possession of a mind a part that enjoys *identity* with the *ruling principle* of the whole. Thus the other aspect of man's proper relation to the universe is that of *adequating* his own existence, confined as it is as a mere part, to the essence of the whole, of reproducing the latter in his own being through understanding and action. . . . [I]n achieving this knowing relation, human reason assimilates itself to the kindred reason of the whole, thereby transcending the position of a mere part.[1]

Something very close to this constitutes the dominant image governing much deep ecological writing. Deep ecologists, many of them, are asking us to submit to the modern equivalent of the divine logos, to discern that logos at work in the ecosystem and identify with it. There are a number of disquieting aspects of this tendency within Deep Ecology which come to mind once the affinity between it and Stoicism becomes apparent.

(2) *Methodology.* I am not claiming point-by-point similarity between ancient Stoicism and Deep Ecology; nor am I claiming that deep ecologists have been influenced by ancient Stoicism. It is certainly true that ancient Stoicism differs in many ways from Deep Ecology. What I *am* claiming is (a) that there is a certain sensibility present in Stoicism, a theme which can be read there as a subtext when Stoicism is considered in the social and political context of its rise to prominence in the ancient world, and (b) that this same sensibility can be read as a subtext in the deep ecological literature I will be considering.

Reading the subtexts underlying social and historical events as well as those underlying human behavior is nothing new and is no more (or less) problematic than the hypothesizing of theoretical entities in science. Freud's reading of the family romance, the Marxist understanding of liberalism as capitalist ideology, Kuhn's reading of the social dimension of scientific progress, and Rorty's deconstruction of modernist epistemology are all instances of such subtextual analysis. The reading of a subtext is the discernment of a hidden agenda underlying the text itself, an agenda which stands in complex relationships to the agents participating in the construction of the text. Subtextual analysis is not the reduction of text to subtext. Text and subtext interact with one another, and individuals who participate in one participate in some measure in the other, regardless of conscious intent. I do *not* want to claim that the subtext I discern underneath particular texts I consider is consciously ascribed to by the authors of those texts; *nor* do I wish to claim that they are the subtexts of those authors' lives. What I *do* want to claim is that the subtexts are *there*, and that the fact that they are there (and are being played out) has social, political, and personal implications. Just as a man who is not in any obvious or explicit way sexist may contribute, by the very accumulation of small and seemingly insignificant actions,[2] to structural sexism, so deep ecologists may—and, I think, *do*—contribute, in certain of their theoretical formulations, to the continuation of the very alienation from nature that they consciously oppose.

(3) *Focus of the Analysis.* My concern is with certain *tendencies* within what is often referred to as the *Deep Ecology movement*. As case studies in Neo-Stoicism I take the recent monograph by Warwick Fox, *Approaching Deep Ecology: A Response to Richard Sylvan's Critique of Deep Ecology*, Bill Devall and George Sessions' *Deep Ecology: Living as if Nature Mattered*, and several papers by Fox, Sessions, and Arne Naess.[3]

These deep ecologists make a distinction between what they call the *Deep Ecology platform* (or, alternatively, the *basic principles* of Deep Ecology) and "the fundamental features of philosophies and religions from which that platform (basic principles) is derived."[4] Although accounts of the Deep Ecology platform (basic principles) vary to some extent, those offered by Naess and by Devall and Sessions[5] can be taken as paradigmatic. In those accounts the principle of *biospherical egalitarianism* is stressed. This is expressed in *Deep Ecology* as follows: "The well-being and flourishing of human and nonhuman Life on Earth have value in themselves. . . . These values are independent of the usefulness of the nonhuman world for human purposes."[6] Most of the other elements of the Deep Ecology platform are devoted to a specification of what is needed in order to realize these human and nonhuman values (e.g., richness and diversity of life forms, reduction of the human population, and changes in public policy). In line with this, Fox has delineated "Deep Ecology's positive or constructive focus" in this way: "Deep ecology is concerned with encouraging an egalitarian attitude on the part of humans not only towards all *members* of the ecosphere but even towards all identifiable *entities* or *forms* in the ecosphere, whether 'living' . . . or not."[7]

*Suitably stated*, I have no difficulty endorsing the Deep Ecology platform.[8] My difficulty is with the "fundamentals" from which deep ecologists tend to derive that platform. "The fundamentals," Naess says, "are Buddhist, Taoist, Christian or of other religious persuasions, or philosophic with affinities to the basic views of Spinoza, Whitehead, Heidegger, or others." He states further that these fundamentals are "mutually more or less incompatible," but that this "does not affect the Deep Ecology [basic] principles adversely."[9] A person could reject *any* of these fundamentals and yet remain a deep ecologist by accepting the Deep Ecology platform. Naess has developed his own set of fundamentals which he calls "Ecosophy T." Ecosophy T is a normative system, the ultimate norm of which is Self-realization. From this norm are derived many of the items which appear in the Deep Ecology platform, such as norms concerning diversity, complexity, exploitation, and egalitarianism. As Sessions puts it, "Naess stresses that Ecosophy T is only *his* version of Deep Ecology, and that many versions need to be worked out."[10]

When we move on to Devall and Sessions and to Fox, however, we encounter an exegetical problem. Prior to stating the basic principles (platform) of Deep Ecology, Devall and Sessions provide an account of the "central intuition" of Deep Ecology—namely, "the idea that we can make no firm ontological divide in the field of existence: That . . . to the extent that we perceive boundaries, we fall short of deep ecological consciousness"—and two "ultimate norms" of Deep Ecology—namely, *Self-realization* and *biocentric equality*.[11] Fox states that Deep Ecology proceeds from two hypotheses and one norm:

> The hypotheses are (i) that "The self is as comprehensive as the totality of our identifications. Or, more succinctly: Our Self is that with which we identify" . . . and (ii) that the self can and does grow/develop/mature (i.e. widen its sphere and intensity of identifications over time). The norm is that the ideal state of being is one that sustains the widest (and deepest) possible *identification* and, hence, sense of Self. . . . [T]his ideal state of being is referred to as "Self-realization" by Naess and as "ecological consciousness" by Devall and Sessions, and its cultivation—effectively a spiritual discipline—is considered to be the "real work" of Deep Ecology.[12]

The exegetical problem concerns the status of these accounts. In relationship to the distinction between the Deep Ecology platform (basic principles) and the "fundamental features of philosophies and religions from which that platform (basic principles) is derived," it seems clear that these accounts step over the line from "platforms" to statements of "fundamentals" (metaphysics). Consider Fox's response to the following passage from Lynn White, Jr.:

> The problem grows if we ask "Do people have ethical obligations toward rocks?" To an ancient Greek, to an American Indian, or perhaps to certain kinds of Buddhists, the question would have meaning. For quite different reasons they would probably reply "Yes". . . .[13]

Fox's response is to claim that White

> . . . does not explore the question of the degree to which these reasons are based on visions of reality that are (i) superficially different [as a function of differing cultural contexts] . . . but otherwise structurally similar in the sense that they are differently described visions of essentially the same reality; (ii) different at both the surface and the structural level (like those of Heraclitus and Parmenides) but nevertheless conducive . . . to similar ways of experiencing reality e.g. conducive to experiencing oneself as bound up with the rest of the world versus separate and alienated

> from it; and (iii) different at the surface, structural and experiential levels.[14]

As an example of an experience of reality which might underlie multiple descriptions or structural accounts of reality, Fox refers to "experiencing oneself as bound up with the rest of the world versus separate and alienated from it." It *may* be that the phrase "experiencing oneself as bound up with the rest of the world" is descriptively neutral enough to do as an account of the experiences underlying certain Greek, American Indian, and Buddhist metaphysics. What transpires in Fox's account, however, is that these experiences of being bound up with the rest of the world become equated with the central hypothesis of Fox's account, mentioned above—namely, that "the self is as comprehensive as the totality of our identifications," that "our Self is that with which we identify." At the experiential level, the claim is that Greek, American Indian, and Buddhist responses are essentially similar to Fox's views concerning Self-realization. In other words, Fox's view captures the experiential core of these various traditions and is therefore privileged; it is not just one more philosophical position—a "fundamental" view—from which the Deep Ecology platform can be derived. In this way, it comes to pass that American Indians, Greeks (some of them), and Buddhists (some of them) are deep ecologists. But in passing from talk of "experiencing oneself as bound up with the rest of the world" to Fox's account of identification we have passed from neutral description—if that is what it is—to an anything-but-neutral metaphysics. What we have is a metaphysics claiming to be a privileged account of the experiences underlying a great multiplicity of widely differing voices.[15]

We need to detach "Ecosophy S," as I will now call these accounts,[16] from the Deep Ecology platform and treat it as a statement of "fundamentals." The platform, suitably stated, is laudable and should be saved; the subtext underlying Ecology S, however, makes the adoption of this position extremely problematic; detaching the one from the other might make it possible to achieve the aim Naess seems to have had in mind in distinguishing "fundamentals" from "platform"—namely, the creation of a broad-based Deep Ecology movement.

## Stoicism and Ecosophy S

> The tragedy of the West is rooted in number: one is too few and two are too many. To be one should mean to be unified, whole; that should be enough, yet it is lonely. But all human community involves difference. Difference is a challenge to autonomy, to wholeness. Memory, always about the origin, is about a lost oneness.... The telos or goal is about perfect union. The origin and the purpose, then, both are about the desire to be One.... This is the essential structure of

> masculinist eros so deeply embedded in Western stories, including the natural sciences.
>
> Donna Haraway[17]

Ecosophy S places itself in a tradition that conceives of correct environmental practice as involving, in some fundamental way, an understanding of the cosmos and humanity's place in the wider scheme of things. *Ethics*, as this term is used in the context of Ecosophy S, is to be understood "in the sense of the development of an experientially grounded state of being that issues in 'an organic ethos a way of life.'"[18] The central concept of this approach is that of "ecological consciousness" (or "Self-realization" or "sustaining the widest (and deepest) possible *identification* and, hence, sense of Self").[19] Devall and Sessions' statement that "The foundations of Deep Ecology are the basic intuitions and experiencing of ourselves and Nature which comprise ecological consciousness"[20] expresses the centrality of the notion of ecological consciousness for Ecosophy S. There are two normative dimensions to ecological consciousness. First, as stated by Fox, "the ideal state of being is [ecological consciousness or Self-realization or] one that sustains the widest (and deepest) possible *identification* and, hence, sense of Self."[21] Second, ecological consciousness figures into an understanding of "correct behaviour" in the world. As stated by Fox, supporters of Ecosophy S

> ... see correct behaviour as a natural consequence of aligning ourselves with "the nature of reality and out place ... in the larger scheme of things." And by "aligning" is not meant the following of certain rationally constructed rules so much as an "empathizing" of this vision or reality i.e. an incorporation of it that goes beyond (as distinct from substituting for) intellectual assent....[22]

It is this intimate relationship between ecological consciousness and practice, a relationship mediated by the notion of identification, which constitutes the tendency toward Neo-Stoicism inherent in Ecosophy S.

(1) *The Salvational Project of Ecosophy S.* The cultivation of ecological consciousness is often presented as salvational. Theodore Roszak is quoted approvingly by Devall and Sessions as claiming that "the essential business of life ... is to work out one's salvation with diligence."[23] Sessions places Ecosophy S in the company of Pythagoras, for whom

> Purification of the soul ... was to be got through *philosophy* (love of wisdom), the intellectual activity consisting in *theory* (looking-at, contemplation) of the *cosmos* (order—that is, the world regarded as a patterned whole). ... Pythagoras's point was that awareness is liberation; to understand the world is to rise in some measure above it.[24]

In so doing, Sessions criticizes the "separation of the 'theory of the cosmos' from a search for self-knowledge or spiritual autonomy."[25] Ecosophy S is viewed as part of the "perennial philosophy" tradition, instances of which

> ...characteristically began with a metaphysical account of the world or reality which placed humans in the wider scheme of the things. Human psychology was then understood in terms of adjusting to this larger reality. Finally, a system of ethics or a way of life was the end result of the perennial philosophy approach. Spiritual growth and human maturity developed from the "enlightenment" of realizing that our narrow isolated socialized self is an illusion....[26]

Among the "sources of the Deep Ecology perspective" Devall and Sessions cite the poet Robinson Jeffers:

> I believe that the universe is one being, all its parts are different expressions of the same energy, and they are all in communication with each other, therefore parts of one organic whole.... The parts change and pass, or die, people and races and rocks and stars; none of them seems to me important in itself, but only the whole. This whole is in all its parts so beautiful, and is felt by me to be so intensely in earnest, that I am compelled to love it, and to think of it as divine. It seems to me that this whole alone is worthy of the deeper sort of love; and that there is peace, freedom, I might say a kind of salvation, in turning one's affections outward toward this one God, rather than inwards on one's self, or on humanity, or on human imaginations and abstractions....[27]

Jeffers refers to this sensibility as a "tower beyond the reach of tragedy."[28] This phrase is revealing: in its sensibility it is thoroughly Stoic.

Stoicism is characterized by its attempt to overcome the alienation resulting from the "tragic" loss of the Greek city-state by construing the *cosmos* as its home. The theme of the development of ecological consciousness as a means of moving beyond alienation is explicit in the writing of John Seed:

> When humans investigate and see through their layers of anthropocentric self-cherishing, a most profound change in consciousness begins to take place.
>
> Alienation subsides. The human is no longer an outsider, apart.... [Y]ou start to get in touch with yourself... as a species only recently emerged from the rain forest....

> "I am protecting the rain forest" develops to "I am part of the rain forest protecting myself. I am that part of the rain forest recently emerged into thinking."[29]

What I suggest is that underneath this Neo-Stoic text of alienation and salvation is a subtext: the building of ecological consciousness as the building of a "tower beyond the reach of tragedy." The notion that correct practice is intimately connected with the empathic internalization of, or alignment with, a vision of reality and our place in it has peculiar attraction to the alienated soul searching for wholeness, health, home.[30]

But why would it occur to one that alienation is to be overcome by, of all things, metaphysics, by the empathic internalization of a highly abstract, humanly constructed *vision* of wholeness, connectedness, and health? It is instructive, in this regard, to notice that Stoicism is a response to a particular cultural situation. Hans Jonas provides us with an account of Stoicism as a response to the passing of the classical polis and of the fragility inherent in the Stoic response:

> Grand and inspiring as [the Stoic] conception is, it must not be overlooked that it represented a position of retreat inasmuch as its appeal was addressed to a human subject that was no longer a part of anything *except the universe*.... According to classical doctrine, the whole is prior to the parts, is better than the parts, and therefore that for the sake of which the parts are and wherein they have not only the cause but also the meaning of their existence. The living example of such a whole had been the classical *polis*, the city-state, whose citizens ... not only were *dependent* on the whole for their being but also *maintained* that whole with their being.... Thus this whole, making possible first the very life and then the good life of the individual, was at the same time entrusted to the individual's care.... Now it was the *cosmos* that was declared to be the great "city of gods and men," and to be a citizen of the universe, a *cosmopolities*, was now considered to be the goal by which otherwise isolated man could set his course. He was asked, as it were, to adopt the cause of the universe as his own.... Wisdom conferred inner freedom in shouldering the tasks, composure in facing the whims of fortune besetting their execution, but did not set or revise the tasks themselves.... A role played is substituted for a real function performed....[31]

According to Andrea Nye's description of the passing of the classical polis, "Gone was the closed self-congratulatory security of the small city-state community whose upper class shared the same ideals.... Alexander's world was full of differences...."[32] But the need for some analogue of the security of the city-state did not disappear. Nye goes on to tell of the Stoic attempts to maintain the hegemony of the Athenian world, establishing small polises within indigenous cultures—this within the context of the Stoic notion that "to adopt the cause of the universe as his own" was "the goal by which otherwise isolated man could set his course." It is this seeming contradiction that allows us to see the Stoic sensibility for what it is, the creation of a "tower beyond the reach of tragedy," the continuance of isolation and alienation *by the very forging of a metaphysics of embedment in the cosmos*. The (or *a*) subtext of Stoic philosophy is the desire for embedment in some greater whole coupled with a refusal to give up the hegemony that went with Athenian supremacy. The result was alienation from the real world, the Alexandrian world of difference, and the creation of an *abstract* embedment which provided for an analogue of the security of the polis. Stoic logic, metaphysics, and ethics provided a unifying framework in which otherness could be encountered without risk; it provided a way of containing the Alexandrian world of difference.

What has disappeared for *us*, creating the vacuum in which Ecosophy S has appeared, is the self-congratulatory security of *modernism* and (most important for deep ecologists) its declared anthropocentricity. Ecosophy S (with its central concept of ecological consciousness) functions in a manner analogous to Stoicism, but within a different context, that of the demise of modernism, its shattering into a world of difference, the postmodern world. It expresses a yearning for embedment coupled with a refusal to forgo the ultimate hegemony so characteristic of modernism. This is the *subtext* underlying the concepts of Self-realization, identification, and ecological consciousness. Subtextually, the central operative idea at work in these concepts in the idea of *containment*, containment of the other, of difference, rather than genuine *recognition* of the other, genuine acknowledgement and embracing of the other.

(2) *The Bootstrapping of Ecosophy S.* Nye explores in detail the mechanism by which Stoicism maintained cultural hegemony within the context of a metaphysics which proclaimed that one was ultimately a citizen, not of this or that city-state, but of the universe. That mechanism was Stoic logic: "Logic would be... the grammar of the cosmos, a divine law that keeps the physical world in place, a universal will to which a virtuous man must conform."[33] Access to this grammar of the cosmos by the wise man constituted a new way of dividing the human world into an analogue of the old division into Greeks and barbarians. It would not be the particularities of the Greek language which effected the division, but rather a particular kind of perception, of which only rational men were capable. "With Zeno... logic... is to be the scaffolding for a ruling culture, a way to order its thought and perpetuate its most deeply held

convictions. This is accomplished in Stoic Metaphysics and Ethics." Stoic wisdom is powerful indeed:

> Once rational thought [defined by access to the "grammar of the cosmos"] is firmly instituted a man can do no wrong. His actions will follow naturally, not requiring any deliberation or conscious choice. He has achieved the state of Stoic a-pathy and has driven passion from his life. He is not without emotion, but his emotions are rational. . . .[34]

But what guarantees that Stoic access to the "grammar of the cosmos" is in the truth? The epistemic privilege of the Stoic wise man requires

> . . . a radical revision in the notion of truth. Originally the Greek concept of truth was closely related to the credibility of the logos, or account, of a witness of an event. Could he or she be trusted? . . . Was his narrative of the event likely to be true? . . . Gone [in Stoicism] is the complex determination of motives and events and purposes that characterize truth in its communicative setting.[35]

In other words, truth is no longer to be negotiated in its social and historical setting; truth is no longer contextual[36] and historical—it is ahistorical and eternal. The new Stoic logic achieved this "eclipse of the speaker" by the introduction of the idea of a "meaning" which stands between sign and referent, and which allows for a distinction between perception and grasped or rational presentation, the latter of which is the guarantor of truth. Unfortunately, the arguments involved were circular: ". . . the new concept of truth and the idea of a 'meaning' are conceptually realted: each is defined in terms of the other."[37] The effect of this introduction of the notion of a "rational presentation" is to insulate Stoic wisdom from cultural negotiation by lifting it beyond the reach of time and circumstance. Access to "rational presentations" distinguishes the wise man from the barbarian.

Corresponding to the "grammar of the cosmos" in Stoic logic we have in Ecosophy S what Fox calls the "central intuition" of Deep Ecology, "the idea that there is no firm ontological divide in the field of existence,"[38] otherwise stated as "a vision of the world as a 'unity in process.' "[39] This central intuition has the same epistemic status as the Stoic "grammar of the cosmos." It is, strictly speaking, beyond the reach of the science of ecology, not deducible from it; it is an *intuition* partly constitutive—along with the norms of Self-realization and biospherical egalitarianism—of ecological consciousness.[40] We have seen that Fox argues that the "vision of reality" of Ecosophy S constitutes the experiential core of a wide range of religious and philosophical traditions, that it constitutes, as it were, the "deep structure" of those traditions, underlying their cultural and [surface] structural differences. Ecological consciousness, tuned to the "grammar of the cosmos," also puts one in an *ethically* privileged place: "... Deep Ecology supporters see correct behaviour as a natural consequence of aligning ourselves with 'the nature of reality and our place ... in the larger scheme of things.' "[41] Of central significance to Ecosophy S is the salvational dimension of ecological consciousness; the parallels between Stoicism and Ecosophy S in this respect are striking.

Most importantly (and most problematically), ecological consciousness provides a new way of effecting a modern equivalent of the distinction between Greeks and barbarians. As with the Stoics, the distinction—this time between those *with* and those *without* ecological consciousness—is not made along cultural lines; nor is it made with respect to what Fox calls [surface] "structural" features of religions or philosophies. It is made at a deeper level, analogous to that of access to the "grammar of the cosmos" for the Stoic wise man. The distinction generates a contradiction similar to that in Stoicism when it tries to maintain the hegemony of the Athenian world within the context of its philosophy of adopting the cause of the universe as its own. The key is the centrality of metaphysics in the form of an internalized vision of reality and the linking of ethics to this metaphysics. Ecological consciousness with its internalized vision of reality has a non-contingent relationship to Deep Ecology: it is definitive, for proponents of Ecosophy S, of what it is to be a deep ecologist, as I have argued. Metaphysical *systems* (even Spinoza's) are detachable form Deep Ecology, according to Ecosophy S, in the sense that one can be a deep ecologist without subscribing to any metaphysical system. One cannot be a deep ecologist, however, without sharing the ecological consciousness delineated in Ecosophy S. So far this is not *philosophically* problematic (although, as I have argued, ecological consciousness should *not* be made definitive of Deep Ecology if what one is after is broad-based support of the Deep Ecology platform). What *is* problematic is the way in which ecological consciousness in insulated from contextual negotiation.

Part of this insulation is achieved, I have argued, by claiming that the internalized vision of reality partly constitutive of ecological consciousness expresses the experiential core of a wide range of religious and philosophical views. More important, however, is the "bootstrap" nature of more direct arguments in support of the rationality of Ecosophy S's vision of reality. The argument presented by Fox goes like this. First, Fox subscribes to Naess' view that "the norms and tendencies of the Deep Ecology movement are not derived from ecology by logic or induction. Ecological knowledge and the life-style of the ecological field-worker have *suggested, inspired, and fortified* the perspective of the Deep Ecology movement." The "unity in process" vision of reality is then affirmed as an "intuition" (the *central* intuition) which goes beyond the findings of the science of ecology (or any science). Acceptance of such intuitions is said to be a matter of trusting one's inner voice. Fox then says that "While I refer to this view as the central *intuition* of Deep

Ecology, I do not in any way mean that it is irrational or ungrounded." It has, Fox claims, "a basis *in ecology*"; "this cosmology is what ecology directly implies." (This claim must be understood in a way that is consistent with the earlier claim that the vision of reality is an "intuition" and "not derived from ecology by logic or induction." What Fox seems to be after is something a bit stronger than Naess' terms *suggest, inspire* and *fortify* afford.) After referring to Roger Walsh's "hypothesis . . . that we can enhance our perceptual sensitivity by the augmentation of normal sensory perception (as in science), by intellectual conceptual analysis (as in philosophy), or by direct perceptual training (as in meditation)," Fox assents to Walsh's claim that

> . . . no matter how it is obtained, (perceptual) enhancement of sufficient degree may reveal a different order of reality from that to which we are accustomed. Furthermore, the properties so revealed will be essentially more fundamental and veridical than the usual. . . .

Fox then returns to the claim that, nonetheless, "the central vision of deep ecologists *is* a matter of intuition. . . . That is, it is a matter of trusting one's inner voice in the adoption of a value stance or view that can not itself be proven or disconfirmed." (Curiously, Fox then undermines this distinction between ecological consciousness and science by noting—citing Mary Hesse—that there is nothing unusual in his use of the expression *intuition*, since scientific theories are similarly constrained, but underdetermined, by facts.) Fox concludes by claiming that in the end we decide between competing theories or world views by adopting an evaluative stance.[42]

With this last statement Fox avoids the particular kind of circularity found in the Stoic argument, for in the end he does not claim to have shown that the vision of Ecosophy S is the truth. There is, however, a curious "bootstrap" quality to the argument which has the effect of placing the deep ecological vision of reality and its associated norms beyond the reach of cultural negotiation, and hence produces the same effect as circularity.[43] What happens is that the vision of Ecosophy S is bootstrapped beyond the social, historical, and personal conditions of its origin, placed outside time and circumstance. The "inner voice" which is to be trusted is one which has been purified[44] by elimination of self-centered thoughts and bootstrapped to a new level of veridicality. By this technique we penetrate to the deep structure, the experiential core, underlying contingent, cultural encrustations.

For Sessions, Spinoza is the main *metaphysical* inspiration for his adoption of Ecosophy S. As such, Sessions regards Spinozism as optional to the adoption of Deep Ecology. Nonetheless, his attraction to Spinoza tells us something important about the concept of ecological consciousness in Ecosophy S.

> But the highest level of knowledge, for Spinoza, is direct intuitive knowledge of individual things (modes) and this is clearly a mystical kind of knowing. The subject/object distinction disappears—actually, one goes beyond all conceptual knowledge, and experiences the "union that the mind has with the whole of Nature." And only at this level is there *understanding*. . . .
>
> While scientific knowledge is a kind of "seeing things under the aspect of eternity," it ultimately is only a higher and more objective form of imagination than mere opinion, but it can lead to direct understanding. Given that all rational conceptualization is a form of imagination, what would Spinoza have to say of his own rational-logical metaphysical system? Like theoretical science, it too is a higher form of imagination which can help guide one to intuitive understanding. Wittgenstein seemed to understand his metaphysical system in the *Tractatus* in a similar way (6.54): "My propositions serve as elucidation in the following way: anyone who understands me eventually recognizes them as nonsensical, when one has used them—as steps—to climb up beyond them. . . . He must transcend these propositions, and then he will see the world aright. What we cannot speak about we must pass over in silence."[45]

There is certainly no negotiating *this* kind of knowledge; it has left the realms of discourse altogether. The kind of understanding that is claimed is quite ahistorical and quite Stoic in temperament; it is not positioned with respect to the personal, social, or historical contingencies within which that understanding arises. Naess' acceptable statement that the *norms* of Deep Ecology go beyond the claims of the *science* of ecology seem to be the basis for the bootstrapping of ecological consciousness into eternity by other proponents of Ecosophy S. This is the intent of Stoic logic as well.

What subtext is at work her? What could possibly motivate such an extreme response, the contradictory mixing of deep ecological call to embedment in this world with the insulation of the vision of reality and norms of Ecosophy S from that very embedment, the bootstrapping of Ecosophy S into eternity? The answer I have suggested is that the subtext of Ecosophy S parallels that of Stoicism: the yearning for a home in the wake of the breakup of the polis while refusing to give up the privileges of the polis which the description of the new home seems to imply. Read *modernism* for "the polis" and we have a description of the condition of Ecosophy S. The seeming need to bootstrap its account into eternity is one of the characteristics of modernism which Ecosophy S refuses to forgo. The result is a continuing, perhaps deepening, alienation wedded to a strong need to overcome that alienation.

(3) *Autonomy and Control in Ecosophy S.* This tendency toward bootstrapping, with its insulating effect, goes hand in hand with foundationalism,

both in Stoic logic and metaphysics as well as in modernism. The move toward foundationalism is an attempt to generate a kind of knowledge which is autonomous in the sense of not being subject to social negotiation. Subtextually, this need for an autonomous realm of knowledge often goes hand in hand with a need for autonomy on the part of the knowing subject—autonomy in the static sense of "separation and independence."[46] The detailing of this connection in modernism (and the demonstration that the conception of autonomy at work is a peculiarly masculinist one[47]) cannot be undertaken here, but one instance of this must be mentioned since it is germane to my analysis.

A long series of works in the philosophy of science, beginning with the work of Thomas Kuhn, has demonstrated that science is socially constructed and that integral to this construction is the notion that, as Sandra Harding puts it, science is "sacred":

> The project that science's sacredness makes taboo is the examination of science in just the ways any other institution or set of social practices can be examined.... To what other "community of natives" would we give the final word about the causes, consequences, and social meanings of their own beliefs and institutions?[48]

Questions which can be asked of any other social activity cannot be asked with respect to the practice of science for fear of committing the genetic fallacy. Modern science has constructed itself in such a way that it has insulated itself from social negotiation.

The ideology of "sacred" science has been connected with an alienating form of autonomy by Evelyn Fox Keller:

> *Autonomy*, at one end of its range [of meanings], connotes a radical independence from others, mapping closely onto an interpretation of *objectivity* that implies a reductive disjunction of subject from object.... [T]he same interpretation of autonomy also correlates with a conception of power as power over others, that is, with power defined as domination.[49]

What is at work here is the need to control the other on the part of the knowing subject which has become radically disjoined from the object of its inquiry, from the other from which it has become separate, "autonomous." For the most part, the literature, which has dealt with this conception of autonomy as separation and independence, has dealt with the social and psychological conditions that foster its development and with an understanding of how this drive for autonomy manifests itself in such diverse forms as the ideology of modern science, liberal political ideology, the domination of women, and the degradation of the environment.

What have been given less attention are the ways in which this need for autonomy and control can be discerned in behaviors which seem to eschew the desire for domination and radical separateness.[50] This can be explored by examining distorted forms of connectedness in which connectedness is achieved either by forgoing the development of a genuinely autonomous self or by an assimilation of the other by the statically autonomous self, an assimilation which constitutes a subtle form of control or domination.

One form that the (pseudo-) connectedness achieved by the forgoing of genuine autonomy takes is submission in "ideal love":

> The fantasy of submission in ideal love is that of being released into abandon by another who remains in control. Here I would argue that the freedom and abandon called forth by this powerful, controlling other represents an alienated version of the safe space that permits self-discovery, aloneness in the presence of the other.[51]

The "controlling other" is the precise complement to this "submission in ideal love" and the ideology of such submission is ideally suited to (and fostered by) a context in which domination is part of the subtext determining practice, not part of an explicit personal or social ideology of the "controlling other." As Jessica Benjamin and others have noted, the subtext which calls for static autonomy, for separation and independence, also calls for control:

> Since the child continues to need the mother, since man continues to need woman [and nonhuman nature], the absolute assertion of independence requires possessing and controlling the needed object. The intention is not to do without her [woman or nonhuman nature] but to make sure that her alien otherness is either assimilated or controlled, that her own subjectivity nowhere asserts itself in a way that could make his dependency upon her a conscious insult to his sense of freedom.[52]

When conscious ideology is one of interconnectedness and equality, but the subtext is one of alienation wrought by a need for separation and independence, control must be exercised in more subtle ways. A king of control is required which in some sense recognizes the real existence of the other as well as the need for, or dependence upon, the other. The preeminent form such control takes is that of exercising control over discourse, control over "naming the other." One form that understanding can take is that of the construction of a totalizing theory designed to assimilate the other into a unifying conceptual framework. The control that is exercised is primarily psychological in nature, but, of course, it seeps into practice in myriad ways, all of which can be rationalized as consistent with the ideology of interdependence and equality. One of the functions of totalizing theory is to rationalize the "colonizing" of the other, the control of the other by means of control over naming reality.

In contrast to the alienated (pseudo-) connectedness achieved by *abandonment to* the controlling other, this form of alienated (pseudo-) connectedness is achieved by assimilation of the other by means of a totalizing discourse, an instrumental use of discourse designed to guarantee control by, and separateness of, the knowing subject—a subject which knows, not the object of knowledge, but its representation of it, its colonization of it. This is the form that Stoic assimilation of the other takes—namely, an orientation to the world in which a text of connectedness is underwritten by a subtext of alienation and yearning for connectedness. Ecosophy S expresses this same relationship between text and subtext.

Even though the goal in the case of Ecosophy S is an alteration of one's experience of the world, which is to result in a deeper embedment in that world, rather than flight from it, the project is motivated by a deep yearning for wholeness, health, and connectedness coupled with a refusal to give up the safety of alienation. The nature of ecological consciousness, its colonization and assimilation of the world of which it would be "plain member and citizen," guarantees that the venture is without risk. Rather than experiential risk as revelatory of a new understanding of self and world, we have ecological consciousness constructed in a condition of alienation functioning as the vehicle for the overcoming of alienation. Transformation is not accomplished by risking encounter with the real; rather, one's very experience of the real is transformed by a prior metaphysical account. Central to Ecosophy S and its notion of ecological consciousness is a totalizing view which becomes definitive of the self by way of empathic incorporation, Stoicism all the way.

Recently, Nye has raised similar objections to traditional philosophy of language's "attempt to unify the diversity and individuality of spoken language in order to produce a transpersonal intelligibility." She discerns in this project "a willful avoidance of the discourse of others typical of masculine styles of communication"[53] just as I find a willful avoidance of genuine encounter with the world, genuine overcoming of alienation, in the project of Ecosophy S. She sees the "project of the unification of language" as "prompted by a previously undertaken alienation."[54] Whatever the motivations prompting this move toward alienation,[55] the alienated soul finds itself caught in a double bind. Self-alienation is a project undertaken in the interest of autonomy, the consolidation of selfhood. But such a self is lonely; and so a movement is initiated to overcome this self-imposed alienation. This project, however, must be conceived and carried out in such a way that is does not threaten the "autonomous" self. What is constructed is a kind of pseudo-connectedness. As Nye puts it:

> Once it has been decided that we begin from a solitude in which we will not hear what others are saying, will not allow their words to

> touch us or move us, ... the project of the unification of language opens as the only way in which that solitude can be relieved.... Once the grounding of discourse is so laid, any communication becomes a guarded communication, one in which openness is medicated by categories and structures that channel what is to be said into acceptable and bearable idioms.[56]

This totalizing and essentializing voice is that of the colonizer of other people, other minds.[57] Colonization, whether it be the political colonization of one country by anther or the silencing of the other's voice by the imposition of a totalizing discourse, preserves (alienated) autonomy by refusing to hear the voice of the other and achieves (pseudo-) connectedness by articulating a totalizing discourse which subsumes (rather than hears) the voice of the other. "Vision is substituted for voice," and "instead of expressing personal experience, language [is] grounded in authoritarian impersonal truths."[58]

Similar difficulties emerge from the attempt to mediate practice by ecological consciousness and its empathic internalization of a vision of reality. Rather than an ethical voice emergent from existential encounter, from genuine relationship, we get an ethical voice grounded in the authoritarian impersonal truth of metaphysics, of internalized vision. True, proponents of Ecosophy S hold that it "proceeds 'from a certain way of experiencing the world,' "[59] but this experience is the result of an empathic incorporation of a *vision* of reality. The experience is at one remove; it is not so much experience as it is an *alternative* to experience which—grounded as it is in the impersonal, authoritarian voice of metaphysics—is safe, insulated, alienated. "Mother Earth," having been safely wrapped in totalizing vision, can now be safely eulogized. Diversity is in this way colonized and contained, controlled by metaphysics. When we are asked to alter our experience of the world in accordance with a totalizing, unifying picture of the world, the fact that assent to the metaphysical vision is to take the form of empathic incorporation hardly helps; it in fact makes matters worse. To make oneself over as a result of experience, genuine encounter, is one thing; to make oneself over in accordance with a totalizing vision of reality is quite another.

It will surely be objected that this characterization of Ecosophy S must be misguided because its proponents so often emphasize the importance of direct experience of nature as a means to the achievement of ecological consciousness. Granted this emphasis, the question remains: why is the apparatus of *Ecosophy S* necessary in order to achieve the right relationship to nature? What requires explanation is the fact that one might need or desire to back up the Deep Ecology platform with either totalizing vision or the project of Self-realization of Ecosophy S. Although Fox, for example, does speak of the "soft boundaries" between ecological consciousness and practice such that the two inform one another, it remains true that the function of ecological consciousness within Ecosophy S is such that the other is *always* contained within, colonized by, the central unifying intuition and the salvational project of Self-realization. Moreover, as with Stoicism, every effort is made to bootstrap the intuitions and norms of Ecosophy S outside the realm of social negotiation. The effect of all this is not to deny the existence of otherness, of difference, but to control it, to contain it within a unifying framework of understanding so that it may be enlisted in the service of a particular kind of salvational project: the project of one who yearns for connectedness, but who can loose the bonds or boundaries of the self only with the promise of opening up into a larger Self.

(4) *The Other and a Politics of Difference.* Genuine encounter with, and acknowledgment of, the other is not fundamental within Ecosophy S; it is controlled by unifying vision and subservient to the salvational project.[60] What makes it difficult to see this is that it conflicts utterly with the stated views of Ecosophy S. Yet, as with Stoicism, the subtext is one of domination and control. What is called for, in contrast, is an environmental version of a "politics of difference." Such an environmentalism would *begin* with an acknowledgment of otherness, of difference, would acknowledge the ethical centrality of difference rather than subsume it within a totalizing vision or salvational project. Some of the pieces of such an environmentalism are emerging, for example, in the work of Tom Birch. His view is that

> The essence of an ethical world view is to establish ties or relationships. An ethical world view requires the refusal to gain control over others.
>
> What's necessary in order to have a situation that makes it possible to have ties or maintain them? There has to continue to be an other, another person or nature. What's necessary in order for there to be an other? That's where wildness comes in. Wildness is the center of otherness. . . . When you attempt to control people, or when you attempt to control nature, the result is to inevitably alienate yourself from them or it.

This notion of the ethical centrality of otherness and, hence, wildness is connected to the notion of wild love:

> But the wild lover knows that nothing can be secured with guarantees—there can be no final certainties in loving—and the wild lover loves this otherness of the world also, this otherness that denies the lover's own desire for guarantees, as, simultaneously, the world beneficently sponsors the wild lover's own participation in desiring and loving, with all their attendant uncertainties. Wild love is not altruistic or selfless in a manner that implies self-denial, for the continued participation of the self is just as indispensable as that of

> the other to the maintenance and furtherance of the relations of loving. Thus wild love requires a love (a wild love) of love itself, and consequently of all that it presupposes, all the parameters of its wildness. Wild love is thus love of the world as it has been given to us.

If there *is* a metaphysical vision at work here it is a metaphysics of otherness—that is, one which provides a mechanism, an epistemology and an ethic, for the systematic deconstruction of totalizing and colonizing visions. The place of ethics is specified in relationship to the central notion of otherness. The term Birch uses is *deontic experience*, referring to the binding claims (not the claim of *duty*) that otherness can make upon us. Deontic experience can be had anywhere when we are open to the claims of otherness; hence, Birch takes ethical considerability "all the way down." This, for me, is Birch's analogue (or rereading) of biospherical egalitarianism.[61]

In contrast, Ecosophy S relativizes its version of universal considerability (biospherical egalitarianism or biocentric equality) to the norm of Self-realization:

> ... I would add here that the second norm [biocentric equality] is actually derivative from the first [Self-realization]. Devall and Sessions acknowledge this at one point despite referring to two "ultimate norms." Naess also realizes this since his elaborated eco-philosophy ("Ecosophy T") proceeds from the single "ultimate norm" of "Self-realization."[62]

Where is the other in the epistemic and ethical space of Ecosophy S? "Correct behavior," Fox says, making use of Devall and Sessions, is "a natural consequence of aligning ourselves with 'the nature of reality and our place ... in the larger scheme of things.' "[63] This alignment is in the service of (or constitutes) the salvational project of Self-realization (i.e., the attainment of ecological consciousness). Closely connected to the totalizing and colonizing vision of reality of Ecosophy S, then, is something very close to an antinomian ethical position. Knowledge of the "other" at the "highest level of knowledge" is experienced as the "union that the mind has with the whole of Nature."[64] Ethics is the antinomian behavioral correlate of this epistemic union, what one naturally does once this "union" is experienced. There is no genuine ethics, metaphysics, or politics of *difference* here; the other is there, certainly, but colonized.

As with the Stoic notion of the "Unity of Man," there is within Ecosophy S a unifying tendency which can easily block genuine recognition of difference. What cannot be readily assimilated to its normatively charged central intuition of "unity in process" might easily be stigmatized as "other" in the same sense in which women have been branded as "other" by masculinist understandings of mankind which take men as the norm for what it is to be human; within the framework of such accounts of humanness women become unintelligible, aberrant. My sense is that the notion of ecological consciousness might—to some extent does now—function analogously within Ecosophy S. The problem is that to the extent that ecological consciousness as defined by Ecosophy S is taken as normative for what it is to have deep ecological concerns (rather as an optional, personally or culturally idiosyncratic way of arriving at an endorsement of the Deep Ecology platform), to that extent those who do not share this ecological consciousness are cast as "other," aberrant. This tendency is inherent in any discourse which leans toward totalization and, hence, colonization. It is inherent in Stoicism and it is inherent in Ecosophy S. It has made itself manifest in a tendency to divide the world into those who have ecological consciousness on the one hand and those thinking is anthropocentric on the other, and in a tendency to blur real and important distinctions between the members of each group. This breeds what Karen Warren has called a "unity in sameness" movement defined by references to the possession of ecological consciousness. What is needed, however, what is even *implied* by a view which calls for biocentric equality, is a *politics of differences*:

> . . . [T]he logic of domination must be abolished both to make possible a meaningful notion of *difference which does not breed domination* and to prevent [radical environmentalism] from becoming a "support" movement based on shared experiences [i.e., ecological consciousness]—a "unity in sameness" movement—rather than a "solidarity movement" based on shared beliefs and attitudes [i.e., the Deep Ecology platform].[65]

The move made by proponents of Ecosophy S is from the intuition of interconnectedness and the salvational project of Self-realization to the conclusion that what is of fundamental importance for *all* people is the gaining of ecological consciousness. To fail to have such consciousness is to fail in a fundamental way to be a (healthy) human being. But what if nonwhite people of color, for example, don't see things in this way? What if they nonwhite people of this stance and, instead, see colonizing domination and oppression of themselves as a people and as individuals as the central issue, and view Ecosophy S not only as a luxury possible only for the dominant culture, but also as part of the very fabric of domination, a subtle mind game for those who see the injustice of grosser forms of domination but who cannot, in the end, give up their colonial ways? Within the framework of Ecosophy S (as within the framework of any colonizing discourse), such a view is easily read as pointing to an issue which is certainly of concern—the domination of one class by another—but as expressing (understandably) the limited (and anthropocentric) concerns of people living under the thumb of oppression. That is, a paternalistic and condescending attitude is easily adopted in which the anthropocentrism of some people is "understood," though regretted. In this way the concerns of these people are colonized, deflected, ultimately delegitimized. *Difference* is not made a central dimension of a revolutionary politics of difference which might hope to eliminate *all* forms of domination, including the domination of nature; rather, difference is colonized, assimilated into a "higher," more "enlightened" discourse. It is in this sense that the very structure of Ecosophy S militates against the recognition of genuine otherness.

Central features of Ecosophy S which militate against its genuine recognition of difference are its ahistorical and essentializing tendencies. Both of these are inherent in what I have called its technique of bootstrapping itself into eternity. This technique is closely related to the *salvational* (as opposed to *liberational*) mood of Ecosophy S. (For Ecosophy S the *liberation* of nature is subsumed under its *salvational* project.) A key feature of *liberational* movements is their focus on the historical positionality of the subject of oppression in relationship to the oppressor. A key feature of *salvational* movements, on the contrary, is their tendency to ahistoricity, their desire to create a safe place outside time and circumstance. Liberation is from the chains of oppression and does not deny the historical positionality of the subject; salvation is from time itself, from the claims that others make upon us that truth and justice be *negotiated*. It is precisely this that salvational movements have in common with the totalizing and colonizing tendencies of dominant cultures. To stand outside time and circumstance (in my sense) it is not necessary to literally leave this temporal sphere; it is only necessary to gain control over it with an ahistorical, totalizing discourse.

There is virtue, certainly, in the salvational project of Ecosophy S. It is a gospel of redemption from an anthropocentric ideology which has been the correlate of a ruinous alienation from, and destructiveness of, nature. Its limitations are that, as a salvational project, it speaks only to a small set of humans—namely, those imbued with the masculinist consciousness[66] of an imperialist culture—and that, as I have argued, it is not radical enough, since it perpetuates the very alienated consciousness it would escape.[67]

The proper context for addressing the issue of the liberation of nature is not within the salvational project of Ecosophy S but within the context of the project of the elimination of all oppression—and what is required for this is to place at center stage not the project of the creation of a timeless ecological consciousness, but the historically positioned voices of oppression in all its forms. Though it claims to acknowledge and value difference, the effect of the salvational project of Ecosophy S is to *colonize* difference and use it as an instrument of salvation. We begin to discern this subtext at work when we start to ask questions such as: what are the politics—personal and cultural—behind the view that "we can make no firm ontological divide in the field of existence . . . [no] bifurcation in reality between the human and the non-human realms"[68]—as opposed, say, to the politics at work behind the view that other beings are my sisters, brothers, grandmothers, grandfathers, aunts, nephews, daughters? What is going on when it is claimed that the "experiential core" underlying the latter view—typical of Native American cultures—is that expressed by Ecosophy S and its vision of "unity in process"? Is it merely the particular focus of an *Ecosophy* which leads it to say that there is no "bifurcation in reality between the human and the non-human realms" and *not* to say, for example, that there is no bifurcation in reality between white, middle-class environmentalists and the residents of Spanish barrios or black ghettoes? What would it mean to take the latter claim seriously? *Might* we listen with the same ear to the residents of Harlem (or to the corporate executives engaged in the destruction of old-growth forests) with which we listen to the voices of a tall-grass prairie in southern Wisconsin? What would such a listening be like?[69]

To listen in such a way Ecosophy S would have to take the notion of *human* ecology as seriously as it has taken the notion of ecology in connection with the nonhuman world. By its procedure of reading out of the conversation all voices which it deems "anthropocentric," however, it forgoes a deep understanding of the dynamics of oppression and, therefore, forgoes the possibility of a truly revolutionary practice. Its stance toward those it deems anthropocentric is thoroughly modernist in its dualistic opposition of anthropocentric and ecological consciousness; and its techniques of disenfranchisement are also thoroughly modernist—including the techniques of essentialism and totalization. It wants to conduct its revolution from the top down. Difference is tolerated only within the limits of those certified as having true ecological consciousness. With the exception of tribal peoples, the voices of oppressed people are disenfranchised by Ecosophy S. Their wisdom and strategies for living within the confines of the dominant culture are not taken seriously and understood ecologically[70] and so cannot be used to construct an understanding of what a true revolution would be like, how it might come about. Of course, if these comments are to be taken seriously, this line of reasoning must be applied both to *itself* as well as to the elements of Ecosophy S of which I have been critical. To be sure, this is simply to call for a *full* contextualization of ethical and political thought and practice, privileging no position *in the manner in which* modernism privileges certain of them with its foundationalist and essentializing assumptions. To forgo *this* sort of privileging, however, is not to say that distinctions cannot (or should not) be made.

What is needed is what Linda Alcoff (following Teresa de Lauretis and others) calls an "identity politics" and a "conception of the subject as positionality":

> If we combine the concept of identity politics with a conception of the subject as positionality, we can conceive of the subject as

> nonessentialized and emergent from a historical experience and yet retain our political ability to take gender [or difference in general, even though culturally constructed] as an important point of departure.[71]

Even the strategies of the colonizers must be understood ecologically. They are not to be understood or condemned using allegedly timeless and ahistorically "true" criteria. They are to be understood and critiqued contextually in relationship to (among other things) the identity politics of historically positioned voices of oppressed people. The debate between proponents of Ecosophy S and others should be understood as occurring within nature and should be understood ecologically. It cannot sensibly be understood as occurring outside time and circumstance; to do this is to subscribe to the very modernism which Ecosophy S condemns. Ecosophy S, though it tries to part ways with the anthropocentrism of modernism, still underwrites, in its mode of discourse, the totalizing and colonizing discourse of modernism which underlies its domination of nature. Ecological consciousness is the consciousness of modernist alienation seeking reconnection with that from which it has lost touch. To that extent it is going in the right direction. But it speaks form alienation and does not escape it. The remedy is not thc salvational politics in which Ecosophy S is ensconced but a politics of difference.

What must be explored are the necessary conditions—both cultural and individual—for the emergence of healthy voices. What Ecosophy S does not provide is precisely this understanding. Its concern is not with the construction of a human community which will provide a context in which such voices will emerge. What it provides, rather, is a picture of the cosmos to which we are to conform. The vision tells us how the whole show fits together. Correct behavior is in an important sense derived from the vision—that is, even though direct experience of wild nature is very important in the generation of ecological consciousness, the otherness of nature is, when all is said and done, *colonized* by ecological consciousness, a consciousness which is the result of a salvational project which colonizes nature through its program of Self-realization carried out within the context of a vision of unity. This has it backwards. The real work is in providing a context, a practice, in which conversation with the coyote is made possible. The question is whether ecological consciousness is to function for us as a totalizing consciousness which, while praising diversity in nature, often seems quite insistent on listening to only one voice—that shaped by the internalization of its preferred vision of reality—*or* whether metaphysical visions, the stories we tell of the cosmos and our place in it, must make do catch as catch can in a cultural setting in which primacy is given to the voices which emerge in encounters with the coyote.[72] The ragtag of such a method of doing "metaphysics" sacrifices unity to the authenticity of the voices heard in its ragbag of tales, myths, stories; but this is all to the good—totalizing accounts have a way of achieving unity at the cost of silencing[73] those voices which cannot be heard in the official jargon. What is needed is a *grass-roots* conception of metaphysics, a notion of metaphysics as bioregional narrative, a metaphysics directly responsive to genuine encounter.[74]

(5) *Self-realization and Identification.*[75] The concept of identification which plays such a crucial role in Ecosophy S is intimately tied to the notion of the expansion of the self to the boundaries of the whole. One of the two basic hypotheses included in Fox's version of Ecosophy S is that "The self is as comprehensive as the totality of our identifications. Or, more succinctly: Our Self is that with which we identify." The norm of Ecosophy S, on Fox's reading, is a state of being in which one "sustains the widest (and deepest) possible *identification* and, hence, sense Self,"[76] Self-realization is the ultimate norm in Naess' Ecosophy T and Devall and Sessions include it (along with biocentric equality) as one of the two ultimate norms of Deep Ecology.

As I have argued elsewhere,[77] there is a superficial similarity between this notion of an expanded self achieved through identification and the notion, explored by Carol Gilligan,[78] of defining relationships which, while acknowledging dependencies and bonds of care and responsibility, leave selves intact; however, because the former notion leaves in place a fundamentally atomistic notion of the self which is absent in the latter, Ecosophy S defines itself in relation to a false dichotomy: we are either atomistically defined selves who are strangers to one another or each of us is (potentially, ideally) a gigantic Self inclusive of all the rest.[79] Rightly rejecting the first alternative, Ecosophy S opts for a position which, in effect, tries (impossibly) to combine the notion of defining relations with a "safe" but alienating atomism. The concept of an expanded Self is what mediates the two (in fact incompatible) notions. What makes the notion of an expanded Self atomistic and ultimately alienating is the fact that the identification which makes possible the expansion of the self is achieved by the empathic incorporation of a totalizing vision. It is this vision which in fact produces the unity of the expanded Self, makes it atomistic, alienating, "safe." This is the same move made by classical Stoicism: the intimacy and unity of the polis is reproduced in alienation and isolation by internalizing the logos of the cosmos.

That this unity is the unity of the logos is evident, for example, in the following passage from John Seed: ". . . as the implications of evolution and ecology are internalized and replace the outmoded anthropocentric structures in your mind, there is an identification with all life. . . ."[80] Identification is mediated by internalized metaphysical vision. This, again, is very much in the spirit of Stoicism. Identity, for Stoicism, is identification with the logos of the whole. Identification is with particulars only in the derivative sense that the logos of the cosmos threads its way through the cosmos, binds it together as a totality. Identification, within Ecosophy S, involves, not seeing or hearing the other *as* other,[81] but seeing the other *sub specie aeternitas*:

> And seeing all entities, including one's "small" self, as aspects of one's wider or "big" self is, of course, also captured in Spinoza's ideal of living *sub specie aeternitas* i.e. under the view or aspect of eternity, which is to say "seeing the universal in all the particulars of one's existence."[82]

The problem inherent in the notion of identification can be clarified by introducing Jessica Benjamin's account of the difference between masculine and feminine modes of desire and the differing forms that self-other recognition takes in each.[83] The fact that "the salient feature of male individuality is that it grows out of the repudiation of the primary identification with and dependency on the mother" leads to a "one-sided autonomy" requiring possession and control of the needed other. This control need not be physical; the control is a psychic need which may in fact be satisfied by a form *understanding* which contains and controls the other in a purely symbolic mode. In such understanding, very close attention may be paid to the other—penetrating perception, in fact—but the overriding factor governing that attention is the unifying understanding which is its goal, the achievement of which satisfies the intertwining needs for "autonomy" and control. (This contrasts with the "wild love" of the otherness of the other and the kind of understanding which such love can bring.) Benjamin calls this control by symbolic means—the creation of totalizing and colonizing discourse in the service of domination and control—"fantasy."[84]

Genuine self-other recognition "occurs when the other is truly placed outside the control of fantasy." This is not to say that fantasy has no legitimate place in our lives.[85] The important question is whether fantasy—a general term for what are also called "phallic symbolic processes"—is organized or controlled by self-other recognition or whether, on the contrary, self-other recognition is organized or controlled by fantasy. My worry about Ecosophy S is that the centrality of the project of Self-realization within the context of the "intuition of unity" in process functions as fantasy, organizing and controlling self-other recognition.[86] The colonization of nonhuman nature which this brings about is in many ways benign.[87] Where it is potentially most problematic (in a *direct* way) is in relationship to other humans (and, therefore, *indirectly* in relationship to nonhuman nature). By not making the perception and recognition of difference central, the tendency of Ecosophy S, as I have argued in the previous section, is to understand the human world dichotomously—as divided into those who *have* and those who *lack* ecological consciousness. Those who lack ecological consciousness are understood as other in the negative sense of lacking that which is definitive of full humanity. That is, the others are colonized by the notion of ecological consciousness and labelled "anthropocentric"—the *lack* of ecological consciousness becomes, *politically speaking*, their defining characteristic, the characteristic around which the politics of Ecosophy S is organized. I have argued for the necessity of a politics of difference and have urged that this is the only politics which makes *ecological* sense. It is also the only politics which is ultimately consistent with the stated deep ecological notion of biospherical egalitarianism.

Benjamin (drawing on the work of Winnicott) offers a noncolonizing alternative to the notion of Self-realization in which self-other recognition not controlled by fantasy is articulated in spatial metaphors "by describing a space that contains and a space in which we create":

> This space begins between mother and baby—[Winnicoot] calls it the holding environment—and expands into what he calls the transitional area, the child's area of play, creativity, and fantasy.... The central experience to which Winnicott refers is being and playing *alone* in the presence of the other; to be truly alone with oneself paradoxically requires this sense of the other's being there.... It is in this way, through the unobtrusive mediation of the other, that drives become one's own desire.... "On the seashore of endless worlds children play".... This image suggests something that both forms a boundary and opens up into endless possibility; it evokes a particular kind of holding, similar to the first bodily holding by the mother. It refers to ... a sense of self that evolves through relationships that validate what we can do for ourselves. The confidence in one's inner resources is also founded in the experience of the other's integrity and separateness.... The awareness of one's own intentions, the ability to express them through action, and the confidence that they are one's *own*, evolve through the flow of recognition between two persons.[88]

Paul Shepard has expanded this notion of the holding environment in a way which constitutes a viable alternative to the colonizing notion of Self-realization.[89] He refers to the nurturing "matrix" within which maturation takes place. What he adds to Benjamin and Winnicott is the notion of a transition from the "maternal matrix" provided by the first nurturer to the "earth matrix." This transition is critical in order that a highly charged ambivalence toward the first nurturer[90] does not solidify and become the subtext of later adult behavior. Such a subtext can play itself out in various ways. Neo-Stoicism simply buries this subtext under a layer of reaction formation not present in the explicit domination present in the modern "rape of nature." As Shepard puts it:

> Fear and hatred of the organic on the one hand, the desire to merge with it on the other; the impulse to control and subordinate on

> the one hand, to worship the nonhuman on the other; over-differentiation on the one hand, fears of separation on the other: all are two sides of a coin.[91]

What Benjamin and Shepard offer are images of connectedness and the growth of the self which have at their very *heart* the recognition of difference, genuine otherness, Benjamin associates desire with a *space* (in contrast to the association of desire with incorporation, penetration, control)—the notion of the other as providing and defining the space in which self-discovery or creation takes place (as opposed to the swallowing or colonizing of the other as a means of identity-creation). This space allows for the creation/discovery of a genuine (and genuinely autonomous) self, because it does not involve a fantasizing of the other, a casting of the net of totalization over all of existence as a peculiar way of guarding the autonomy of the self while achieving (pseudo-) connectedness.

Identification, in Ecosophy S, is mediated by the internalization of metaphysical vision, and identification, in turn, mediates the relationship between ecological consciousness and practice. John Seed, who, as we have seen, makes the connection between ecological consciousness and identification, is quoted with approval by Fox as claiming that "It is only by identification with the *whole process* that correct values will emerge."[92] Once again we have a superficial resemblance between Ecosophy S and the work of Carol Gilligan. She observes that ethical decisions or values emerge for her female subjects from highly contextual and narrative embedment in quite specific communities or networks of relationships. For proponents of Ecosophy S, value (or correct behavior) emerges contextually also, but the process is different. The context out of which values emerge for Ecosophy S is not that of narrative embedment in a specific set of relationships—they do not emerge out of that *space* in which genuine self-other recognition occurs—but that of a unified and totalizing understanding of the whole process. That is, the main determinant is not genuine encounter with the other, but rather metaphysical vision with which one identifies by empathic incorporation. *Specific* valuations or decisions *do* emerge from embedment in concrete situations, but only from those situations as colonized by abstract vision.[93] In *that* sense valuations and decisions are *deduced* from abstract vision, not emergent from narrative embedment. The valuation *appear* not to be deductions because the proponent of Ecosophy S has identified with the logos from which the deductions are made. As a result, they have the feeling of valuations emergent from genuine encounters. The response, however, is not, say, to the coyote, but to the *logos* of the coyote-in-the-ecosystem.

That the response emerges not from the abstract understanding of this logos, but from its empathic incorporation into the self makes the situation worse, not better. Totalizing theories have such a strong grip on us precisely because of their ability to colonize the mind, making it impossible to listen to voices not unified, codified, stripped of singularity by the theory. It may seem as though we are listening to the coyote when we hear her through the self which has empathically incorporated a vision of the coyote-in-the-ecosystem, but we do not hear her; we hear out own isolation and longing for wholeness and connectedness with that which we have shut out and are willing to readmit only on the condition that it conform to a unity, a logos, of our own making. It matters little that the unity includes the coyote, for it includes her only as a "logosized" coyote. What we need is not so much the internalization of a total vision of the whole, but rather the nuanced narrative of myth and story, the articulation of local encounters with the world. Total visions have their uses,[94] certainly, but only when held accountable to the subtly shaded voices of an ethical vernacular.[95]

The parallels between ancient Stoicism and Ecosophy S are pervasive. What is needed, however, is not a rebirth of Stoicism, but an account of how it can come to be that a people lives in such a way that it stays within the carrying capacity of the land, lives in a way consistent with its health, and maintains and deepens it *own* health as part of the health of the land. What is needed is not the *salvational* project of Ecosophy S, but a politics of difference, an account of how a people educates itself to those virtues necessary for such health and how ethical decisions are made in such a community, an account of the social and communal conditions under which decisions which promote the health and well-being of individual, human community, and ecosystem *emerge*. Instead of being transfixed by totalizing vision, we need to hear the multiple voices of this Earth and learn to speak forth that listening.

## Notes

1 Hans Jonas, *The Gnostic Religion: The Message of the Alien God and the beginnings of Christianity* (Boston: Beacon press, 1963), pp. 246–47.
2 See Marilyn Frye, "Oppression," in *The Politics of Reality: Essays in Feminist Theory* (Trumansburg, N.Y.: Crossing Press, 1983).
3 Warwick Fox, *Approaching Deep Ecology: A Response to Richard Sylvan's Critique of Deep Ecology*, Environmental Studies Occasional Paper 20 (Hobart, Australia: Board of Environmental Studies, University of Tasmania, 1986) (this monograph is especially important inasmuch as it purports to be explicating the work of Naess, Devall, Sessions, and other deep ecologists). Bill Devall and George Sessions, *Deep Ecology: Living As If Nature Mattered* (Salt Lake City: Peregrine Smith Books, 1985). When "Devall and Sessions" appears in this paper I am to be understood as referring to this book. I take Naess' work to be a far less paradigmatic instance of Neo-Stoicism than that of the other authors I have mentioned.
4 Devall and Sessions, *Deep Ecology*, p. 225. See also Arne Naess, "Intuition, Intrinsic Value and Deep Ecology," *The Ecologist* 14 (1984): 201–02; Warwick Fox, "On Guiding Stars to Deep Ecology," *The Ecologist* 14 (1984): 203–04.
5 Arne Naess, "The Shallow and the Deep, Long-Range Ecology Movement. A Summary," *Inquiry* 16 (1973): 95–100; Devall and Sessions, *Deep Ecology*, pp. 69–73.

6 Devall and Sessions, *Deep Ecology*, p. 70.
7 Warwick Fox, "The Deep Ecology-Ecofeminism Debate and its Parallels," *Environmental Ethics* 11 (1989): 5–25.
8 The "suitably stated" caveat is of some importance. The particular formulation in the literature have been the source of no end of misunderstandings, as even deep ecologists sometimes admit (see, for example, Fox, *Approaching Deep Ecology*, pp. 37–41, on the misleading us of the language of *rights* in the deep ecology literature).
9 Quoted in Devall and Sessions, *Deep Ecology*, p. 225.
10 Ibid., p. 227.
11 Ibid., p. 66. The quotation is drawn from Warwick Fox, "Deep Ecology: A New Philosophy of Our Time?" *The Ecologist* 14 (1984): 196.
12 Fox, *Approaching Deep Ecology*, pp. 86–87. The quotation Fox uses is from Arne Naess, "Identification as a Source of Deep Ecological Attitudes," in Michael Tobies, ed., *Deep Ecology* (San Diego: Avant Books, 1985), p. 261.
13 Lynn, White Jr., quoted in Fox, *Approaching Deep Ecology*, p. 72.
14 Ibid., pp. 75–76.
15 In the introductory paragraph of the section ("Entering Deep Ecological Territory") from which I have taken Fox's account he asks, rhetorically, whether the various supporters of deep ecology have anything in common at the philosophical level (ibid., p. 53). It is clear that he regards his account as the *philosophical* common ground of deep ecology. It is also clear that he does not regard his account, or something *very* like it, as optional, as one that can be rejected by one claiming to be a deep ecologist: "I have argued that the work of Naess, Devall, Sessions and others of like mind . . . suggests that the *appropriate framework of discourse for describing and presenting deep ecology* is . . . one that is fundamentally to do with the nature and possibilities of the Self . . . " (ibid., p. 85; my emphasis). We encounter similar difficulties in Devall and Sessions, *Deep Ecology*, pp. 65–66.
16 *S* for "Self-realization." The appropriateness of this label will emerge in due course. See Arne Naess, "Identification," p. 269, n. 1.
17 Donna Haraway, "Primatology is Politics by Other Means," in Ruth Bleier, ed., *Feminist Approaches to Science* (New York: Pergamon Press, 1984), p. 82.
18 Fox, *Approaching Deep Ecology*, p. 42. See also pp. 78–84.
19 Ibid., p. 87. These are alternate terms for the central concept in what Fox refers to as the "ultimate norm" of deep ecology.
20 Devall and Sessions, *Deep Ecology*, p. 65.
21 Fox, *Approaching Deep Ecology*, p. 87.
22 Ibid., p. 34. The internal quote is from Devall and Sessions, *Deep Ecology*, p. 69.
23 Devall and Sessions, *Deep Ecology*, p. 10.
24 George Sessions, "Spinoza and Jeffers on Man in Nature," *Inquiry* 20 (1977): 483. Sessions is quoting Wallace Matson on Pythagoras.
25 Ibid., p. 484.
26 Devall and Sessions, *Deep Ecology*, p. 80.
27 Ibid., pp. 101–02. This quotation from Jeffers also appears in Sessions, "Spinoza and Jeffers," p. 512.
28 See Sessions, "Spinoza and Jeffers," p. 513.
29 Quoted in Devall and Sessions, *Deep Ecology*, p. 243.
30 See Jim Cheney, "Eco-Feminism and Deep Ecology," *Environmental Ethics* 9 (1987): 115–45; and Ariel Kay Salleh, "Deeper than Deep Ecology: The Eco-Feminist Connection," *Environmental Ethics* 6 (1984): 339–45.

31 Jonas, *The Gnostic Religion*, pp. 247–49. I would argue that the environmental activism engaged in by various proponents of Ecosophy S does not serve as a refutation of the claim that Ecosophy S embodies a fundamentally Stoic sensibility. Suffice it to notice, for now, that much of this activism is not on the part of "plain members and citizens" (Aldo Leopold's term) of the biotic community. "I am part of the rain forest protecting myself" is a thoroughly *stoicized* version of environmental activism. The "plain member and citizen" has disappeared and activism is mediated by *identification* with the entire biotic community. Stoicism is not inconsistent with activism; it is simply the activism of one who is no longer "plain member and citizen" of the biotic *community*.

32 Andrea Nye, "Logos Spermatikos," unpublished chapter of a work in progress on the history of logic.

33 Ibid.

34 Ibid.

35 Ibid. This move is analogous to the foundationalsim characteristic of modernist philosophy; it privileges ahistorical philosophical inquiry into the "foundations" of truth, knowledge, and obligation. An analysis of the subtext of modernist philosophy concerns itself, in part, with the motivation underlying this move to such foundationalist conceptions. See Sandra Harding, *The Science Question in Feminism* (Ithaca: Cornell University Press, 1986). I want to claim that Ecosophy S continues this them of modernism.

36 A contextualist account views truth as what has emerged from social and political negotiation. Such a view doesn't mean that anything goes: which views are to be accepted is subject to *negotiation*, and among the items negotiated are criteria of acceptability, which include "the complex determination of motives and events and purposes." The historical and social "positionality" of claimants to truth is an important factor in such negotiations. This is precisely what Stoicism (and modernism) tries to read out of the picture.

37 Nye, "Logos Spermatikos."

38 Fox, "A New Philosophy," p. 196.

39 Fox, *Approaching Deep Ecology*, p. 74.

40 See, for example, Fox, "A New Philosophy," pp. 195–97.

41 Fox, *Approaching Deep Ecology*, p. 34.

42 Fox, "A New Philosophy," pp. 195–98.

43 When an argument is circular it closes in on itself, thereby making it immune to all criticism except that which points out the uselessness of an argument which closes in on itself and thereby cuts itself off from the world of discourse in which it claims to function. Bootstrap strategies have the same effect of insulating the view to which one is bootstrapped from criticism of the usual contextual sort.

44 On the relationship between purification and what I have called bootstrapping within modernism, see Susan R. Bordo, *The Flight to Objectivity: Essays on Cartesianism and Culture* (Albany: State University of New York Press, 1987), esp. chap. 5.

45 George Sessions, in Devall and Sessions, *Deep Ecology*, pp. 239–40. As has been noted about this passage from Wittgenstein, if it can't be said, it can't be whistled either. Sessions' endorsement of the "positivistic/mystical" early Wittgenstein is clear. His rejection of the "postmodern," late Wittgenstein seems clear enough in his appendix H to *Deep Ecology* as well as in his earlier "Spinoza and Jeffers."

46 The definitions of static and dynamic autonomy used here are from Evelyn Fox Keller, *Reflections on Gender and Science*. (New Haven: Yale University Press, 1985), p. 97.

47 A partial listing of the places in which this work has been undertaken is as follows: Harding, *The Science Question;* Keller, *Gender and Science*; Bordo, *The Flight to Objectivity*; Nancy Chodorow, *The Reproduction of Mothering: Psychoanalysis and the Sociology of Gender* (Berkeley: University of California Press, 1978).

48 Harding, *The Science Question*, p. 39.

49 Keller, *Gender and Science*, p. 97.

50 I have attempted some of this in my "Eco-Feminism and Deep Ecology."

51 Jessica Benjamin, "A Desire of One's Own: Psychoanalytic Feminism and Intersubjective Space," in Teresa de Lauretis, ed., *Feminist Studies/Critical Studies* (Bloomington: Indiana University Press, 1986), p. 97.

52 Ibid., p. 80.

53 Andrea Nye, "The Unity of Language," *Hypatia: A Journal of Feminist Philosophy* 2 (1987): 95.

54 Ibid., p. 108.

55 Object relations theory—as exemplified in the work of Nancy Chodorow, *The Reproduction of Mothering*, and Dorothy Dinnerstein, *The Mermaid and the Minotaur: Sexual Arrangements and Human Malaise* (New York: Harper & Row, 1976)—constitutes but one of many attempts to get at the motivational roots of this *undertaken* alienation. For some limitations of this approach, see Elizabeth Fee, "Critiques of Modern Science: The Relationship of Feminism to Other Radical Epistemologies," in Ruth Bleier, ed., *Feminist Approaches to Science*, p. 85.

56 Nye, "The Unity of Language," p. 107.

57 See Harding, The Science Question, chap. 7, and numerous references contained in that book. See also Jim Cheney, "Postmodern Environmental Ethics: Ethics as Bioregional Narrative," *Environmental Ethics* 11 (1989): 117–34.

58 Nye, "The Unity of Language," pp. 96 and 99.

59 Fox, *Approaching Deep Ecology*, p. 34.

60 Proponents of Ecosophy S often refer to Oriental religions to illustrate, and draw parallels with, Ecosophy S. It might be pointed out, however, that the tendency is to opt for the more totalizing positions among these traditions, such as the statement found in the *Upanishads* that "Thou (*atman*, *pursusa*) art That (*Brahman*)." Even when Buddhism is mentioned, those interpretations or versions are preferred which most closely approximate their Hindu cousins. That side of Buddhism which stresses the importance of *not* clinging to the grand metaphysical flights of Hinduism, which stresses that enlightenment consists in experience of the *emptiness* of all *dharmas* (where the notion of emptiness is not itself reified), does not appear. The agnosticism expressed in the Buddha's "four-cornered negation" and the relentlessly "deconstructive" technique employed, for example, in Nagarjuna's dialectic are at odds with the Orientalism preferred by proponents of Ecosophy S, the point of that agnosticism and deconstruction being that the Hindu notions of *Brahman* and *purusa and* the relationship between the two colonize the mind, providing points of attachment for the clinging of desire and, in so doing, prevent release (*moksa*, *satori*). On Nagarjuna, see Frederick J. Streng, *Emptiness: A Study in Religious Meaning* (New York: Abingdon, 1967). I am taken by Gary Snyder's description of Taoism, Hinduism, and Buddhism as the "younger brothers" of the "old ways," "slightly confused because passing through the temporary turbulence called civilization" (quoted in Devall and Sessions, *Deep Ecology*, p. 84). Whatever *Snyder* means by this, I take it to mean that all these traditions fundamentally represent salvation techniques, not for humanity in general, but for the alienated masculine ego. As such, I think the deconstructive versions (typified by Nagarjuna) are the only ones which offer much real hope, and then only when

combined with what has been called a "politics of difference," a genuine acknowledgment of otherness outside the scope of totalizing discourse.

61 The quotations have been culled from various unpublished manuscripts provided me by Tom Birch.

62 Fox, *Approaching Deep Ecology*, p. 50. See Devall and Sessions, *Deep Ecology*, pp. 225–28 for Ecosophy T.

63 Fox, *Approaching Deep Ecology*, p. 34.

64 George Sessions on Spinoza, in Devall and Sessions, *Deep Ecology*, p. 239.

65 Karen Warren, "The Power and Promise of Ecological Feminism," delivered at the American Philosophical Association meeting in New York City, December, 1987. "Radical Environmentalism" was inserted in place of *feminism*; the other bracketed phrases are additions to the text.

66 On the masculinist nature of salvational projects claiming to be concerned with the salvation of *all* humans, see Judith Plaskow, *Sex, Sin and Grace: Women's Experience and the Theologies of Reinhold Niebuhr and Paul Tillich* (New York: University Press of America, 1980).

67 Ecosophy S might take a cue here from the Oriental traditions it so admires. Zen Buddhism, in particular, is quite clear about the myriad ways in which religious views can themselves be used as devices for resisting enlightenment. It is occasionally noted that the philosopher is the *last* to achieve *satori* (enlightenment) precisely because the logical acumen of the philosopher makes it possible to stave off the final dissolution of the craving and defensive self, using the teachings of enlightenment themselves for this purpose. I believe this to be the case with Ecosophy S.

68 Fox, "A New Philosophy," p. 196.

69 It will be noted that Ecosophy S does seem to include humans in its "central intuition." Devall and Sessions (*Deep Ecology*, p. 67), for example, say that "Spiritual growth . . . begins when we cease to understand or see ourselves as isolated and narrow competing egos and begin to identify with other humans and friends to, eventually, our species. But the deep ecology sense of self requires . . . an identification which goes beyond humanity to include the nonhuman world." This passage would lead one to believe that proponents of Ecosophy S, on their way to identification with old-growth forests, *first* identified with (as part of the project of Self-realization) the corporate executives bringing about the destruction of old-growth forests and the residents of Harlem. Two points are problematic here. First, if we take the actual words seriously, the corporate executives and Harlem are elided, summed up in the word *eventually*, and colonized with the word *species*. In fact, for the purpose of articulating the nature of Self-realization, they are erased as something of an embarrassment to the notion of *inclusive* identification. The second difficulty is that wider identification, in any sense in which that term refers to actual human experience, does not proceed from family, *to the human species*, and *then* to the nonhuman world. Only within the abstractions of metaphysics does that happen. The widening of identification is in accordance with what one comes to include as part of one's *community*. This most often includes nonhuman residents in the area before it comes to include "distant strangers." (Black inmates at a prison at which I taught had little difficulty in arriving at the perception that the rats who had chewed on their fingers and those of other they knew were their "brothers"—they saw them as members of their community with whom they had to deal *as such*, that is, as members of the community with claims of their own. They did *not* (by and large) see whites as their "brothers"—they were oppressors and colonizers who stood *outside* the moral community. I would also note that this use of *brother* is but one indication

of the fact that, in general, colonized people think in terms of moral *community*; it is perhaps a mark of ethical *imperialism* that it uses organic images such as "unity in process" in its ethical deliberation. The organic model is totalizing and colonizing in this connection. These observations call for a reading of identification linked with the notion of "community" rather the Self-realization.)

70 See Carol B. Stack, *All Our Kin: Strategies for Survival in a Black Community* (New York: Harper & Row, 1974).

71 Linda Alcoff, "Cultural Feminism versus PostStructuralism: The Identify Crisis in Feminist Theory," *Signs* 12 (1988): 433.

72 Tom Birch, taking it that the reference is to the Trickster so prevalent in American Indian tales, notes that *Coyote* is (often *disgustingly*) male and that my references to "her" (later in the paper) might be revised. Noting his point, I prefer to retain the lower-case *c* and the pronoun *her* and to let the Trickster resonate through the passage as may be.

73 I have noted this silencing earlier in my discussion of Fox's claim that Ecosophy S "captures" the experiential core of a wide variety of religious and philosophical traditions.

74 This is my reading of Vine Deloria, Jr.'s observation of a central difference between Christianity and Native American religions—namely, that Native Americans did not regard their religious as *exportable*. That is, they were not instruments of colonization. Vine Deloria, Jr., *God is Red* (New York: Grosset & Dunlap, 1973).

75 Some of the themes discussed in this section are further discussed by Marti Kheel in her unpublished "Ecofeminism and Deep Ecology: Reflections on Identity and Difference."

76 Fox, *Deep Ecology*, pp. 86–87. This "ideal state" is equated, by Fox, with Self-realization (Naess) and ecological consciousness (Devall and Sessions).

77 Cheney, "Eco-Feminism and Deep Ecology."

78 Carol Gilligan, *In a Different Voice: Psychological Theory and Women's Development* (Cambridge: Harvard University Press, 1982).

79 The notion of *identification* at work here is not be confused with that of *identity*. Entities are held to have their own relative autonomy. See Fox, *Approaching Deep Ecology*, pp. 56–57.

80 Quoted in Devall and Sessions, *Deep Ecology*, p. 243. Also quoted in Fox, *Approaching Deep Ecology*, p. 59.

81 Again, this is not to deny that one might begin on the path toward ecological consciousness through establishing relationships with particulars. The point is that ecological consciousness, as this is delineated within Ecosophy S, is a totalizing consciousness.

82 Fox, *Approaching Deep Ecology*, p. 57.

83 Benjamin, "A Desire of One's Own." Masculine and feminine modes of behavior are cultural constructs and, therefore, the connection between the modes of desire delineated in Benjamin's account and actual men and women is a matter of cultural contingency. The point in calling them masculine and feminine is to mark the *pervasiveness* (not the *necessity*) of this connection.

84 Ibid., p. 80. As Keller argues in *Gender and Science*, this is the kind of control sometimes sought by those interested in the theoretical sciences. What this shows is that the control of the other by means of *fantasy* described by Benjamin—what I call a totalizing and colonizing discourse—is not inconsistent with that fantasy being *true* according to the canons of modern science. The point is that fantasy, in such a case, has its place in a subtext of domination and control. Keller does not argue that science done in such a setting is false or illusory, just that a different

kind of science would be done outside that setting. On this last point see "A World of Difference," in *Gender and Science*, pp. 158–76.

85 Benjamin ("A Desire of One's Own," pp. 93–94) mentions its legitimate (nondominating) role in love when it does not control self-other recognition. Its role in what Thomas Kuhn calls "normal science" might be mentioned as well.

86 Tom Birch's appropriate and graphic term for this is "Pac-man-ism."

87 But see, for important caveats, Tom Birch, "The Incarceration of Wildness: Wilderness Areas as Prisons," unpublished manuscript.

88 Benjamin, "A Desire of One's Own," pp. 93–94.

89 Paul Shepard, *Nature and Madness* (San Francisco: Sierra Club Books, 1982).

90 Described by Dorothy Dinnerstein in *The Mermaid and the Minotaur*.

91 Shepard, *Nature and Madness*, p. 123.

92 Fox, *Approaching Deep Ecology*, p. 63; emphasis added.

93 Here, in particular, I think that Naess's view might constitute an exception to my characterization of Ecosophy S. See Naess, "Self-Realization in Mixed Communities," *Inquiry* 22 (1979): 231–41.

94 As, for example, in adjudicating the claims made upon us by "distant strangers." Modernist ethical theory has as its primary focus the development of "rules of the road" for civilized commerce among relative strangers. In a world in which most of the people with whom we have dealings *are* relative strangers, we cannot hope to do without such an ethic. A satisfactory contextualist ethic must, therefore, accommodate this fact. It must also make provisions for the specific problems to which it itself is prone, most obviously the dangers of loss of autonomy and degeneration into fascism. Such a contextualist ethic has yet to be developed. Central to the safeguarding of a contextualist ethic from degenerating into fascism and to the development and preservation of the genuine autonomy of which Keller speaks is the development of a politics of difference. These depend as well upon the fostering of the kind of desire of which Benjamin treats, one which develops in the space provided by genuine recognition of the other, a non-totalizing and noncolonizing recognition. A politics of difference is, of course, conducive to such recognition.

95 See Cheney, "Postmodern Environmental Ethics." Carolyn Merchant has suggested the wonderful term *ethical vernacular*. For an instance of the use in environmental ethics of the notion of contextual embedment see Holmes Rolston, III, "The Human Standing in Nature: Storied Fitness in the Moral Observer," in Wayne Summer, Donald Callen and Thomas Attig, eds., *Values and Moral Standing* (Bowling Green, Ohio: Applied Philosophy Program, Bowling Green State University, 1986). Judith Plaskow, in *Sex, Sin and Grace*, notes that several female authors—she focusses on Doris Lessing—have observed that women trapped by patriarchal definitions of what they are, or what they should be, feel free only in natural surroundings. This freedom is spoken of as a merging with nature, a feeling of unity or oneness. Is this an instance of a Neo-Stoic attitude of identification with the logos of the universe? I don't think so. These women are not alienated males longing for the disappearance of the masculine ego or the self-aggrandizement of the ego made possible by identification with the ecosystem. They are women in bondage longing *for release* from the endless oppression of culturally imposed self-sacrifice. The oneness with nature that they feel is not be understood, I think, as self-immolation, but rather as the recognition of a context which avoids *debilitating* dependence and permits *facilitating* dependence, a dependence which opens out onto health and well-being: that is, dependence which makes possible the realization, the actualization, of a self, a self which, nonetheless, stands in defining relations to various elements of the human and

nonhuman world and conceives of itself in that way. (In Benjamin's terms, nature provides a *space*, a holding environment for self-discovery.) This sense of facilitating dependence is made possible by extending our sense of contextual embedment to include nature, by ceasing to regard nature as merely a background against which we play out out lives. See also Cheney, "Eco-Feminism and Deep Ecology."

# Part 6

# ASIAN PHILOSOPHY AND THE ENVIRONMENT

# 94

# THE JAPANESE APPRECIATION OF NATURE

*Yuriko Saito*

Source: *British Journal of Aesthetics*, 25:3 (1985), 239–49.

ONE OF the characteristic of Japanese culture is often said to be the close and harmonious relationship between man and nature. Accordingly, the Japanese attitude towards nature is described as 'man in harmony with nature' or 'man in nature' while the Japanese appreciation of nature invokes such phrases as 'the Japanese traditional love of nature' or 'the great Japanese love of nature and sense of closeness to it'.[1]

Commentators on Japanese culture point to phenomena principally taken from the aesthetic realm as evidence for this unique attitude towards and appreciation of nature in Japan. There are important seasonal festivals that celebrate the beauty of nature (such as cherry blossom-viewing, moon-viewing, snow-viewing festivals); the Japanese often attempt to bring nature into the proximity of their daily lives by designing patterns in kimono fabric after natural objects and phenomena; they construct gardens even in confined spaces, or reduce nature into miniaturized presentations by arranging flowers in an alcove, cultivating a dwarfed pine tree (*bonsai*) or creating a miniature landscape on a tray (*bonkei*). Traditional Japanese architecture is designed to harmonize with, rather than dominate, its natural surroundings. The Japanese gardens are designed without references to the kind of abstract geometrical forms employed in European formal gardens. Various Japanese folk arts, crafts and even packaging often express the respect for the qualities inherent in the natural materials. Similarly, Japanese cooking is noted for preserving as much as possible the natural qualities (not only taste, but flavour, texture, colour, and shape) of the material. Perhaps most importantly, the subject-matter of Japanese art and literature is predominantly taken from natural objects and phenomena.[2]

Such a predominance of natural themes in art, cooking, literature and architecture appears to imply that the Japanese have a very intimate relationship with and a special love for nature. However, a precise analysis of the Japanese aesthetic appreciation of nature is needed to examine the ways in which the Japanese relationship between man and nature is considered to be harmonious, since there are several possible ways in which this is rendered harmonious. Man can conceive of nature as a contrasting force and still consider the relationship to be harmonious and unified through a balance of conflicting elements. Or man can consider nature as essentially identical with man and the relationship between the two to be harmonious because of their identity. In this essay, I shall show that the traditional Japanese love of nature is based upon the conceived identity between man and nature and this conception of nature forms an important basis for their aesthetic appreciation of nature.

## 1 The lack of sublime objects in Japanese appreciation of nature

Many commentators have noted that the Japanese appreciation of nature is directed exclusively towards those objects and phenomena which are small, charming and tame. This characteristic becomes conspicuous especially when we compare it with the Western and other Oriental (such as Chinese and Korean) traditions which appreciate not only those small, tame objects of nature but also gigantic or frightful aspects of nature.

Citing various short ancient poems which are perhaps the best record of the traditional Japanese appreciation of nature, Hajime Nakamura points out that 'the love of nature, in the case of the Japanese, is tied up with their tendencies to cherish minute things and treasure delicate things'.[3] Even when a grand landscape is appreciated, it is not the grandeur or awesome scale of the scene but rather its composition compressed into a compact design that is praised.

Consider, for example, how mountains are described and appreciated by the Japanese. The following well-known poem from the eighth century illustrates that the poet's appreciation of Mount Fuji is due to its graphic, compositional aspect rather than its soaring height or voluminous mass.

> When going forth I look far from the Shore of Tago,
> How white and glittering is
> The lofty Peak of Fuji,
> Crowned with snow![4]

We do not find, in this kind of appreciation, the Chinese taste for the grandiose, 'a broad prospect from the top of a tall-peak . . . craggy mountains and the dim vastness of waters'.[5] Neither do we detect a sense of fright, mystery or darkness exuded by the mountain.

A graphic illustration of the Japanese appreciation of mountains as friendly and warm rather than hostile and formidable can be found in some of the wood block prints of the Edo period. Consider, for example,

Andō Hiroshige's depiction of Mount Hakone from the *Fifty-Three Sceneries of Tōkaidō*. While successfully conveying the difficulty of passing this steep mountain by fantastically exaggerating its profile, this print does not give the viewer an impression that the mountain is hostile, or that it challenges us to conquer it. Moreover, despite its steep shape, the size of the mountain relative to the size of men in procession is rather reduced, avoiding the impression that the mountain is overbearing. In addition, the colour used for the men in procession and the mountain are almost in indistinguishable, again avoiding a stark contrast between the two.

The same observation can be made of the Japanese depiction of the sea. When the Japanese appreciate the ocean, it is never 'great oceans' but 'bays where boats passed to and fro between islands'; never 'the mighty ocean over which the great ships to T'ang China sailed', but rather 'the sea . . . somewhere small used by tiny boats and pleasure craft'.[6] The following poems are representative of such appreciation of the ocean.

As the tide flows into Waka Bay,
The cranes, with the lagoons lost in flood,
Go crying towards the reedy shore.

As we row round the jutting beaches,
Cranes call in flocks at every inlet
Of the many-harboured lake of Omi.[7]

Even when a rough sea is depicted in visual art (which is not frequent), it never gives the impression of ferociousness. Take, for example, the famous wood block print by Katsushika Hokusai of a gigantic wave almost swallowing boats, with Mount Fuji seen at a distance. While the represented state of affairs might be horrifying, the work does not convey such a feeling at all. Although highly evocative of dynamic movement, because of a fairly contrived and calculated composition with a distant Mount Fuji as a static focal point, this print gives us a feeling which is neither insecure not dreadful.

Likewise, creatures depicted by the Japanese are often small, harmless ones such as butterflies, warblers, copper pheasants, cuckoos. On the other hand, ferocious, life-endangering animals such as tigers are frequently objects of appreciation in other traditions. Indeed, in the Japanese tradition we do not find a praise for 'forests filled with wild beasts'; instead there is a constant appreciation of things which are 'small, gentle and intimate'.[8]

Some thinkers ascribe this conspicuous absence of the sublime in the Japanese appreciation of nature wholly to Japan's relatively tame landscape and mild climate.[9] Tall cliffs, unbounded landscapes and soaring mountains may indeed be lacking in Japanese topography. However, the lack of appreciation of the sublime in the Japanese tradition cannot be wholly accounted for by reference to this factor. The fierce and awful aspects of nature such as annual autumn typhoons, earthquakes and rough seas are fully experienced by the Japanese, perhaps most eloquently documented by a mediaeval Buddhist recluse, Kamo no Chōmei, in his *An Account of My Hut* (1212).

In spite of the frequent occurrences of devastating typhoons, however, it is noteworthy that the morning *after* a typhoon, not the typhoon itself, is praised for its aesthetic appeal in three major classics in the Japanese tradition, *The Pillow Book* (c. 1002), *The Tale of Genji* (c. 1004) and *Essays in Idleness* (c. 1340). For example, in *The Pillow Book*, a series of anecdotes and essays concerning the Heian period court life, Sei Shōnagon praises the beauty of the morning after the storm without describing her experience of the storm itself during the previous night. The only reference made to the storm is her amazement at recognizing the arrangement of leaves 'one by one through the chinks of the lattice-window' is the work of 'the same wind which yesterday raged so violently'.[10]

How do we then account for the fact that the grand and fearful aspects of nature, while experienced by the Japanese, are not acknowledged as objects of aesthetic appreciation in their tradition? It may be helpful here first to examine, as a point of comparison, the Western notion regarding the appreciation of the sublime. Perhaps the most theoretical discussion of man's appreciation of the sublime in the West can be found in Kant's aesthetic theory. In his theory of the sublime, Kant proposes that man's appreciation of the sublime in nature (either vast or powerful parts of nature) is based upon the fundamental contrast between nature and man. The contrast is twofold. First, man is contrasted with nature because of his apparent inadequacy to grasp the magnitude of a vast part of nature or to have dominion over its powerful part. However, second, man is also contrasted with nature because of his ultimate dominion and superiority over nature. That is, in experiencing the vast parts of nature, the feeling of pleasure is generated because we recognize that our rational faculty is capable of *thinking of* infinity in spite of the inability of our sensible faculty to grasp it. Our appreciation of the powerful aspects of nature is brought about by a similar recognition. While the power of nature may have dominion over our physical being we have ultimate dominion over nature due to our super-sensible faculty of reason which is free from those causal laws governing the phenomenal world. Indeed, Kant describes the play of mental faculties (imagination and reason) involved in our experience of the sublime to be 'harmonious *through their very contrast*'.[11]

The lack of appreciation of the sublime in the Japanese appreciation of nature, then, is explained by the Japanese view of nature in its relation to man. Rather than conceiving the relationship between man and nature as contrasting, I shall argue that the Japanese appreciated nature primarily for its identity with man. As Masaharu Anesaki observes:

> Both Buddhism and Shintoism teach that the things of nature are not essentially unlike mankind, and that they are endowed with

> spirits similar to those of men. Accordingly awe and sublimity are almost unknown in Japanese painting and poetry, but beauty and grace and gentleness are visible in every work of art.[12]

In what way then is nature considered to be essentially the same as man in the Japanese tradition? There are two ways in which the Japanese have traditionally identified with nature. One may be called emotional identification and the other is identification based upon the transience of both man and nature.

## 2 Emotive identification with nature

There is a long tradition in Japanese literature of emotional expression in terms of natural objects or phenomena. Lament and love, two strong emotions which constitute the major subject-matters of Japanese literature, are often expressed not directly but in terms of or by reference to nature. This tradition goes as far back as the oldest anthology of Japanese poems, *Manyōshū*, compiled in the eighth century. Shūichi Katō explains that this anthology indicates 'the court poets of the *Manyōshū* expressed their profound feelings in terms of their daily natural surroundings'.[13]

Perhaps the most explicit expression of this tradition in Japanese literature is found in Kino Tsurayuki's preface to *Kokinshū*, another anthology poetry, compiled in 905. In this preface Tsurayuki explicitly defines the nature of poetry as expression of emotion *in terms of* nature. Its opening paragraph states:

> Japanese poetry has the hearts of men for its seeds, which grow into numerous leaves of words. People, as they experience various events in life, speak out their hearts *in terms of* what they see and hear. On hearing a warbler chirp in plum blossoms or a kajika frog sing on the water, what living thing is not moved to sing out a poem?[14]

This identification of man and nature through emotive affinity is developed into an important aesthetic concept by an Edo-period philologist and literary critic, Motoori Norinaga (1730–1801) in his theory of *mono no aware.* Variously translated as 'pathos of things' or 'sensitivity of things', and sometimes compared to the Latin notion of '*lacrimae rerum*' ('tears of things'), *mono no aware* refers to the essential experience of sympathetic identification with natural objects or situations.

The experience and appreciation of the identification with natural objects or situations occur in two ways. Sometimes we intuit the *kokoro* (essence, spirit) of the object or situation and sympathize with it: this results in an aesthetic experience of the object based upon *mono no aware.* Hence, with respect to situations (*koto*), Norinaga claims,

> What does it mean for one to be moved by knowing *mono no aware*? It is, for example, when one is confronted by some situation which is supposed to be happy, one feels happy. One feels happy because one apprehends the *koto no kokoro* which is happiness in this case. By the same token, one feels sad when confronted with what is supposed to be sad; because one apprehends its *koto no kokoro.* Therefore, to know *mono no aware* is to apprehend the *koto no kokoro*, which is sometimes happy and sometimes sad, depending upon the situation in question.[15]

Regarding natural objects (*mono*), he makes a similar point:

> To see cherry blossoms in full bloom and to see them as beautiful flowers is to know *mono no kokoro.* To recognize their beauty and to be moved by feeling that they are deeply beautiful is to know *mono no aware.*[16]

Some other times, when we are possessed with a strong emotion, we experience an identification with natural objects and events by colouring these objects with our emotion. Norinaga claims that the aesthetic appeal of many classical Japanese literary works is derived from descriptions of this kind of emotional identification with nature. He agrees with Ki no Tsurayuki that '*mono no aware*, which is so intense that any verbal expression seems inadequate, can be expressed in a profound manner if one expresses it by what one sees or hears' such as 'the sound of wind or crickets, . . . the colour of flowers or snow'.[17] Indeed, according to Norinaga, the most important aesthetic appeal of *The Tale of Genji*, the work he praises for its expression of *mono no aware*, is its description of nature which has affinity with the characters' emotive states.

Some natural objects and phenomena are associated with certain emotive content in Japanese literature so frequently that they have been established as symbols for expressing particular emotions. For example, cherry blossoms (especially when they are falling) are often associated with sorrow in classical Japanese literature because they epitomize the transience of beauty. The autumn evening is a favourite symbol among mediaeval poets for expressing desolation and loneliness.

Whether the emotive identification between man and nature is rendered primarily as a result of man's intuitive grasp of the essence of a natural objects or as a result of the imposition of feeling onto the outward reality, this appreciation of nature for being emotionally charged constitutes an important aspect of the Japanese aesthetic appreciation of nature: the appreciation of nature for its expressive quality.

While the notion of expression relevant in the aesthetic sense has become a point of dispute, I believe that George Santayana's 'two-term' account of expression is generally correct. According to him, the aesthetic expression takes place when the following 'two terms' are united or fused together: 'the first is the object actually presented, the further thought, emotion, or image evoked, the thing expressed'.[18] Only when these two terms are 'fused' or 'confounded' in our experience is the object said to be expressive of the idea or emotion. Many instances of our aesthetic appreciation of nature are based upon this 'fusion' between the object's sensuous surface and various associated facts such as scientific facts, historical or literary associations, or practical values. For example, we may appreciate the tremendous geological age *manifested* by the weathered surface and many layers of a geological formation; the way in which the fierceness of a battle is *reflected* in a disfigured landscape with poor vegetation; or the manner in which the danger of an animal is perceptibly *realized* in its fierce appearance. The above are all *aesthetic* appreciation of these respective objects, distinct from mere appreciation of *the fact* that the landscape is aged, the battle took place on the site, or the animal is dangerous to us.

Emotion is also often associated with a natural object or phenomenon. Emotion can be said to be aesthetically expressed by a natural object when we can *see* the landscape as emotionally charged. If the emotive content remains distinct from the object and the viewer's experience is dominated by the emotion he experiences, then the aesthetic component in the appreciation diminishes. In other words, if the appreciation is directed merely towards the feeling of loneliness, the appreciation does not seem to be aesthetic; but if it is directed towards the way in which the feeling of loneliness is embodied by the actual landscape, then the appreciation is aesthetic. Which this mode of appreciating nature as a mirror of one's emotion is not limited to the Japanese tradition, it constitutes an important aspect of the Japanese aesthetic appreciation of nature.

## 3 The Japanese appreciation of the transience of nature

The Japanese appreciation of nature for its affinity rather than contrast with man has another basis. In addition to appreciating the relatively small and gentle objects and phenomena of nature, the Japanese are also known for their appreciation of the transitory aspects of nature. This fact is most significantly reflected in the traditional phrase by which the Japanese refer to nature as an object of appreciation—*kachōfūgetsu*, flower, bird, wind and moon. Flowers (most notably cherry blossoms) do not stay in bloom forever; the bird song is always changing and passing; wind is literally passing and transitory by definition; and the moon is constantly changing its appearance and location. Indeed these natural objects and phenomena form the favourite subjects for Japanese art. Other natural objects and phenomena frequently referred to in Japanese art are also short-lived: rain, dew, fog, insects, and various seasonal flowers.[19]

The Japanese preoccupation with the change of seasons should be understood in this regard. In many instances of appreciation of nature from the earliest record, the Japanese have been most sensitive to the characteristics of each season and the transition from one to the other. Consider the following examples. *The Pillow Book* begins with the famous description of the best of each season; the first six volumes of *Kokinshū* is organized according to the four seasons;[20] a famous passage in *Essays in Idleness* (a well-known series of essays by a fourteenth-century retired Buddhist monk, Yoshida Kenkō) also is directed towards appreciating the transition of season.[21]

The Japanese sensitivity towards seasonal change is even today manifested in the following aesthetic phenomena. Some natural objects or phenomena are celebrated for their symbolic presentation of their respective seasons. This symbolic import has been established throughout the long tradition of the required use of the season word, *kigo*, in haiku poetry and the celebration of the seasonal festivals mentioned at the beginning of this paper.[22]

The importance of seasons in the Japanese appreciation of nature is not limited to the symbolic import vested in various individual objects and phenomena. Just as emotion often organizes various components of nature into a unified expressive whole, seasons are also used as an aesthetic organizing principle. In other words, sometimes a composition made up of various natural objects and phenomena is praised for the 'fittingness' of the objects which creates a unified whole suggestive of a particular season. For example, one of the norms of flower arrangement is to suggest the mood of a season. Master Sennō advises: 'when there comes the season for autumn flowers like chrysanthemums and gentians, your work must suggest the desolation of a withered winter moor'.[23] Japanese cooking also reflects the Japanese appreciation of seasonal changes by its emphasis on seasonable dishes and the incorporation of appropriate materials for garnish and decoration. Accordingly, many contemporary Japanese cookbooks are arranged by season.

What is the basis for this Japanese appreciation of the transitory nature of natural objects and phenomena? There is an immediate aesthetic appeal of something which does not last for long. Psychologically we tend to cherish and appreciate objects or events more if we know that they will never be the same. Hence, commentators discussing the notion of Japanese wisdom point out that the Japanese appreciation of the flower, moon and snow is based upon 'regret for the transience of phenomena' which compels them to cherish 'those rare occurrences fitting to each season and time'.[24]

A contemporary Japanese painter, Higashiyama Kaii, indicates that his experience of viewing the full moon in the spring against the foreground of drooping cherry blossoms in full bloom in the Maruyama district of Kyoto is intensified by the recognition of the transitory and non-recurring nature of the phenomenon.

> Flowers look up at the moon. The moon looks at the flowers... This must be what is called an encounter. flowers stay in their fullest bloom only for a short period of time and it is very difficult for them to encounter the moon. Moreover, the full moon is only for this one night. If cloudy or rainy, this view cannot be seen. Furthermore, I must be there to watch it...
>
> If flowers are in full bloom all the time and if we exist forever, we won't be moved by this encounter. Flowers exhibit their glow of life by falling to the ground.[25]

Higashiyama, therefore recommends that we 'think of the encounter with a particular landscape occurring only once'.[26] Such advice would have us avoid the fatigue factor which is detrimental to the aesthetic experience of any object. It is not yet clear whether transitoriness itself directly contributes to the aesthetic quality of the object, but it is aesthetically relevant in the sense that it predisposes the viewer to attend very carefully to the object and fully savour whatever the object has to offer at the moment.

Another appeal of the transitoriness of natural objects and phenomena is also aesthetic. It is based upon the pleasure we derive from imagining the condition of the object before or after the present stage and comparing them. This aspect of the appeal of the transitory and changeable nature of natural objects and phenomena is discussed by Yoshida Kenkō. In a well-known passage in *Essays in Idleness* he claims that natural objects such as flowers or the moon are best appreciated before or after their full state.[27] 'Branches about to blossom or gardens strewn with faded flowers are worthier of our admiration' than blossoms in full bloom. As in a love affair between a man and a woman, 'in all things, it is the beginnings and ends that are interesting' because such stages of the phenomena are more stimulating to one's imagination. In particular, we appreciate the exquisite contrast between the present condition and the imagined condition of the previous or following stage. Even when an object or phenomenon is at the peak of its beauty, the appreciation is deepened by pathos based upon the apparent contrast between its present appearance and what will become of it later on.

The Japanese taste for such natural objects as cherry blossoms and moon, therefore, can be explained from the aesthetic point of view: these objects most eloquently exhibit to one's senses the transience of nature in general. Cherry blossoms are more effective than other flowers for symbolizing transience because they look most fragile and delicate, they stay in full bloom for only a short period of time, and they drift down slowly petal by petal, giving an impression that they regret falling.[28] But why is such sensuous manifestation of transience so cherished and appreciated? Why not appreciate the (apparent) permanence and stability of a rock, for example?

After all, isn't transience of everything, including ourselves, considered a primary source of man's suffering?

The Japanese traditional appreciation of the transient aspect of nature stems from a further metaphysical consideration. One of the most important ideas spread by the introduction of Buddhism in the sixth century was the impermanence of everything. Everything, both nature and man, will sooner or later change through modification, destruction or death. Transience of human life was often considered a source of people's suffering and an object of lament. Youth and beauty pass. Wealth and power do not last. And, of course, no one avoids death.

Lament over these facts is the subject-matter of major literary pieces in Japan. *An Account of My Hut*, for example, presents in the first chapter the following observation on the human condition.

> It might be imagined that the houses, great and small, which vie roof against proud roof in the capital remain unchanged from one generation to the next, but when we examine whether this is true, how few are the houses that were there of old. Some were burnt last year and only since rebuilt; great houses have crumbled into hovels and those who dwell in them have fallen no less. The city is the same, the people are as numerous as ever, but of those I used to know, a bare one or two in twenty remain. They die in the morning, they are born in the evening, like foam on the water. (197)

The same theme is expressed in the beginning paragraph of perhaps the most famous tale from the Japanese mediaeval period, *The Tale of the Heike*:

> Yes, pride must have its fall, for it is as unsubstantial as a dream on a spring night. The brave and violent man—he too must die away in the end, like a whirl of dust in the wind . . . [29]

What interests us here in these two passages is not merely their rather pessimistic outlook on man's life. What is noteworthy is that the description of the transience of human life is compared to the transience of natural phenomena. Kamo no Chōmei's passage is preceded by: 'The flow of the river is ceaseless and its water is never the same. The bubbles that float in the pools, now vanishing, now forming, are not of long duration: so in the world are man and his dwellings' (197). The passage from *The Tale of Heike* is also preceded by the famous beginning: 'The bell of the Gion Temple tolls into every man's heart to warn him that all is vanity and evanescence. The faded flowers of the sala trees by the Buddha's deathbed bear witness to the truth that all who flourish are destined to decay'.

This practice of comparing the transience of human life to the transience of natural phenomena abounds in Japanese literature. Consider, for example, a well-known poem by Ono no Komachi, a ninth-century poetess renowned for her beauty, in which she laments the passing of her youth and beauty by comparing them to the passing of flowers:

> The flowers withered,
> Their colour faded away,
> While meaninglessly
> I spent my days in the world
> And the long rains were falling.[30]

This frequent association between transience of nature and transience of human life stems from the conviction that nature and man are essentially the same, rooted in the same principle of existence. As Higashiyama remarks, referring to his discussion of viewing the full moon against the cherry blossoms at Maruyama,

> Nature is alive and always changing. At the same time, we ourselves, watching nature change, are also changing day by day. Both nature and ourselves are rooted in the same fated, ever-changing cycle of birth, growth, decline and death.[31]

This belief concerning the co-identity of man and nature is the ground of the Japanese appreciation of the evanescent aspects of nature. Grief experienced at the transience of human life is transformed to aesthetic pathos when it is compared to the transience of nature. By identifying human life with nature, the Japanese find a way to justify the transience of life. That is, since *everything* is in constant flux there is no escaping change and this recognition leads to resignation and finally to an acceptance of the sorrow of human existence.[32]

As a psychologist, Hiroshi Minami, commenting on Japanese psychological characteristics, suggests, this preoccupation with the co-identity of man and nature and the appreciation of the transient are based upon 'the perception of nature and life as one and the same, and the ascription of unhappiness and misfortune to the transiency and evanescence of nature and things impermanent'.[33] This identity between man and nature leads to resignation before the facts of life and then to acceptance of life, with all its sorrow and suffering. 'Unhappiness in life is expressed through the guise of nature; because of the evanescence of nature, man realizes that it is senseless to grieve and should become reconciled to fate'.[34]

Many contemporary thinkers, in particular those concerned with ecological matters, often praise the Japanese attitude of 'man in harmony with nature' for being ethically more desirable than the Western tradition of 'man over nature' or 'man against nature'. I believe that a further critical study is needed to determine whether their praise of the Japanese attitude towards nature for its ecological implication is justified. However, the preceding discussion does suggest that the Japanese regard man and nature as fundamentally identical and appreciate nature for its unity with man. The content of this unity and co-identity between man and nature should be understood in the sense that both man and nature share the most important principle of existence in common: transience.[35] The Japanese appreciation of the evanescent aspects of nature is rooted in the psychological benefit the Japanese derive from them: justification of the impermanence of human existence.

## References

1 For discussion of the Japanese harmonious relationship to nature, see Ian McHarg, *Design with Nature* (New York: Doubleday/Natural History Press, 1971), pp. 27–8; Ian McHarg, 'The Place of Nature in the City of Man', in *Western Man and Environmental Ethics* (Reading: Addison-Wesley Publishing Company, 1973); Ian Barbour, 'Environment and Man: Western Thought', in *Encyclopedia of Bioethics*, ed. Warren Reich (New York: The Free Press, 1978), I: 336–73; Lynn White, 'The Historical Roots of Our Ecological Crisis', *Science*, 155 (March 1967): 1203–7; 'Towards an Ecological Ethic', Editorial, *New Scientist*, 48 (December 1970): 575. As for the Japanese appreciation of nature, the first phrase comes from Hajime Nakamura, *Ways of Thinking of Eastern People* (Honolulu: The University Press of Hawaii, 1981), p. 355. The second phrase comes from Edwin O. Reischauer, *The Japanese* (Cambridge, Mass.: Harvard University Press, 1982), p. 148.

2 In addition to various sources which discuss each specific item (such as Japanese architecture, kimono, or Japanese cooking), the above observation is presented in the following writings on the general characteristics of Japanese culture. Masaharu Anesaki, *Art, Life, and Nature in Japan* (Tokyo: Charles E. Tuttle, 1973), Chapter I; Tatsusaburō Hayashiya, Tadao Umesao, Michitarō Tada and Hidetoshi Katō, *Nihonjin no Chie* (*The Japanese Wisdom*) (Tokyo: Chūōkōronsha, 1977); Hajime Nakamura, 'Environment and Man: Eastern Thought', in *Encycolpedia of Bioethics*; Hajime Nakamura, *Ways of Thinking*, Chapter 34; Masao Watanabe, 'The Conception of Nature in Japanese Culture', *Science*, 183 (January 1974): 279–82.

3 Nakamura, *Ways of Thinking*, p. 356.

4 By Yamabe no Akahito, included in *Manyōshū*. Translation is taken from Nakamura, *Ways of Thinking*, p. 356.

5 Sōkichi Tsuda, *An Inquiry into the Japanese Mind as Mirrored in Literature*, trans. Fukumatsu Matsuda (Tokyo: Japanese Society for the Promotion of Science, 1970), p. 282.

6 Shūichi Katō, *A History of Japanese Literature*, trans. David Chibbett (New York: Kodansha International, 1979), I; pp. 65–6.

7 The translation of the first poem is taken from Nakamura, *Ways of Thinking*, p. 356. The second poem is taken from Katō, *History*, p.65.

8 Katō, *History*, p. 66. And Kamo no Chōmei explains that his appreciation of the mountain life with the hooting of the owl and various other elements is dependent upon the fact that 'it is not an awesome mountain' (209). *An Account of My Hut*, trans. Donald Keene, included in *Anthology of Japanese Literature*, ed. Donald Keene (New York: Grove Press, 1960). The page reference from this work will be indicated within parentheses.

9 Anesaki, *Art, Life, and Nautre*, pp. 7–8.
10 *The Pillow Book of Sei Shōnagon*, trans. and ed. Ivan Morris (Harmondsworth: Penguin Books, 1982), p. 194. Another famous passage on the typhoon in *The Tale of Genji*, while describing the storm itself, does not express a sense of awe felt by being wholly overpowered by nature's brutal force. This is partly because the experience during the storm is narrated from a house, looking out to a small enclosed garden rather than in the midst of a vast moor or a thick forest where the effect of the storm is felt to the utmost degree. See the chapter entitled 'Typhoon' in *A Wreath of Could: Being the Third Part of 'The Tale of Genji'*, Trans. Arthur Waley (Boston: Houghton Mifflin Company, 1927). See also section 19 of *Essays in Idleness*, trans. Donald Keene (New York: Columbia University Press, 1967).
11 Kant, *Critique of Judgment*, trans. J. H. Bernard (New York: Hafner Press, 1974), section 27, p. 97, emphasis added.
12 Anesakim, *Art, Life, and Nature*, p. 10.
13 Katō, *History*, p. 66 (my translation). Chibbett's translation reads rather inaccurately: 'the court poets of the *Manyōshū* were profoundly influenced by nature and both their love poems and their elegies reflect this'.
14 Translation is taken from Makoto Ueda in *Literary and Art Theories in Japan* (Cleveland: The Press of Western Reserve university, 1967), p. 3. This practice of expressing one's emotion by reference to nature gives rise to the important aesthetic effects of 'constraint' and 'suggestion' in Japanese literature. See donald Keene's discussion of these effects in *Japanese Literature: An Introduction for Western Readers* (New York: Grove Press, 1955), Chapter I 'Introduction'.
15 Motoori Norinaga, *Isonokami Sasamegoto* (*My Personal View of Poetry*) in *Complete Works* (Tokyo: Chikuma Shobō, 1968), II, pp. 99–100, my translation.
16 Motoori Norinaga, *Shibun Yōryō* (*The Essence of The Tale of Genji*) in *Complete Works*, IV, p. 57, my translation.
17 Norinaga, *Isonokami*, pp. 110–11.
18 George Santayana, *The Sense of Beauty* (New York: Dover Publications, 1955), p. 121.
19 Barbara Sandrisser points out that rain is often an aesthetically celebrated phenomenon in Japanese literature and visual art. She correctly attributes the Japanese finding attraction in rain to their preoccupation with impermanence: 'Rain is inspiring. Perhaps its impermanence encourages this feeling . . . Rain appears and disappears. The experience of rain, although similar, is never the same'. 'Fine Weather—the Japanese View of Rain', *Landscape*, 26 (1982), p. 47.
20 Volumes I and II of *Kokinshū* are devoted to spring poems, volume III to summer poems, volumes IV and V to autumn poems, and volume VI to winter poems. The rest (volumes VII–XX) is devoted to poems concerning love, travel, farewell, etc., but it is commonly regarded that the best poems are found in the first six volumes.
21 Sei Shōnagon further observes in section 245 that 'things which pass quickly' are 'boat with a hoisted sail, people's age, spring, summer, fall and winter'. Also see *Essays in Idleness*, section 19.
22 The bond between cherry blossoms and spring is so firm and entrenched in the Japanese mind that an Edo-period thinker felt it necessary to remind people that cherry blossoms blooming after the spring is over are also appreciable. (Referred to by Tsuda, *Inquiry*, p. 265.)
23 Translation is by Ueda in his *Literary and Art Theories*, p. 81.
24 Hayashiya, *et. al., Nihonjin*, pp. 60–1, my translation.
25 *Nihon no Bi o Motomete* (*In Search of Japanese Beauty*) (Tokyo: Kodansha, 1977), pp. 26–7, my translation. The same point is made by Yoshida Kenkō in *Essays in Idleness* in which he claims that 'if man were never to fade away like the dews of

Adashino, never to vanish like the smoke over Toribeyama, but lingered on forever in the world, how things would lose their power to move us!', p. 7.

26 Ibid., p. 27.

27 Yoshida Kenkō, *Essays*, pp. 115–18.

28 This aspect of cherry blossoms can be contrasted with camellia flowers, for example, which Natsume Sōseki describes 'will live in perfect serenity for hundreds of years far from the eyes of man in the shadow of the mountains, flaring into blossom and falling to earth with equal suddenness'; they 'never drift down petal by petal, but drop from the branch intact'. *Kusamakura* (*Grass Pillow*) in *Complete Works* (Tokyo: Chikuma Shobō, 1971), II, p. 188, my translation.

29 *The Tale of Heike*, trans. Hiroshi Kitagawa and Bruce T. Tsuchida (Tokyo: University of Tokyo Press, 1975).

30 Translated by Donald Keene, included in Keene, *Anthology*, p. 81. Yoshida Kenkō similarly describes the changeableness of this world 'as unstable as the pools and shallows of Asuka River' (25) and transitoriness of people's affection 'like cherry blossoms scatering even before a wind blew' (27). Kenkō, *Essays*.

31 Higashiyama, *Nihon no Bi*, p. 27.

32 The Japanese tendency towards the absolute acceptance of the phenomenal world is thoroughly discussed by Nakamura in *Ways of Thinking*, Chapter 34.

33 Minami, *Psychology of the Japanese People*, trans. Albert R. Ikoma (Toronto: University of Toronto Press, 1971), p. 63.

34 Ibid., p. 60. I changed Ikoma's translation here because he translates the passage incorrectly as: 'because of the evanescence of nature, man should realize that it is senseless to grieve and should become reconciled to fate'.

35 I thank the editor for his helpful comments and advice on this paper.

95

# ON THE ENVIRONMENTAL ETHICS OF THE *TAO* AND THE *CH'I*

*Chung-ying Cheng*

Source: *Environmental Ethics*, 8 (1986), 351–70.

How the *Tao* applies to the ecological understanding of the human environment for the purpose of human well-being as well as for the harmony of nature is an interesting and crucial issue for both environmentalists and philosophers of the *Tao*. I formulate five basic axioms for an environmental ethic of the *Tao*: (1) the axiom of total interpenetration; (2) the axiom of self-transformation; (3) the axiom of creative spontaneity; (4) the axiom of a will not to will; and (5) the axiom of non-attaching attachment. I show that each axiom generates important consequences for environmental ethics and that together they provide a necessary foundation for environmental ethics.

## I Methodological considerations

Although environmental ethics is one of applied ethics arising from contemporary interest in applying and exploring certain ethical concepts and positions in relation to a set of concrete situations which human persons confront in their life world,[1] a close reflection on this applied ethics leads to a metaphysical critique of certain basic ethical positions concerning relationships of human beings to nature, to other human beings, and to themselves. The fact is that ethics cannot be applied until we have a clear understanding of the underlying concepts of the human person and his/her end-values as well as a clear understanding of the objects or situations to which the applications pertain. Both understandings require a disclosure of presupposed reality and, therefore, a resolution on the order or scheme of things in which human persons find themselves.

Methodologically speaking, we could treat problems of applied ethics at three levels: the metaethical level, where meanings of ethical terms are clarified; the metaphysical level, where the fundamental premises of the nature of reality are examined; and finally, the normative level, where ethical norms of actions and attitudes are formulated. Environmental ethics apparently cannot be clearly understood until we come to grips with these three levels of understanding. This approach is particularly appropriate in consideration of the fact that environmental ethics is not a well-formed system of ethical concepts, and that no system of norms for this ethical concern has been fully formulated and agreed upon. The openness of this field is a feature not to be deplored, but to be commended, because inquiry into identifying its problems, not even to say resolving its problems, is a worthy methodological and metaphysical exercise and will be rewarding in terms of possible creative insights, not only into environmental-ethical issues, but also into a deeper foundation for ethics in general. We may indeed consider the three levels of understanding mentioned above as pertaining to the analytics of ethics, the teleology of ethics, the teleology of ethics, and the deontology of ethics. In dealing with environmental ethics we must be concerned with its analytical, its teleological, as well as its deontological dimensions.[2]

## II Analytical considerations

One central question for environmental ethics which must be raised before any other questions is what the term *environment* means or stands for. *Environment* is derived from *environs*, meaning "in circuit" or "turning around in" in Old French.[3] It is apparently a prepositional word, indicating an external relation without a context, also certainly devoid of a relationship of organic interdependence. Yet when we reflect on the experience of environment, we encounter many different things and different processes in the context of organic interdependence. We might say what we experience presupposes the existence of life and the living processes of many forms. This experience of environment is better expressed by the Chinese philosophical paradigm, *sheng-sheng-pu-yi*[a] ("incessant activity of life creativity").[4] We must, therefore, make a distinction between a surface meaning and a depth meaning for environment. Without understanding life and the living process of life, we cannot understand the depth meaning of environment. On the other hand, without understanding the constituents and conditions of life, we cannot understand life and the living process of life. Hence, the very essence of environment requires an understanding of reality and the true identity of life in both its state and process aspects. This means we have to understand the *Tao* content and the *Tao* process in the environment, whereas Tao indicates the way of life-creativity in ceaseless movements and in a multitude of forms.

With the above analysis of the meaning of environment, it is clear that the essential depth meaning of environment was lost in modern man's conception of environment. The modern man's conception of environment is founded on the surface meaning of environment, which is typified by technology and science; with its underlying philosophy of modernday materialism, Cartesian dualism, and mechanistic naturalism, the concept of environment of modern man was very much objectified, mechanized, rigidified, dehumanized, and possibly even de-enlivened, and so de-environmentalized.[5] Environment is no longer an environment at all; environment becomes simply "the surroundings," the physical periphery, the material conditions and the transient circumstances. The environment is conceived as a passive deadwood, and very often as only visible and tangible externalia. In fact, as the depth meaning of environment suggested above, environment is active life; it is not necessarily visible or tangible, and certainly it cannot be simply a matter of externality. Hence, it cannot be treated as an object, the material conditions, a machine tool, or a transient feature. Environment is more than the visible, more than the tangible, more than the external, more than a matter of quantified period of time or a spread of space. It has a deep structure as well as a deep process, as the concept of *Tao* indicates.

The distinction between the surface meaning and the depth meaning of environment also suggests a distinction between the Western and the Chinese approach to environment. Whereas the West focuses on the external relation of man to his surroundings based upon a qualitative separation and confrontation between the human and nonhuman worlds, the Chinese focus on the internal relation of man to his surroundings based upon an integrative interdependence and a harmony between man and the world. For modern Western man after Descartes, the nonhuman world is to be rationally studied, researched, and then scientifically manipulated and exploited for the maximum utility of serving man. This will to conquer and dominate nature is, of course, premised on the externality of nature to man, but there are two other rational principles or assumptions involved in exercising this will to conquer and dominate.

First, it is assumed that nature is a completed work of mechanical forces with one-dimensional natural laws controlling its workings. The one-dimensional natural laws are revealed in the physical sciences and the reductionistic methodology of physicalism. Hence, biological laws are very often reduced to laws of physics and chemistry; no other laws are permitted to stand on their own. Yet the relationship between various forms of life in the totality of nature cannot be said to be fully captured by physicalistic laws; nor can the relationship between man and the world of things be said to be regulated by these laws. The very fact of the breakdown of the environment in industrialized societies, as reflected, for example, in the problems of water-air-noise pollution, precisely points to the lack of understanding of the relationship between various forms of life and man and his environment by way of modern science and technology.

There is a second assumption of the modern mechanical sciences: everything in the world forms an entity on its own as a closed system, and therefore can be individually and separately dealt with. This isolationist and atomistic assumption in the problem-solving methodology of modern science is strongly reflected in Western medical diagnostics and treatments. It was not until recent times that modern medical and health care researchers became aware of the potential limitations of this isolationist and atomistic approach, and became awakened to a holistic approach.

In contrast with the Western externalistic point of view on environment, the Chinese tradition, as represented by both Confucianism (with the *I Ching* as its metaphysical philosophy) and Taoism (with Chuang Tzu and Lao Tzu as its content), has developed an internalistic point of view on the environment. The internalistic point of view on the environment in Chinese philosophy focuses on man as the *consummator* of nature rather than man as the conqueror of nature, as a participant in nature rather than as a predator of nature. Man as the consummator of nature expresses continuously the beauty, truth, and goodness of nature; and articulates them in a moral or a natural cultivation of human life or human nature. This is paradigmatically well expressed in Confucius' saying, "Man can enlarge the Way (*Tao*) rather than the Way enlarging man."[6] It is also expressed in Chuang Tzu's saying, "The *Tao* penetrates and forms a Unity."[7] As part and parcel of nature, man does not stand opposite nature in a hostile way. On the contrary, man has profound concern and care for nature at large, as befitting his own nature. For his own growth and well-being, man has to cultivate the internal link in him between himself and Mother Nature. To conquer nature and exploit it is a form of self-destruction and self-abasement for man. The material consequence of the conquest and exploitation must be forestalled by an awakening to what man really is or in what his nature really consists.

In contrast with the two Western assumptions about the environment, Chinese philosophy clearly asserts that nature, and therefore man's environment, is not a complete work of production by a transcendent God, but rather is a process of continuous production and reproduction of life. In Bruno's words, nature is *Natura naturans*, not merely *Natura naturata*. In other words, nature is an organism of continuous growth and decay, but never devoid of internal life. With this understanding men cannot treat nature as an isolated and atomic part without regard for the totality involving a past and a future. This leads to the second understanding contrary to the Western methodology of atomism: man has to interact with nature in a totalistic manner, realizing that there is no single linear chain of causality. There is always a many-to-many relationship between cause and effect. Hence, man has to consider a many-to-many level approach to relate the potential needs of man to nature. Man has to naturalize man as well as to humanize nature, treating nature as his equal and as a member within the family of the *Tao*. This approach to nature is reflected in the holistic approach of Chinese medicine in both its diagnostic and medical/health care aspects.

The modern mandarin translation for *environment* is *huan-chin*, meaning "world of surroundings." This translation apparently reflects the surface meaning of *environment* correctly. But when embedded in the contexts of Chinese philosophy and Chinese cultural consciousness the "world of surroundings" does not simply denote individual things as entities in a microscopic structure; it also connotes a many-layered reality such as heaven and earth in a macroscopic enfoldment. This "world of surroundings" is generally conceived as something not static but dynamic, something not simply visible but invisible. It is in this sense of environment that we can speak of the *tao* as the true environment of man: the true environment of man is also the true environment of nature or everything else in nature.

When asked about the presence of the *tao*, Chuang Tzu had this to say: "(The *tao* is) nowhere not present." Pressed as to where exactly the *tao* lies, Chuang Tzu replied that the *tao* is in the ants, in the weeds, in the ruins, and in the dungs.[8] The import of Chuang Tzu's message is that the *tao* embraces everything, large or small, in the universe and imparts a unity of relationships in our environment, and that the *tao* is a totality as well as a part of the totality pervading everything beyond our perception so that we cannot ignore what is hidden in our understanding of the environment. If understanding is the basis for action, this understanding of the environment in terms of the *tao* is very essential for formulating an ethic of the environment, namely for articulating what human persons should do or attitudinize toward their world of surroundings. Two more observations have to be made in order to explicate the philosophy of the *tao* for the purpose of the formulating an environmental ethics of the *tao* or an ethics of the environment based on an understanding of the *tao*.

The first observation concerns the *Tao* as the *tzu-jan*. *Tzu-jan* means "doing-something-on-its-own-accord," or natural spontaneity. In the *Tao Te Ching* it is said that "Man follows earth; earth follows heaven; heaven follows the *Tao* and the *Tao* follows *tzu-jan*."[9] But *tzu-jan* is not something beyond and above the *Tao*. It is the movement of the *Tao* as the *Tao*, namely as the underlying unity of all things as well as the underlying source of the life of all things. One important aspect of *tzu-jan* is that the movement of things must come from the *internal life* of things and never results from engineering or conditioning by an external power. That is why the life-creativity nature of the *Tao* is the only proper way of describing the nature of the movement of the *Tao*. However, to say this is not to say that only the *Tao* can have the movement of *tzu-jan*. In fact, all things can follow *tzu-jan* insofar as they follow the *Tao*, or in other words, act and move in the manner of the *Tao* and in unison and in accordance with the *Tao*. Perhaps a better way of expressing this is: things will move of their own accord (*tzu-jan*) insofar as they move by way of the *Tao* and the *Tao* moves by way of them. One has to distinguish between, on the one hand, *Tao*-oriented or *Tao*-founded movement and, on the other hand, thing-oriented or thing-founded movement. Only when the movement of a thing comes from the deep source of the thing—the *Tao* and its harmony with the totality of the movements of all other things—will the movement of things be genuinely of its own accord and, therefore, be spontaneous. Spontaneity (*tzu-jan*) is a matter of infinite depth and infinite breadth in an onto-cosmological sense.

One can, of course, speak of different degrees of *tzu-jan* in view of the different degrees of depth and breadth in harmonious relating and self-assertion among things. Just things have their own histories and defining characteristics in form and substance, things also have their relative freedom of self-movement and life-creativity. Things, in fact, can be considered as conditions or preconditions of various forms of *tzu-jan* (spontaneity): insofar as things preserve their identity without destroying the identities of other things, and insofar as things change and transform without interfering with the process of change and transformation of other things, there is *tzu-jan*. This explains the mutual movement, rise and decline, ebb and flow, in things of nature.

For human beings, *tzu-jan* finds its rationale not only in the internal movement and life-creativity of human activity, but in the principle of least effort with maximum effect. Whatever produces maximum effect by minimum effort in human activity manifests natural spontaneity. One may, therefore, suggest that only in following natural spontaneity is there least effort and maximum effect. This can be called the *ecological principle of nature*.[10] Using this principle we can correctly interpret the most important point ever made about the nature of the *Tao*: "The *Tao* constantly does nothing and yet everything is being done" (*Tao-chang-wu-wei erh wu-pu-wei*).[11] That the *Tao* constantly does nothing means that the *Tao* does not impose itself on things: the *Tao* only moves of its own accord. This also means that all things come into being on their own accord. The constant nonaction of the *Tao* is the ultimate cosmological principle of life-creativity and the only foundation for the evolution of the variety of life and the multitude of things. The nonaction of the *Tao* in this sense is an intrinsic principle of ultimate creativity; this intrinsic principle of ultimate creativity consists in an unlimitedness and an unlimitation of expression of life forms and life processes in a state of universal harmony and in a process of universal transformation.[12] In this ultimate sense of creativity, there is no effort made by the *Tao*, and yet there is an infinite effect, achieving life-creativity. The ecological principle reaches its ultimate limit in the principle of *chang-wu-wei*. Hence, we can conceive of the principle of least effort with maximum effect as an approximation to the *tzu-jan* of the *Tao* on the human plane.

With this principle correctly understood, we can resolve the dilemma and predicament arising from civilization and knowledge. The Taoist questions the value of knowledge and civilization, since they lead to greed, lust, and evil (tricks and treachery) in human society. In the same spirit, we can question the value of science and technology. In resolving many problems of man, do science and technology create more problems for man? Do science and technology seem to lead man to a purely pessimistic future? The Taoistic criticism here is that without an understanding of the *Tao* it is indeed possible and necessary that knowledge and civilization, science and technology, will doom man to self-slavery and self-destruction. Man simply falls into the bondage of his own conceptual prison and becomes a victim of his own desires. The Taoistic criticism of *wu-wei* is supposed to awaken man to self-examination and self-doubt; in this way man is awakened to a quest for self-surpassing and self-overcoming in an understanding of the totality of reality and its secret of creativity through *wu-wei* and reversion (*fan*).[13] With this awakening, man can still proceed with his knowledge and civilization, science and technology, if he is able to neutralize and temper his intellectual and intellectualistic efforts with a sense of the *Tao*. This means that man has to develop knowledge and civilization, science and technology, not out of pace with his efforts to relate to things, other humans and himself. His knowledge and civilization, science and technology, have to contribute to his relating to and integrating with the world of his surroundings. To do this he has to keep pace with his own growth as a sentient moral being, having regard and respect for his own identity and dignity as well as the identity and dignity of other beings, including his fellow man. Furthermore, he has to use his knowledge and, hence, science and technology, in keeping with the order of things, with his best interests conceived and deferred in harmony with life and in preservation or promotion of universal creativity. He also has to closely follow the principle of least effort, if not the principle of no effort, with maximum effect, if not infinite effect in terms of life and creativity—preservation and promotion—for his intellectual/scientific/technological/organizational activities.

As man is part and parcel of the *Tao*, it is only when man loses the sense of the *Tao* and respect for the *Tao* in his actual life that man becomes alienated from the *Tao* and his activities become a means of self-alienation which will inevitably result in losing the true identity of man by way of self-destruction. This is the natural and spontaneous reaction of the *Tao* to the self-alienation of man in his intellectual/scientific/exploitative engrossment and obsession with himself. Hence, the remedy for knowledge and civilization, or for science and technology, is not more knowledge and more civilization, or more science and more technology, but a constant relating and integrating of these with the *Tao*. To do so is to naturalize as well as humanize knowledge and civilization, science and technology. It is to make these a part of the *Tao*. Although knowledge and civilization, science and technology, are man's forms for the appropriation of nature (the *Tao*), these forms should not remain apart: man should also let nature reappropriate them by integrating them into nature (the *Tao*). This is the essential point of an ethics of man's relation to the environment. To understand the *Tao* and to follow the *Tao* is the essence of the ethics of the environment; it is also the way to transform the artificiality and unnaturalness of knowledge and civilization, science and technology, into the spontaneity and naturalness of the *Tao*.

In light of this understanding, the conflict between the *Tao* and knowledge/civilization/science/technology can be resolved; the true ecology and life-creativity of nature can be restored with knowledge/civilization/science/technology. They can be seen as enhancing rather than obstructing, complementing rather than opposing, the actual spontaneity and harmony of the creativity of the *Tao*. This is the true wisdom of the Taoist critique of knowledge and civilization, science and technology. It is called *hsi-ming*, "hidden light," by Lao Tzu, and *liang-hsing*, "parallel understanding," by Chuang Tzu.[14] In this wisdom lies the most profound principle of both the ecology of nature and the ethics of the environment.

My second observation concerns the *Tao* as a process of the ramification and differentiation of the *ch'i*. Before I explain the meaning and reality of *ch'i* in Chinese philosophy, it is important to appreciate the significance of bringing in *ch'i* as an explanation of the depth structure and depth process of the environment. We have seen that the depth structure and depth process of the environment has been explained in terms of the *Tao* and its life-creativity (*sheng-sheng*). Even though this explanation is necessary in pinpointing the ontological being and becoming of the environment, it is not sufficient, on the one hand, to illuminate the dynamics and dialectics of the differentiation and ramifications of the *Tao* and, on the other hand, to manifest those dynamics and dialectics of the unification and integration of the *Tao*. In other words, there is a gap between the ontology of the *Tao* and the cosmology of the *Tao* which must be bridged.

It is when the *Tao* is seen in the form and activity of *ch'i* that this bridging takes place. It might be suggested that the *Tao* expresses itself in terms of three perspectives which result in three characterizations in the history of Chinese philosophy. The first perspective is derived from understanding the quality of the activity: it is the perspective of life-creativity as clearly formulated in the texts of the *I Ching*. This perspective has already been discussed above. The second perspective is derived from understanding the patterns of the activity: it is the perspective of the movement of internal spontaneity, reversion and return, as clearly formulated in the texts of the *Tao Te Ching* as well as those of the *Chuang Tzu*. In fact, a concentration on the patterns of the movement of the *Tao* may lead one to see the *Tao* in terms of principles and reasons. The Neo-Confucianist metaphysics of *li* ("principle") is a logical result of this development. This development also leads to an epistemology of the *Tao*. In both the ontology of the *Tao* (life-creativity) and epistemology of the *Tao* (principles of nonaction, etc.), the *Tao* is always conceived as a totality and a unity; the nature of the unity and totality of the *Tao* is stressed above all. In fact, the very concept of the *Tao* carries with it a reference to its unity and totality. Yet the *Tao* is as much a distribution and diversification of being and becoming as a unity and totality of being and becoming. Hence, we need another explanation of this former aspect of the *Tao* which will also serve the purpose of cosmologizing the ontology and epistemology of the *Tao*. This is how the *Tao*-as-the-*ch'i* paradigm comes in. This is also how the concept of *ch'i* based on experience of the *Tao* as *ch'i* develops. We might therefore suggest that to understand environment in its depth meaning, one has to focus on both the totalistic and distributive aspects of the environment. Hence, one must focus on both the *Tao as tzu-jan* and the Tao as *ch'i*.

Another consideration with regard to the importance of the *Tao* as the *ch'i* is that whereas the *Tao* focuses on reality as a passage of dynamic processes, the *ch'i* focuses on reality as a presence of material—stuff which leads to an actualization of things and the concretization of events. Hence, for understanding the formation and transformation of the environment in its substantive structure, one has to understand *ch'i*. It is in understanding *ch'i* that one can see and grasp the subtleties of the environment *vis-à-vis* human beings. It is only on this basis (i.e., understanding the *Tao* as *ch'i*) that one is capable of formulating an ethics of the environment or an ethics of the *Tao* toward the environment.[15] For this reason, we may consider the discussion of the nature of *ch'i* as constituting a metaphysical inquiry into the depth structure and depth process of the environment. As the goal of an ethics of the environment is to understand how human beings should relate to the environment *via* a true understanding of environment, we may see how a metaphysical inquiry into the structure and process of the environment also constitutes a teleological inquiry into the nature of the environment in relation to man. It is only when we are able to understand the nature of the environment in its true identity that we are able to see what the end-values of our thinking about and acting toward the environment are. The end-values are provided by our understanding of reality: to act in accordance with reality and our true nature will be our end and ultimately will be the criterion of value. Hence, the metaphysics of the environment becomes a teleology of the environment, because that is where environmental ethics is grounded. With this foundation we can then rightly speak of an environmental ethics of the *Tao* which is the deontology of the *Tao*.

## III Metaphysical/teleological considerations

There are several passages in the *Tao Te Ching* which justify our attempt to explain the *Tao* in terms of the *ch'i*. In chapter fourteen of the *Tao Te Ching*, the *Tao* is described as something invisible, inaudible, and ingraspable.

> Being invisible is called the Great (*yi*); being inaudible is called the Silent (*hsi*); being ingraspable is called the Subtle (*wei*). The source of these three cannot be fully disclosed. They merge to form Oneness. The Above of this Oneness is not bright, and the Below of

> this Oneness is not dark. As a continuum it cannot be named, and it again vanishes no-thing-ness. It is therefore the form of no-form, the image of no object. It is called "Seeming-To-Be" (*fu-huang*).[16]

This description of the *Tao* is no doubt applicable to *ch'i* as the primary source of being and becoming. Hence, the *Tao* is *Tao*-as-*ch'i,* and the *ch'i* is *ch'i*-as-*Tao*. In chapter twenty-one of the *Tao Te Ching*, the *Tao* is again described as being constantly indeterminate and unformed (*fu-huang*), and obscure and dark (*yao-min*), and yet containing and yielding concrete things.

> The *Tao* is something indeterminate and unformed. Being indeterminate and unformed, there emerges the forms; being unformed and indeterminate, there emerges the things. Being obscure and dark, there emerges the essences (*chin*); these essences are very real, wherein the truth (*hsin*) exists.[16]

The essences (*chin*) are generally explained by commentators as the *chin-ch'i*, or the quintessential *ch'i*, which is the primary or elementary *ch'i*.[17] In this way, it is possible to state how the *Tao* is related to the *ch'i: Tao* in its actual movements becomes the *ch'i*. Or alternatively speaking, the movement of the *ch'i* is the *Tao*. In fact, chapter twenty-five of the *Tao Te Ching* gives an even more vivid characterization of the *Tao* as *ch'i* i.e., as something which is the primary substance for all things.

> There is something which comes into being before heaven and earth. It is silent and formless. It exists without depending on anything else. It moves in circles ceaselessly. I do not know its name, so I just call it the Tao. I also try to call it the Great. The Great is Passing; the passing is distant; the Distant returns to its own.[18]

Although Lao Tzu does not speak straightforwardly about the *Tao* as *ch'i*, it remains clear that the *Tao* must present itself as the *ch'i* for the purpose of cosmos making and thing making. In other words, to speak of the *Tao* as *ch'i* is necessary insofar as there is a need for explaining the constitution of the world of things. Similarly, in order to satisfy the need to explain the totalistic and transformational nature of things, one has to speak of the *ch'i* as *Tao*. It is only when one understands the *Tao* in the form of *ch'i* and *ch'i* in the form of *Tao* that we are able to understand thoroughly the nature of reality which forms our environment, and which defines a teleology of end-values for our ethical demands on our behavior toward the environment.[19] In the following, I give a characterization of *Tao* as *ch'i* by answering the question, "What is *ch'i*?" I follow the classical literature on *ch'i* in *Lao Tzu*, *Chuang Tzu*, *Kuan Tzu* and the *I Ching*, and even more importantly, in the classic of Chinese medicine, the *Nei Ching*.

What is *ch'i*? *Ch'i* is the most fundamental concept in Chinese philosophy as well as in Chinese common sense. *Ch'i* has been variously translated into English as "material force," "vital force," and "energy." Like many fundamental Chinese philosophical terms, no translation captures its full meaning. In fact all the existing translations conceal and obscure the rich experiential structure of meaning in the concept of *ch'i*. It is important to recognize that the concept of *ch'i* already contains a metaphysical theory, an epistemological theory, and a scientific theory. It also contains a structure of levels and implies a process of stages. *Ch'i*, on the natural level, is simply the vapor in the form of steam over the rice paddies and the cloud/mists in the sky. On the human level, with regard to human daily experience, *ch'i* is simply breath. As breath is linked to air, so *ch'i* covers organic life-energy and the energy in nature in various forms. Hence, nothing in human body is not *ch'i* and nothing in nature is not *ch'i*. There are two important conditions involved in understanding *ch'i*: (1) *ch'i* is the vital energy of no form. It moves, rests, goes and comes, condenses and stretches. In other words, it transforms; (2) *ch'i*, nevertheless, forms all things in the universe: it can change into various forms depending on how it moves and rests and how it relates to things formed from *ch'i*. By itslef *ch'i* does not have any determinate form (indeed, no form at all). But it is precisely because of its indeterminateness and inner capacity of creativity that *ch'i* can creatively form into things, and this in turn is the reason why things have no permanent form and substance. Ultimately the life and death of man and other species have to be explained on the basis of the impermanence of the form of *ch'i*.

The onto-cosmology of *ch'i* has been developed to the extent that to think of *ch'i* one also has to recognize three other characteristics of *ch'i*.

1. *Ch'i* is universally present; it is not confined to a single region, just as it is not confined to a simple form. Hence, there exists a universal sympathy and consanguinity among things in the universe. Yet, the difference in kind, structure, and level of things make a difference of degrees of sympathy and consanguinity among things. To discover which things are in sympathetic resonance and which things are not are two important efforts in both traditional applied environmental analysis (*feng-shui*) and diagnosis/healing therapy in Chinese medicine.

2. In the pre-Chin philosophy of *ch'i*, the order, structure, and form exhibited by *ch'i* are inherent in and internal to the *ch'i* itself. As to how different things are formed, the apparent randomness of *ch'i* actually conceals an ordering force of balance, control, cohesion, sustenance, and qualitative elevation or evolution. Hence, *ch'i* is not to be conceived as an arbitrary chaotic existence: it is to be conceived as leading to seasoned and proportionate evolution and sustenance of life. In fact, reason and principle (*li*) can be explained as the essence of *ch'i*, i.e., as that which constitutes the nature of *ch'i* insofar as *ch'i* behaves in accordance with them. Hence, one

should not speak of *li* apart fom or external to *ch'i*. *Li* belongs to *ch'i*. This explains why there are orderly principles in the formation of the natural world and human physiology, and similarly why nature in itself exhibits regularity and produces *laws* and *patterns* open to human understanding. The possibility of human understanding can indeed be explained as the sympathetic activities of *ch'i* inside one person in resonance with *ch'i* externally in the world.

3. The basic order inherent in *ch'i* is exhibited in the distinction between *yin* and *yang*. *Yin* and *yang* refer to two aspects of *ch'i* as a whole; yet in the process of change and transformation one aspect could become dominant. As such the dominant aspect of *ch'i* becomes perceptible and is experienced. Yet, at the same time, the other aspect continues to exist or subsist, although it takes a subtle mind to experience it, and it takes a period of time to see its functioning. *Yin* and *yang* originally referred respectively to the shady and the lighted (the bright) under the sun. Both the shady and the bright have light as their central subject. Similarly, the *yin* and *yang* have *ch'i* as their central subject.

It terms of the *yin/yang* distinction we can speak of two polarities of *ch'i*. In doing this, we must remember that the *yin/yang* concepts have primary experiential roots. They are not arbitrary inventions of man, but instead reflect deep features of the nature of *ch'i*. Thus, we can speak of *yin-ch'i* and *yang-ch'i*. In light of the primary experience of *ch'i*, one can immediately see how *yang-ch'i* can be interpreted and perceived in motion, brightness, and firmness (*Kang*), and *yin-ch'i* in rest, darkness, and softness (*jou*). The reason for this interpretation is experiential: *yang-ch'i* or *yin-ch'i* is not experienced as a definite single quality, but rather is experienced as an indefinite mass phenomenon, the nature of which reveals itself in different ways under different ways under different circumstances. Hence, in terms of the *yin/yang* distinction, and indefinite subcategorization for either *yin* or *yang* can be introduced by way of sympathetic understanding and in spontaneous revelation. In addition to the *yang* natures of motion/light/firmness and the *yin* natures of rest/shade/softness, a long list of other qualities or states of things can also be divided naturally into *yin/yang* groups. In each of these divisions, an identification of the *yin* and the *yang* will enrich and make more definitive the *yin/yang* experiences and their corresponding concepts. Yet no complete distinction and exhaustion of the *yin/yang* totality can be given. Perhaps, we can speak of an inductive and implicit definition of the *yin/yang*: *yin/yang* are those natures or qualities compatible with and strengthening those natures or qualities already known as *yin/yang*.

Philosophically speaking, both the *I Ching* and Tzu's *Tao Te Ching* have conveyed a general criterion for distinguishing between *yin/yang*. *Yin* is passive, receptive, close-in, downward, soft, resting, and background-like, whereas *yang* is active, creative, open-ended, upward, firm, moving, and foreground-like. *Yin/yang* become more clearly described when we look at them relative to concrete things on different levels. Therefore, we can speak of heaven/earth as *yin/yang*, man/woman as *yin/yang*, life/death as *yin/yang*, etc. In each case, of course, the substantive reference to *ch'i* is presupposed, for apart from *ch'i* how can the *yin/yang* become discovered or revealed? In reference to *ch'i*, we can establish the following list of *yin/yang* distinctions:

For *yin*: shade/rest/softness/cold/the weak/closure;
For *yang*: light/motion/firmness/hot/the strong/openness.

There are five principles governing *yin/yang* distinctions:

1. *The principle of universality.* *Yin/yang* exist in all things and for all things on different levels. Indeed, it is because of the *yin/yang* distinction that things come to exist in different kinds and on different levels. This is related to the principle of creativity established by the *I Ching*.

2. *The principle of relationality.* The *yin* and the *yang* always exist in close relation to each other, which should explain how things are always related in *yin/yang*. Insofar as there are relations, there are *yin/yang* distinctions and unities underlying all things. The relationships are established in accordance with the next principle.

3. *The principle of opposite complementarity.* *Yin/yang* form a unity in which *yin* and *yang* are opposite to each other, and yet depend on each other, and therefore complement each other for their being distinguished as opposites. What makes opposite complementarity possible is the pervading unity I spoke of earlier. The opposite complementarity is opposite because of difference, and complementary because of unity. Unity and difference exist at the same time as internal natures of the *ch'i* or as internal dynamical tendencies in the *ch'i*. As such, we can speak of the unifying and differentiating aspects of *ch'i* which give rise to *yin/yang* and their natural bond of oneness in difference and opposition. This is again made very clear in the *I Ching*. In light of this principle, we may indeed speak of *yin/yang* as polarities of *ch'i*.

4. *The principle of relativity.* All states of *ch'i* and hence all things, can be both *yin* and *yang* relative to different states of *ch'i* or things. As individual things *ch'i* reveals a manifold of context in which things are related. Hence, no single *yin/yang* distinction is sufficient to characterize things in their relation to many other things. A wife who is an employer can be *yin* relative to her husband, but can be *yang* relative to her employees. In fact, she can be both *yin* and *yang* to the same object, depending on the contexts of relationships which give rise to differences of activities and positions. A network of *yin/yang* develops for a single state of *ch'i* or an individual thing. One can see how complex the human body or the whole universe can be presented in terms of a network of the *yin/yang*.

5. *The principle of creativity.* With the polarity of *yin/yang* aspects inherent in *ch'i*, activities involving the disturbance of the balance or harmony in *yin/yang*, the development of a new level of balance or harmony, or the restoration of a given balance and harmony are possible. In fact, all things in the universe are results of the creative activities of *yin/yang*; and all these creative activities are actual results of *yin/yang* movements of opposition and unity. By *yin* alone or by *yang* alone there can be no continuation of life creativity. Since there are life and creativity in the evolution of new entities or new generation of things, *yin/yang* can be perceived as ever present among all things. In regard to this creative capacity, the *yin/yang* unity of differences is specifically called the *ta'i-chi* (the great ultimate). It is clear that the *ta'i-ch'i* is *ch'i* in its totality, the ever-renewing activities of producing and maintaining the order of things in the environment of man.

Given these principles, it is easy to see how the Chinese notion of the natural environment, insofar as it is developed on the basis of *ch'i*-philosophy, and hence in terms of the *yin/yang* distinction, is totalistic, phenomenalistic, and organismic. The reason is simple. *Ch'i* with its internal *yin/yang* activities is totalistic, phenomenalistic, and organismic, since *ch'i* is a totality, dynamically presenting itself on different levels of reality, and yet well-related with reference to all its parts. These characteristics of *ch'i* not only lead to the development of a conception of the natural environment as a *ch'i* differentiation without loss of its unity, but also as a process, on the one hand, constantly involving the internal interaction of its parts and circulation within the *ch'i*, and, on the other, constantly involving the human *ch'i* with external things in the world. The purpose of environmental care is to maintain both the internal and external harmony of the dynamic interaction of *ch'i*, just as the purpose of medical care is to restore balance and harmony to either the disrupted internal state of *ch'i* or to the disrupted state of *ch'i* in reference to the external world. Chinese traditional environmental studies is precisely and intimately founded and developed on the basis of the *ch'i* philosophy.

With *ch'i* as the basic substance-process of all things in the world, we have the *yin/yang* activities of *ch'i* providing both a structure and a process of creative transformation among all things in the world. With regard to concrete things in the context of the *yin/yang* interchange, we have to introduce another structural distinction, the concretizing activities of the *ch'i* based on *yin/yang*. This is the distinction of *wu-hsing* (five powers). The "five powers" theory dates back to ancient times when people started to wonder about how things were constituted and related. Because *ch'i* philosophy recognizes *ch'i* as a dynamics power, the "five powers" are conceived as agencies rather than as elements, as processes rather than as states. Hence, the *wu-hsing* can be seen as five agencies or even five functions of *ch'i* realized in the concrete processes of things. The *wu-hsing* are powers of metal, wood, water, fire, and earth. The basic characteristics of these five powers are as follows: metal is sharp and cutting; wood is firm and growing; water is soft and penetrating; fire is bright and burning; and earth is solid, yet adjustable. In light of these basic characteristics, two orders of relationships can be established between them: the generative order and the destructive order. These two orders can be described as follows: the generative order consists of wood generating fire, fire generating earth, earth generating metal, metal generating water, and water generating wood again; the destructive order consists of wood destructing earth, earth destructing water, water destructing fire, fire destructing metal, and metal destructing wood.

At the level of the five powers, *yin/yang* are still the background force, and indeed they can be considered the motivating force of the *wu-hsing*. Although each power can be analyzed into *yin/yang* components or aspects, they cannot be described as exclusively *yin* or exclusively *yang*. Thus, water is more *yin* than *yang*, fire more *yang* than *yin*, wood can be either more *yin* (*yin* wood) than *yang* or more *yang* (*yang* wood) than *yin*; similarly metal can be either more *yin* (*yin* metal) than *yang* or more *yang* (*yang* metal) than *yin*. Similarly, one can speak of a *yin* earth (more *yin* than *yang*) or *yang* earth (more *yang* than *yin*). This analysis is possible in terms of the light trigrams (*pa-kua*) of the *I Ching*, a generative-empirical (*hou-t'ien*/late heaven) arrangement, which precisely produces a generative order of the *wu-hsing* with more detailed *yin/yang* identification of the *wu-hsing*.

The significance of the *yin/yang* analysis of the *wu-hsing* theory for the Chinese conception of environment is that we can consider the opposite and complementary relations among *ch'i* structures, in the human being or between the human being and the external environment, not only in the second-order relations of generation/destruction of *wu-hsing*, but also in the primary-order relations of differentiation, unification, balancing, and harmonization of *yin/yang*. This means that the *I Ching* philosophy of the unity of difference and identity and its dialectics of transformation based on "one-dividing-into-two" and "two-unifying-into-one" are always in the background of the formation of views on human life, the environment, and their relationship in the Chinese tradition.[20]

## IV Deontological/transdeontological considerations

What I have said above constitutes a metaphysical picture of reality on the basis of which our conception of the environment can be built. As reality is conceived in terms of *sheng-sheng* (life-creativity), *Tao* (the way of transformation), and *ch'i* (vital force), so environment can also be understood under these three categories, and thus can be seen as being intimately related to human beings in a relationship of part and whole: man is part of the environment; the environment is a part of man; and they both are a part of the same whole. It is evident that these propositions are meaningful in light of the *Tao* and *ch'i* metaphysics of reality as developed in the Chinese philosophical tradition. Although man and his environment may form different orders of their own, they also belong to the same totality, which is the universal process of transformation which is called the *Tao*. Within this totality of transformation, man and his environment are interdependent, interacting as well as interpenetrating elements. Hence, man cannot treat his environment as a mere object for knowing, controlling, and exploitation. The environment consists of many processes, involved with and related to the process exemplified and embodied in the *Tao*, life-creativity, and the *ch'i* properties. Hence, the environment should always preserve a unity, a harmony, and a balance inherent in the relationship of things and processes on an appropriate level. Man is part of this unity, harmony, and balance, and should contribute to their continuous sustenance and growth. This is the supreme categorical imperative for man's behavior toward the environment.

With the environment understood as *Tao* and *ch'i*, we can also deduce important conclusions regarding the misuse or abuse of the environment. The environment can react in a harmful way if it is itself harmed. This is a natural consequence of the intrinsic movement of reversal in the *Tao*. On the other hand, the environment can preserve life and enhance health, if life-preserving and health-enhancing measures are adopted by man in relation to the environment. The environment is a perfect mirror of man. Man's benevolence, justice, and integrity are reflected in the life-generating and health enhancing powers of the environment, just as human selfishness, inconsiderations, and cruelty are reflected in catastrophes and disasters in the environment. From this point of view, benevolence, justice, and integrity are not just simply human virtues toward one's fellow man and the human community, but are also virtues toward things and nature in general, since they can be conceived as life-preserving and health-enhancing powers of nature. On the basis of a metaphysical understanding of the environment, human ethics can be transformed into environmental ethics.

The environmental ethics of the *Tao* begins with an understanding of the *Tao*. After the *Tao* is understood, the *Tao* becomes the supreme end for man. Man has to treat nature as if there is an intrinsic end in nature. But the proper way of treating nature in this sense is to share and embody the *Tao* in nature; in other words, to achieve the supreme end of the *Tao* is to be with the *Tao*, in the *Tao*, and of the *Tao*. The environmental ethics of the *Tao* becomes an art of harmonizing with the *Tao*, as well as the art of self-realization in nature by way of the *Tao*. Here we may indeed say that our deontological duties toward nature or the environment come from nature or the environment itself. It may appear to be an imposition of nature on man independent of his understanding of nature. But when man comes to understand nature, the deontological considerations in the environmental ethics are no longer deontological. This is because the deontological becomes part of the teleology of nature; they represent what nature really is. It is in this sense that the environmental ethics of the *Tao* is not only founded on the basis of the environmental metaphysics of the Tao, but becomes one with it. Of course, we still can recognize that the metaphysics of the environment can produce a true picture of the nature of reality as an end.

We can now formulate four axioms of a metaphysics of the environment on the basis of our understanding of the reality as *Tao* process and *ch'i* structure. Then we can establish four corresponding principles of the deontology of environmental ethics on the basis of these four metaphysical axioms.

1. *The axiom of Self-Transformation.* The *Tao*, or reality as the *Tao*, presents itself in a process of temporal, spatial, material, immaterial, and relational transformations. There is no simple linear relation of cause and effect, but a manifold of levels and dimensions in organic relation.

2. *The axiom of Creative Spontaneity.* The creativity of reality consists in a natural process toward the emergence of life. Even though proper physical conditions for emergence of life are required, the emergence of life and its nourishing have to be conceived as inherent in a self-transforming movement of the *Tao* and *ch'i*. Life is not only part of the whole nature of *ch'i*, but it is vested in every particle and wave of the *ch'i*

3. *The axiom of Interpenetration.* All elements in nature are interdependent, interacting, and interpenetrating. There can be no real separation among them. Hence, there is the effect of the whole on the part and the effect of the part on the whole. No life process or creative effort can be achieved without taking into consideration the interpenetrating relationships of things.

4. *The axiom of Harmonization.* All elements in nature can maintain a relationship of harmony and balance on relevant levels of being and becoming. There is always an inherent harmony pertinent to a given level of relationship. But to achieve a proper harmony on the proper level, one has to relate to all elements in a proper way on that level and also to relate to all elements in a proper way on lower levels. Otherwise, a natural return to a lower level of harmony would appear as an ecological breakdown of harmony on a higher level.

Corresponding to these four axioms of nature there are four corresponding deontological principles in man's relationship to nature or his environment.

1. *The Principle of Harmonization.* Man has to preserve a sense of harmony with regard to nature. This sense of harmony should be articulated in his conscious efforts to love life and to appreciate what is good in life. This sense of harmony should not only be expressed in a sound conservations position on the ecology of nature, but also in man's well coordinated

strategies for cultural development. This is because nature and culture can enhance each other's value, and thus together enhance the value of life.

2. *The Principle of Interpenetration.* Man has to develop science and design technology with a totalistic sense of reality. Man must recognize that there are visible and invisible elements of nature, and that there are many concentric circles of *ch'i* structures which all can act and react on the well-being of man. Man has to incorporate his understanding of the interpenetration among things in his technological exploration of nature. This is essential for making technology not only humanized, but naturalized as well.

3. *The Principle of Creative Spontaneity.* Man has to learn to seek creativity in a spontaneous way so that he can contribute to the richness of life and harmony among things while fulfilling his own needs. In becoming creative, following the *Tao* will help insure that man's research and invention do not issue from a desire to dominate or result in a will to control. Many things that men have done in history have been a waste of life and energy. This can be avoided by following the creative spontaneity of the *Tao*.

4. *The Principle of Self-Transformation.* Man eventually has to learn to transform himself in accordance with the totality of reality. He has to see himself as embodying the *Tao*, and acting with a will not to will. It is only with this understanding that he is able to resolve conflicts within himself and conflicts between himself and the environment. His self-transformation consists in constantly interacting with nature so that he can be said to be part of nature and nature can be said to be a part of himself. It is in nature and the *Tao* that man should find his true self.

Whereas the four axioms of the environmental metaphysics of the *Tao* and the *ch'i* exhibit an order of ontogenesis, the corresponding deontological principles of the environmental ethics of the *Tao* and *ch'i* exhibit an order which returns to the source and embraces the totality. It is in this order of going back to the source and embracing the totality that the environmental ethics of the *Tao* and the *ch'i* transforms the metaphysics of the *Tao* and the *ch'i* not only into a teleology, but also into a deontology. The integration and unification of the teleological and the deontological in the metaphysics of the *Tao* and *ch'i* will eventually lead to a maturity of the environmental ethics of the *Tao* and *ch'i*, and it is in terms of this mature form of the environmental ethics of the *Tao* and *ch'i* that we will be able to speak of the embodiment of a new reason in the mind of man.

## Notes

1 J. Baird Callicott has traced the beginning of the origin of environmental ethics in some of his papers. See "Conceptual Resources for Environmental Ethics in Asian Traditions of Thought: A Propaedeutic," *Philosophy East and West* 37, no. 2 (1987) and "The Metaphysical Implications of Ecology," *Environmental Ethics*, this issue.

2 By *teleology* I actually mean the study of the metaphysics or ontology of ends or values for a given ethical concern. Therefore, the teleological corresponds to the metaphysical level of consideration. Similarly, the analytical corresponds to the metaethical level of consideration, whereas the deontological corresponds to the normative level of consideration.

3 Cf. Ernest Weekley. *An Etymological Dictionary of Modern English* (New York: Dover, 1967), p. 516 (*environ*), p. 1583 (*veer*).

4 *Sheng-sheng* is derived from the Great Appendix of the *I Ching*, sec 5. where it is said that "*Sheng-sheng* is called the change." "*Pu-yi*" is derived from the *Book Of Poetry* in "Chou Sung," where it is said that "The mandate of Heaven is indeed profound and incessant (*pu-yi*)." This stanza is quoted in *Chung Yung* to describe the depth and width of the reality of Heaven and Earth. It is quite clear that the incessant activity of life-creativity is precisely what the *Tao*[b] is. As a life-experience based concept of reality. *Tao* was university conceived by ancient philsophers as a universal process of change and transformation as well as the fountainhead of all forms of lift in the world. Therefore, *Tao* can be said to be in-depth foundation, background, and context of the so-called environment for any sentient being. We shall see a more metaphysical consideration of the *Tao* in the writings of Lao Tzu.

5 When I use the word *objectified*, I mean "being treated as an object"; when I use the word *mechanized*, I mean "being used merely as a machine"; and when I use the word *rigidified*. I mean "being placed in the state of *rigor mortis*; and when I use the word *de-enlivened*, I mean "being depleted of life and the living process"; when I use the word *dehumanized*, I mean "being given no consideration of human feeling and care"; finally, when I use the word *de-environmentalized*, I mean "being devalued as environment."

6 *Analects*, 15: 28.

7 *Chuang Tzu*, Chi Wu Lun.

8 See the Chih Pei Yu chapter of *Chuang Tzu*.

9 *Tao Te Ching*, 25.

10 *Ecology* originally meant the economy of nature: when nature acts, it acts ecologically. The production of life and all things in nature can be said to come from the ecological movement of nature. In understanding the ecology of nature, one would naturally understand the *Tao*; but only when one independently sees the universality, unity, and life-creativity of the *Tao*, will one truly understand the ecology of nature. Hence, the *Tao* can be said to be the metaphysical foundation of the ecology of nature, whereas the ecology of nature is one principle of movement manifesting the *Tao*, corresponding to its spontaneity.

11 Cf. *Tao Te Ching*, 37.

12 This principle can be indeed expressed as the following equivalence: *cheng-wu-wei* = *tzu-jan* = *sheng-sheng* = *wu-pu-wei*, i.e., the constant self-restraining of externality = spontaneity = life-creativity = the natural harmony of all things.

13 *Jan* ("reversion") refers to the fact that the *Tao* reverses what is done against the nature of things. But *jan* is also *fu* ("return"). If things are done according to their nature and of their own accord, things will return to their origin and their identity will be recurrently assured. See the distinction in the context of the *Tao Te Ching*, 40. "Reversion is the movement of the *Tao*." 25, " The distant is the reversion," and 16. "All ten thousand things take place concurrently. I observe their recurrence (*fu*) (*via* their origin)."

14 Cf. *Tao Te Ching*, 2; *Chuang Tzu*, Chi Wu Lun.

15 *Tao Te Ching*, 14.

16 *Tao Te Ching*, 21.

17 Cf. Ren Zhi-Yu, *Lao Ze Jin Yi* (Hong Kong: 1955). See also *Zhong Guo Zhe Xue Shi Ze Liao Suan Zhi, Xian Chin Zhi Bu Zhong* (Beijing: 1959), p. 451.
18 *Tao Te Ching*, 25.
19 In the *Tao Te Ching* there are two direct references to the term *ch'i*. Chapter 10 speaks of "concentrating on *ch'i* to reach softness," and in chapter 55 we find: "when heart-mind controls the *ch'i* it is called strength." It is clear that *ch'i* is the primary substance for the concretion of life, mental effort, and things. In the case of *Chuang Tzu*, although no full theory of *ch'i* was developed. Chuang Tzu has implicitly referred to *ch'i* in describing the piping sound of wind and nature in the Chi Wu Lun chapter. He even speaks of nature as "the great mass breathing the *ch'i*" (*ta k'uai yi ch'i*). It is again quite clear that *ch'i* is assumed to be the primary substance of things and the force of all movements in the totality of the reality identified as the *Tao*. The *Tao* and the *ch'i* can be said to refer to the same reality in different aspects: the macroscopic and the microscopic.
20 Cf. my unpublished paper, "On Transformation and Harmony: Paradigms from the Philosophy of the *I Ching*." Also see my paper, "Model of Causality in Chinese Philosophy: A Comparative Study," *Philosophy East And West* 26 (1976): 1–20.

96

# MAN AND NATURE

## Toward a middle path of survival

*David J. Kalupahana*

Source: *Environmental Ethics*, 8 (1986), 371–80.

I highlight the philosophical standpoints of two traditions, one from the East and the other from the West, that seem to avoid any form of reductionism resulting from the search for ultimate objectivity in human knowledge and understanding. I compare the pragmatic teachings of the Buddha and William James in order to show how both accommodate the human perspective as an inalienable part of the philosophical enterprise, and, further, how these perspectives contribute to their humanistic approaches and to the valuing of the environment in a way that is essential for human survival.

The question is being tossed around: "Can the East help the West to value nature?" One of the assumptions behind this question seems to be that in some sense the West has failed to appreciate or value nature. In fact, environmental ethics has become a major issue precisely because Western science has permeated human life on Earth, whether it be in the West or in the East, in the Northern Hemisphere or in the Southern Hemisphere. There is no denying that Western science has made life easier, more comfortable, and in some ways more satisfying. However, this satisfaction is gradually turning out to be a dissatisfaction, a nightmare for some people, especially those who have begun to realize the threat to natural phenomena posed by the increasing demands of the so-called technological world.

Since the development of science and technology in the Western world, although contributing to a more satisfying way of life, has simultaneously generated fear and trembling in our minds regarding the destiny of mankind, it is appropriate to examine the broad conceptual structures or paradigms, serving as the basis for scientific inquiries or emerging as a result of scientific explorations, that have been highlighted by the intellectual community, the so-called "lovers of wisdom." One such paradigm that has dominated classical, and to some extent scientific thinking, is the problem of *paired opposition* in nature. For example, we have the binary oppositions recognized by physicists, biologists, meteorologists, and others. Science, while recognizing such paired oppositions, nevertheless, wants to keep its world view open-ended enough to accommodate future discoveries, especially in the light of its perceptions of other phenomena that demonstrate no such binary oppositions. Thus, problems remain. Even after the complementary opposites are noted and the extreme binary oppositions are resolved, there are still questions regarding the value of these phenomena. For instance, science assumes that its attitude, approach, and methods are so objective and unprejudiced that it can openly claim that nonhuman nature has no value at all.

Neither the Buddha nor his followers can be considered scientists in the strict sense of the term. However, like all scientists and philosophers who formulate their experiences in conceptual terms, the Buddha as well as his followers provided conceptual formulations of what they observed as nature and human life. In these formulations, they carefully avoided three types of solutions to the problem of *paired opposition* referred to earlier, and which were proffered by some of the major philosophers of the East as well as the West.

The first recognizes the substantial reality of the paired opposites such as *yang* and *yin*, male and female, positive and negative, light and darkness, life and death, good and bad, etc. While assuming that these opposites are substantial, it admits the possibility of their reconciliation in a higher synthesis. According to this view, the sense experiences that provide knowledge of the opposites can still be valid. A more optimistic view of life emerges from this perspective. It resembles the position adopted by the Taoists as well as the thinkers of the earlier and more rationalist *Upaniṣads*. Cartesian dualism, with its emphasis on two distinct substances, mind and matter, along with its recognition of the conception of God as a possible unifying ground, may provide inspiration for such a thesis.

The second assumes the opposites to be less substantial and, therefore, *easily* reconciled or dissolved in a higher synthesis. Sense experience yielding knowledge of the opposites is here looked upon with suspicion. It is misleading and enjoys no significant epistemological status. The Advaita Vedānta school of thought, providing one of the philosophical justifications for the *Bhagavadgītā*, can be classed under this category. Advaita Vedāntins in the modern world have not failed to show their enthusiasm for Hegelianism, thereby indicating a similarity between these two traditions. Common to all these solutions is the recognition of an absolutism of one form or another.

The third type of solution to the paired opposition is similar to the first version referred to earlier in that it recognizes the substantiality of the opposites, but, instead of attempting to reconcile them in a higher synthesis, leaves them as inevitable, though intolerable, phenomena. It wholeheartedly advocates a pessimistic view of life by assuming the incapacity of the human being to resolve this problem, leaving its ultimate solution in the hands of an external power. In the Asian context, the dualistic view of Madhva, which has remained the more popular religious expression, seems to come closer to this form of solution. In the Western world, Kierkegaard embodies a similar way of thinking. The Jaina thinkers in India, while retaining a comparable conception of the opposites, replaced the nation of a supreme being with an extremely deterministic notion of karma which likewise led to a totally pessimistic view of life. The extreme notions of nonviolence (*ahiṃsā*) is an offshoot of this very deterministic theory of karma that leaves all responsibility squarely, yet unfairly, upon the individual.

Most interpretations of Buddhism in the modern world are based upon a conscious or unconscious recognition of one or the other of these paradigms. A distorted interpretation of the Buddhist conception of life emerged with the medieval Hindu thinkers like Udayana Ācārya, an interpretation that was responsible not only for the disappearance of Buddhism as a distinctive ideology from the Indian scene, but also for a legacy of misunderstandings and misinterpretations regarding this tradition in the Western world. Schopenhauer was one of the first to be misled by the Hindu interpretation of Buddhism, assuming that Buddhism advocates a notion of universal suffering and hence a tragic view of life. The paradigm that contributes to this absolutistic view about suffering was not examined in detail. The paradigm is the irreconcilable substantiality of paired opposites. It gives rise to an absolutistic view of suffering as embodied in the statement: "Everything is suffering" (*sarvaṃ duḥkhaṃ*). Neither such a paradigm nor a similar explanation of suffering can be found anywhere in the discourses discovered so far.

Confusion about this point continues even today. For example, Walter Kaufmann, who usually has rare insight into the nature of the Buddha's teachings, has made the following remark about Buddhism and Schopenhauer:

> One might have expected Schopenhauer to realize all of this [some points concerning Hume's theory of tragedy], since he stressed the universality of suffering more than any previous philosopher. But at this point he felt a kinship to Buddhism—the universality of suffering is the first of the first of Buddha's "four noble truths"—and Buddhism and tragedy represent two utterly different responses to suffering.[1]

Since the remark is cryptic and Kaufmann does not elaborate upon his observation, it is somewhat difficult to respond. However, the misunderstanding of Buddha's conception of nature and human life, in this case, seems to stem from the wrong identification of his philosophical standpoint with one or the other of the three paradigms referred to above.

The Buddha's conception of nature and human life carefully avoids the substantialist interpretation of paired opposites. Furthermore, the avoidance of substantialist conceptions of opposites does not mean either their negation or reconciliation in a higher synthesis. The philosophical middle path enunciated by the Buddha judiciously avoids the notions of substantial existence and nihilistic nonexistence in favor of an empirical explanation of the notions of arising and ceasing depending upon conditions.[2] In other words, the conception of paired opposites is retained, not reified. For the Buddha, abstract conceptions of masculin-*ity* and feminin-*ity*, light-*ness* and dark-*ness*, positiv-*ity* and negativ-*ity*, etc., have no meaning except in the context of the empirical events or phenomena conceptualized as man and woman, day and night, positive and negative, etc. To the Buddha, non-substantiality (*anatma*) was a methodological instrument, comparable to a surgeon's scalpel, for dissecting experience and conception in order to get rid of cancerous elements and to preserve the healthy ones. It was neither a butcher's knife nor a scaffold for the dismemberment, destruction, or annihilation of all human experience and conception.

To begin with, the Buddhist explanation of experience does not involve a passive intellect, a *tabula rasa*. Neither does it acknowledge a set of *a priori* categories of intellect or mind. A passive intellect is presupposed in order to recognize the possibility of knowing the object of experience "as it is," in its true and uncorrupted form. This is an extremely substantialist or essentialist approach. Similarly, an admission of *a priori* categories sets the stage for the discovery of laws of thought, which again are believed to be universal and incorruptible. The former tips the scale in favor of the object and the latter highlights the subject. Philosophical discourses based upon such assumptions lead to further paired opposites: realism-idealism, realism-nominalism, determinism-indeterminism, eternalism-annihilationism, etc. This is the background in which the search for metaphysical entities either in the object or in the subject has been proceeding. Until such discoveries are successfully completed human beings tend to remain in a state of anxiety and uncertainty. Thus, it is the task of the philosopher to satisfy this curiosity of human beings with well-constructed theories. The Buddha was not unaware of this human search and predicament, as is evident from the following statement:

> In this case, monk, it occurs to somebody: "What was certainly mine is certainly not mine (now): what might certainly be mine, there is certainly no chance of my getting." He grieves, mourns, laments, beats his breast, and falls into disillusionment. Even so, monks, does there come to be *anxiety* (*paritassanā*) about something objective that does not exist.
>
> In this case, monk, the view occurs to someone: "This world is this self, after dying I will become permanent, lasting, eternal, not liable to change, I will stand fast like unto the eternal." He hears the doctrine as it is being taught by the Tathagata or by a disciple of the Tathagata for rooting out all resolve for bias, tendency,

> and addiction to the determination and conditioning of views, for the appeasement of all dispositions, for the relinquishing of all attachment, for the waning of craving, absence of lust, cessation and freedom. It occurs to him thus: "I will surely be annihilated, I will surely be destroyed, I will surely not be." He grieves, mourns, laments, beats his breast, and falls into disillusionment. Thus, monks, there comes to be *anxiety* about something subjective that does not exist.[3]

The conceptualization about something that is not given in experience, merely because it helps us to overcome our anxieties about existence, did not appeal to the Buddha as a long-term solution to the human predicament. This does not mean that the completely discounted all such conceptualizations, for, as a pragmatist, he was not unmindful of some of the immediate benefits that may be enjoyed especially by the "tender-minded" as a result of utilizing some such conceptualization. However, as a "tough-minded" empiricist who was more concerned with immediate than with long-term benefits, he was not willing to admit metaphysical entities either in the objective world or in subjective life as ultimate realities. Substances or essences, permanent and eternal, whether these relate to objectivity or subjectivity could have no lasting value in his empiricist philosophy. He avoided the search for pure and undiluted sense impressions as well as universal and incorruptible concepts. The overcoming of anxiety is not to be achieved by the renunciation of sense experience and conception, but by the relinquishing of the metaphysical search itself.

Relinquishing a metaphysical search does not involve the renunciation of reflection and investigation altogether, for without these, human beings would be like a "sessile sea-anemone on the rock"[4] waiting for the waves to bring its nourishment. They need to abandon the search for mysterious causes and conditions and to understand experience in the *historical present* in order to deal with the future. On the basis of such an epistemological approach, the Buddha formulated his understanding of nature as well as human life.

Reflecting on the basis of past experiences, the Buddha maintained that all experienced phenomena, including human life, have dependently arisen (*pratītyasamutpanna*).[5] The use of the past participle to express this experience is significant. This is not an essentialist enterprise. Rather it is an explanation of experience comparable to that offered by a radical empiricist for whom "the flights and perchings" of a bird are equally relevant to an understanding of its movement.[6] Following upon this understanding of nature and human life, the Buddha formulated an abstract principle of explanation, which he referred to as "dependent arising" (*pratītyasamutpāda*).[7] This latter, expressed in nominal form compared with the previous use of the past participle, does not represent a "real" in its eternal way. On the contrary, it has to be constantly verified, for it is the product of limited experience, not "omniscience." Being the result of limited human experience, and requiring constant verification and revision, this radical empiricist approach needs to be coupled with a pragmatic method. For this reason, Buddha looked upon the conception of the fruit, the effect or the consequence (*artha*) defining the conception of a thing or an event or a phenomenon (*dharma*).[8] Hence, he emphasized the idea that a thing is not simply a thing (*dharma*), but a dependently arisen thing (*pratītyasamutpanna dharma*). Even though the term *dharma* came to be used without this qualification, whenever a definition of a *dharma* was called for or was needed, he immediately utilized this pragmatic definition.

*Dependent arising* is often referred to as *dharmatā*, which is the Buddhist term for *nature*.[9] Thus, the Buddha's discourse begins with an account of the world or the universe (*loka*). It does not presuppose the sharp dichotomy between human life and nature used by others to make the latter an absolutely irresistible force which human beings have to contend with, or an external object created for their pleasure and enjoyment. His was not an absolutistic explanation or theory of nature in which the lines are drawn sharply and distinctly so as to make the human being either a hapless object or the epicenter of the universe. A human being is part of nature. Like everything else in the teeming and dramatic richness of nature, he is dependently arisen or causally conditioned. He comes into being depending upon various conditions, contributes his share to the drama, and makes his exit. He is part of nature, that is, in a constant process of becoming (*bhava*), evolution (*pariṇāma*), and dissolution. This process of becoming or evolution is neither haphazard nor strictly teleological. The world is neither an empirical sand heap nor an ever-elusive absolute. Its dramatic richness lies in its concrete contextual fruitfulness.

This form of qualification has far-reaching philosophical consequences which are at least twofold. In the first place, it enables a person to renounce the search for mysterious substances, subjective as well as objective. Second, it facilities the dissolution of sharp dichotomies and paired opposites through a process of melting the solidified or reified concepts without having to eliminate them altogether. This includes the royal pair of dichotomies that has dominated most essentialist and intellectualist enterprises throughout human history, namely, the dichotomy between fact and value. It is the dissolution of this solidified distinction between fact and value that will enable us to perceive a significant relationship between nature and human life. Nature is not simply a brute fact to be contrasted with otherworldly human values. Abandon this monstrous solidified conceptual thinking, and soon human life, together with its most sublime moral values, will be seen as an inalienable part of mother nature.

In the Buddhist context, the term *dharma* signifies both fact and value.[10] *Dharmatā* is the dependent *nature* not only of the so-called "brute fact," but also of social, political, ethical, or even spiritual "value." It is noteworthy that in the Buddhist discourses the "naturalness" (*dhammatā*) associated with moral and spiritual behavior is illustrated by similes taken from physical nature, such as the healthy growth and development of plant life that receives appropriate nutritive conditions,[10] or the natural behavior of animals,[11] or even the innocence of a little baby lying on its back.[12] These are processes that evolve depending upon various conditions, one of which is "interest." Ignoring the important part played by "interest" in the evolution of nature as well as human life, intellectuals have very often adopted the perspective of an "onlooker," a perspective that leads to the distortion of the nature of mental life of human beings, not only giving rise to psychological fallacies, but also eliminating any possibility of recognizing some form of naturalism in mental life.

William James has the following to say about the perspective of the "onlooker" and the process of evolution:

> Survival can enter into a purely physiological discussion only as an *hypothesis made by an onlooker* about the future. But the moment you bring in a consciousness into the midst, survival ceases to be a mere hypothesis. . . . *Real* ends appear for the first time now upon the world's stage. The conception of consciousness as a purely cognitive from of being, which is the pet way of regarding it in many idealistic-modern as well as ancient schools, is thoroughly anti-psychological, as the remainder of this book will show. Every actually existing consciousness seems itself at any rate to be a *fighter for ends*, of which many, but for its presence, would not be ends at all. Its powers of cognition are mainly subservient to these ends, discerning which facts further them and which do not.[13]

Although James does not hold the view that physical phenomena and animals are possessed of consciousness, what is operative in human consciousness as "a fighter for ends" could not be very different from the so-called "instinct" in animals and "tendencies" in natural events. Another statement of James, furthermore, seems to reduce the sharp dichotomy between animal instincts and human interests:

> Each mind, to begin with, must have a minimum of selfishness in the shape of instincts of bodily self-seeking in order to exist. This minimum must be there as a basis for all further conscious acts, whether of self-negation or of a selfishness more subtle still. All minds must have come, by way of the survival of the fittest, if by no direct path, to take an interest in the bodies to which they are yoked, altogether apart from any interest in the pure Ego which they also possess.[14]

If there is any higher moral nature in human beings which in not found among the animals and physical nature, that moral nature should be founded on the recognition of self-interest, self-instinct, or the fight for survival on the part of man as well as nature. The Buddhist conception of compassion and nonviolence and the Confucian idea of benevolence are both based upon the recognition of such interest or instinct. The excesses of compassion and benevolence, on the contrary, are the results of ignoring the pragmatic outlook in life. The Jaina theory of nonviolence represents such an extremist version of nonviolence.

As mentioned earlier, the Buddha recognized that dependent arising is not the result of omniscience. Rather it is a conceptualization based upon limited human knowledge and understanding. It is the "big blooming buzzing confusion"[15] receiving some form of order and uniformity in the reflective (radical empiricist) and interested (pragmatic) human minds. However, this reflection and interest can be carried beyond their legitimate limits to produce a world that is artificial, rigid and substantial, not only in its conceptualization but also in the actual realization. Thus, more often we find ourselves placed in a world that is "dispositionally conditioned" (*saṅkhata, saṃskṛta*) rather than one in which simple interest prevails. According to Buddhism, it is the former, i.e., the dispositionally conditioned, that is the source of most human suffering (*dukkha, duḥkha*),[16] for even the most powerful and compassionate creator cannot keep up with the demands made by human beings. In James' words:

> ... the best simply imaginary world would be one in which *every* demand was gratified as soon as made.... It would need not only a space, but a time of *n*-dimensions, to include all the acts and experiences incompatible with one another here below, which would then go on in conjunction—such as our spending money, yet growing rich; taking our holiday, yet getting ahead with our work; shooting and fishing, yet doing no hurt to the beasts; gaining no end of experience, yet keeping our youthful freshness or heart; and the like.[17]

The Buddhist view of suffering is based primarily upon the realization that human beings crave such an imaginary world. The way out of this suffering does not consist of relinquishing every human interest and need, every form of human demand and every bit of human satisfaction. It is not the adoption of a life of abject poverty and self-mortification. It is only a call for the renunciation of a life of self-indulgence which is the unfortunate transformation of an attitude of "interest" into one of greed, lust, and thirsting. The middle path recommended is the "appeasement of dispositions" (*saṅkhāra-samatha, saṃskāropaśama*).[18] It is neither the annihilation of interest nor the transformation of interest into a state of craving or thirsting.

Appeasement of dispositions is synonymous with freedom (*nibbāna, nirvāṇa*).[19] In his very first discourse, the Buddha referred to two extreme forms of life adopted by human beings.[20] The first is self-indulgence which he condemned as low, vulgar, individualist, ignoble, and unfruitful. The second is self-mortification which is painful, ignoble and unfruitful. Avoiding these two extremes, he recommended a middle path that is both noble and fruitful. Realizing that what is "actually possible in this world is vastly narrower than all that is demanded,"[21] and utilizing such a realization to appease one's dispositions, the Buddha recognized the possibility of overcoming suffering and attaining happiness and tranquility.

Fruitfulness (*artha*) and nobility (*dharma*)[22] are the foundations of this middle path which is recommended for lay people as well as for those who have opted for a more strenuous religious life. Neither fruitfulness nor righteousness is achieved through the simple renunciation of pleasures of sense or of social life, for if that were the case, then Siddhartha could have attained freedom and happiness on the day he left home. His attainment of enlightenment and freedom came only after he realized the nonsubstantiality not only of all experienced phenomena (*dharma*), but also of views, conceptions, and ideas about such phenomena. Commitment and strict adherence to or grasping after ultimate truths leading to inflexibility in regard to one's beliefs are seen as the primary reason for bondage and suffering. One should be prepared to abandon one's cherished beliefs, whether these relate to physical or social or moral laws, if such beliefs are to cause harm to oneself and others. Revisionism is not a weakness so long as the criterion for it is fruitfulness and righteousness. "Even the ideas about what is good have to be relinquished, let alone those about evil," so declared the Buddha.[23] James struck a similar note when he maintained that "there is always a *pinch* between the ideal and the actual which can be got through by leaving part of the ideal behind."[24]

When abstract ideas are not elevated to the level of absolute truths, when fruitfulness is made a criterion for deciding what is important and relevant, when excessive greed and selfishness are renounced, in short when "possessive individualism"[25] is abandoned, it would not be difficult for human beings to adopt what James called the "moral equivalent to war." If there were to be an element of aggressiveness in nature and human life, it is the responsibility of the most advanced of species, namely, Homo sapiens, to utilize that aggressiveness in order to create an environment where it can survive without destroying itself as well as nature. When the Buddha admonished that "one should take oneself as an example and neither kill nor destroy,"[26] he was not speaking of other human beings only, but was referring to all forms of life (*sabbe*), that is, life in general, life on Earth that is related by way of mutual dependence.

About two and half centuries after the Buddha, the Indian monarch Asoka, who became a Buddhist disciple, adopted this "moral equivalent to war," not in the face of defeat in war, but at the height of success and glory. Even though he gained a convincing victory in the Kalinga war, immediately he renounced his warmongering and pursued a policy that was to lay the foundation for moral reawakening in the Indian continent in which all human beings, irrespective of their caste and creed, and all animal and even plant life were saved from unnecessary and wanton destruction and annihilation. "Live and let live" was the ultimate motto of this monarch. As is well known, it was this monarch that was responsible for the spread of the Buddha's message beyond the confines of the Indian continent leaving lasting impression in South, Southeast, and East Asian countries. In these regions one rarely heard of environmental problems, except those caused by natural physical events, until the introduction of modern commercialism and the resultant "rape of nature."

If the West has failed to appreciate nature, it is because it has made a determined attempt to retain its past conceptual heritage without paying enough heed to more innovative thinkers like those of the pragmatic tradition. Given the obvious similarities between my interpretation of the early Buddhist tradition and American pragmatism, an interesting alternative to borrowing from the East presents itself. Since an appropriate nonabsolutist philosophy has already been formulated in the West, there is really no reason for a Westerner to rush toward the East looking for answers to his environmental problems: such answers are already available to him right at his doorstep.

## Notes

1 Walter Kaufmann, *Tragedy and Philosophy* (Princeton: Princeton University Press, 1968), p. 339.
2 *Saṃyutta-nikāya* (PTS) 2.17.
3 *Majjhima-nikāya* (PTS) 1.139.
4 William James, *Some Problems of Philosophy* (Harvard: Harvard University Press, 1979), p. 39.
5 *Saṃyutta-nikāya* 2.25.
6 William James, *The Principles of Psychology* (New York: Henry Holt, 1908), 1:243.
7 *Saṃyutta-nikāya* 2.25.
8 *Majjhima-nikāya* 1.395.
9 *Aṅguttara-nikāya* (PTS) 5.313.
10 Ibid.
11 *Majjhima-nikāya* 1.324.
12 Ibid.
13 James, *The Principles of Psychology*, 1:141.
14 Ibid., 1.323–324.
15 James, *Some Problems of Philosophy*, p. 32.
16 *Majjhima-nikāya* 1.228; *saṃyutta-nikāya* 3.133; 4.401, etc. See also David Kalupahana, "The Notion of Suffering in Early Buddhism Compared with Some Reflections of Early Wittgenstein," *Philosophy East and West*, 27 (1977); 423–31.
17 William James, "The Moral Philosopher and the Moral Life," in *Essays in Pragmatism,* ed. Albury Castell (New York: Harper, 1948), p. 78.
18 *Majjhima-nikāya* 1.167.

19 Ibid.
20 *Saṃyutta-nikāya* 5.421.
21 *Essays in Pragmatism,* p. 78.
22 *Dīgha-nikāya* (PTS) 1.155.
23 *majjhima-nikāya* 1.135.
24 James, *Essays in Pragmatism,* p. 78.
25 See C. B. MacPherson, *The Political Theory of Possessive Individualism* (Oxford: Oxford University Press, 1962).
26 *Dhammapada* 129–30.

# 97

# "CONCEPTUAL RESOURCES" IN SOUTH ASIA FOR "ENVIRONMENTAL ETHICS"

*Gerald James Larson*

Source: J. Baird Callicott and Roger T. Ames, (eds), *Nature in Asian Traditions of Thought*: *Essays in Environmental Philosophy*, Albany: State University of New York Press, 1989, pp. 267–77.

It is a reasonable simple exercise for a South Asianist to address the subject of this book by setting forth a purely descriptive and/or analytic discussion of traditional South Asian metaethical positions that could be construed as "conceptual resources" for doing "environmental ethics." Such an exercise involves, in my judgment, identifying at least three basic positions.

First, it is possible to identify what a metaethicist might characterize as a perspective of "nonnaturalistic, intuitionist, noncognitivism," or in the indigenous idiom of traditional South Asia, the perspective of Mīmāṁsā, either of the Bhāṭṭa type or of the Prābhākara type. From such a perspective, what is good is not a natural property of the world. The good cannot be defined but is realized directly in intuition, and it cannot be dealt with in terms of the notions of true and false. What is good, rather, has to do with the direct, injunctive prescriptions set forth in the *Veda* about what we should do at certain crucial points in our lives. The Vedas do not tell us anything about the world, but they enjoin us to act in a certain way (or, in other words, the act-deontology of the Bhāṭṭas) or to follow certain rules (the rule-deontology of the Prābhākaras).

Second, it is possible to identify what a metaethicist might characterize as a perspective of "naturalistic, nonintuitionist cognitivism," or, in the indigenous idiom, the so-called "realist" perspectives in traditional Indian philosophy, namely, the Nyāya-Vaiśeṣika, the Jaina, and the early Buddhist positions. From such a perspective, what is good is a natural property of the world. The good can be defined and talked about in terms of the world in which we live, and it can be discussed in terms of what is true and false. Moreover, there are predictable consequences which provide the basic motivation for behaving in a moral way. The cultural trappings for such a position involve psychological hedonism (the *sukha-duḥkha* continuum), *karman* and *saṁsāra*, and *dharma* as a theory of obligation. It is a commonsense point of view given the presuppositions of the South Asian culture frame, and it is the most obvious moral position or moral theory to assume in such a context. As a normative position, it could be tagged as a kind of teleological cognitivism (or what Potter has called "path-philosophy").

Third, it is possible to identify what a metaethicist might characterize as a perspective of "nonnaturalistic, intuitionist a-moralism" (or what I like to call the "wild card" in the Indian deck), or, in the indigenous idiom of traditional South Asia, the perspective of Sāṁkhya, Yoga, Vedānta, and certain varieties of Mahāyāna Buddhist reflection (Mādhyamika and Yogācāra). From such a perspective, there is nothing that is truly of intrinsically good. To be sure, many things appear to be good either in terms of minimizing frustration (*duḥkha*), or in terms of maximizing more favorable rebirths, or in terms of sheer contemplation, or in terms of intellectual reflection. Serious discrimination reveals, however, that all determinate formulations or awarenesses of what is good prove to be temporary, limited, and most important, inextricably allied with dialectical modalities (*sukha-duḥkha-moha*) that undercut the perception, inference, or intuition that *anything* is intrinsically good. There may be some contributory or instrumental (or, in other words, extrinsic) "value" in the world in that it can point one in a certain direction or prepare one for ultimate insight, but, finally, the ultimate experience itself is *not* a moral experience—it is "beyond good and evil," or, putting the matter directly, it is the denial that moral and ethical theorizing has any value at all! As a normative position, one might tag such perspective with (the admittedly barbaric) expression "gnoseological intuitionism" or the claim that the ultimate experience is a non-moral or a-moral intuition that arises through an extraordinary modality of knowing.

Among these three traditional perspectives in South Asian theorizing, the first perspective, namely, the act-deontology or rule-deontology of Mīmāṁsā, is an unlikely candidate for environmental ethics, since its injunctions are based on a corpus of texts, the Vedas, the authority of which is barely relevant even in orthodox communities in modern India, quite apart from nonorthodox South Asian environments or environments altogether outside of South Asia. This leaves us, then, with the second perspective, or, in other words, the "teleological cognitivism" or the Indian realist traditions (Nyāya, Vaiśeṣika, Jaina, and early Buddhist thought), and the third perspective, or what I have called the "gnoseological intuitionism" of Sāṁkhya, Yoga, Vedānta, and Mahāyāna Buddhist thought. In terms of environmental ethical discussions, I am inclined to agree that the South Asian realist traditions (Nyāya, Vaiśeṣika, Jaina, and early Buddhist thought) could be employed fruitfully as a way of undergirding a "pragmatic" approach to environmental ethics, as David Kalupahana has forcefully argued in his paper "Man and Nature: Towards a Middle Path of Survival."[1] I am also inclined to agree that the South Asian "gnoseological intuitionist" views of Sāṁkhya, Yoga, Vedānta, and Mahāyāna Buddhist thought could be employed fruitfully as a way of establishing a somewhat new approach to "natural reverence," as Eliot Deutsch has argued in his paper, "A Metaphysical Grounding for Natural Reverence: East-West."[2] Moreover, J. Baird Callicott's critique of Deutsch has a South Asian answer, in my view. Callicott argues that the South Asian notion of Oneness is "substantive and essential . . . and the experience of it homogeneous and oceanic," whereas "in both contemporary ecology and quantum theory at their respective levels of phenomena the oneness of nature is systemic and (internally) relational."[3] Such a basic difference may be said to be valid in terms of Vendānta and Mahāyāna Buddhist accounts of "gnoseological intuitionism," but, as is well known to South Asianists, such a difference would not hold in Sāṁkhya and Yoga accounts. The notion of *prakṛti* as *triguṇa* (*sattva, rajas, tamas*) is clearly "systemic and (internally) relational" in Callicott's sense, and environmental ethicists could possible find powerful conceptual resources for developing "organic" and/or "holistic" perspectives on nature within the traditions of Sāṁkhya and Yoga in South Asian thought.

In any case, as mentioned at the outset, from a purely descriptive and/or analytic point of view, it is a reasonably simple exercise to proceed in the manner I have briefly been outlining by finding "conceptual resources" in Asian traditions for doing "environmental ethics." To be sure, we could debate whether I have been sufficiently precise in formulating the various South Asian options, and more than that, whether it is possible to identify additional perspective as well—for example, one could, I suppose, suggest "conceptual resources" from South Asian theistic, Tantric, or Śākta traditions, and so forth.

Overall, however, I am increasingly troubled and/or frustrated (*duḥkha*) by such purely descriptive approaches to comparative philosophizing as the ones I have been outlining and which are implicit in our work as a whole, and my frustration is both methodological and theoretical. My methodological frustration can easily be pinpointed. We appear to be using, albeit unconsciously, a particular metaphor that, in my view, is methodologically loaded and seriously misleading. If one substitutes the word "natural" for "conceptual" in the expression "conceptual resources," it becomes immediately apparent that we are using an economic metaphor in our undertaking. Since the eighteenth century, European nation-states have been utilizing Asia to supply a variety of resources: spices, tea, cotton, minerals, oil, natural gas, cheap labor, and hosts of other commodities. Now it seems that we are setting out again, only this time we are on the lookout for "conceptual resources." We appear to be using, in other words, an economic metaphor of raw materials. The needed "ideas" for environmental ethics are presumably in short "supply" in our own environment, but we recognize that there is an increasing "demand" for some new intellectual commodities.

What is methodologically loaded and seriously misleading about such an economic metaphor of raw materials is the corollary component of such a metaphor, namely, that we are not really interested in the raw materials in their natural state. We want, rather, to appropriate the raw materials so that we can use them for making what *we* want. We all know full well that the "ideas" and/or "concepts" that we need are not available directly in Asian contexts. They are deeply embedded in culture frames, kinship systems, traditional institutional frameworks, and so forth, from which they must be detached or "dug out" as it were and then imported into our own frameworks. Moreover, if they are to be utilized profitably, these "resources" will have to be processed, manufactured, mass-produced, and, finally, distributed. Of course, we recognize that the market for the eventual product is worldwide or global, and in that sense we can congratulate ourselves that what we are doing will subsequently benefit not only Asia but all people everywhere. This, of course, is exactly the rationale that the British used in India the Raj. Raw materials were purchased in South Asia at a remarkably cheap price; factories in England then processed and manufactured the raw materials into useful consumer items which were subsequently sold in markets in India and elsewhere for a significant profit (and taxed as well). The rationale for the whole process was that the British were actively developing the Indian economy!

The methodological point for our work is that when we proceed by using an economic metaphor in this fashion, we are committing ourselves to a comparative enterprise of external appropriation. Ideas and concepts come to be construed as "things" or "entities" that can be disembedded from their appropriate frameworks and then processed and made to fit into our own frameworks. Such a method for comparative philosophy is, in my view, one-dimensional, overly selective, forced, anachronistic, sociologically unsophisticated, and, perhaps worst of all, unpersuasive. Surely we can develop more sophisticated methodological approaches in our comparative philosophizing.

As already indicated, however, my frustration (*duḥkha*) is not only methodological but also substantive and/or theoretical. My substantive or theoretical concern can be expressed by referring to some comments made by J. Baird Callicott in his article, "Non-Anthropocentric Value Theory and Environmental Ethics." Callicott suggests that environmental ethics may be thought of in two quite different ways, either (a) a sort of subdivision of applied ethics (on analogy, say, with business ethics), that is to say as "an *application* of well-established conventional philosophical categories to emergent practical environmental problems"; or (b) a creative enterprise that "may be understood to be an *exploration* of alternative moral and even metaphysical principles, forced upon philosophy by the magnitude and recalcitrance of these [environmental] problems." Moreover, says Callicott, if understood in the later, more ambitious sense, the task of the environmental ethicist "is that of a theoretician or philosophical architect (as in Descartes' self-image)."[4] My substantive concern has to do with Callicott's way of putting the problem, namely, the environmental ethicist as a "theoretician or philosophical architect (as in Descartes' self-image)." To me as a comparativist, this way of putting the matter does not go far enough in identifying what is really at issue. It simply is a paraphrase of the basic problem, or, putting the matter somewhat differently, this very idiom is itself a part of the problem. The notion of the philosopher as a "theoretician or philosophical architect (as in Descartes' self-image)" is itself part of the conceptual framework that emerged in the seventeenth century for the first time and which included the rise of quantitative science, technology, manipulative reasoning, and the political economies of the emerging nation-state. It is a conceptual framework or mind-set that divides, classifies, quantifies, and distinguishes discreet "universes of meaning" (to use Thomas Luckmann's idiom).[5] Modern philosophy itself, in other words, as a distinct, separate discipline that analyzes, explores, and applies concepts, is symptomatic of a world view that causes "environmental pollution, the aesthetic degradation of nature, human overpopulation, resource depletion, ecological destruction and ... abrupt massive species extinction," according to Callicott.[6] My point here is not an ethical or moral one. I am not regretting what has happened since the seventeenth century, nor am I seeking to assign some sort of blame. My point, rather, is a theoretical one, and I am inclined to express it in the following Zen-like way: We have not understood the environmental crisis until we realize that there is *no* philosophical answer to it. If we seriously think that we can find "conceptual resources" in Asia and then work them into our own philosophizing, *and* that such an effort would have a serious impact on the environmental crisis, then we really have not understood the environmental crisis at all! Put differently, our effort is itself a part of the problem. We are spinning our wheels, and nothing at all will or can change. It is a bit like the Vedānta of Śaṅkara: cleverly tinkering with concepts that deny everything on one level while allowing everything to remain just as it is on another level.

My substantive and/or theoretical frustration, in other words, is that philosophy (including comparative philosophy) as conventionally construed in the modern world since Descartes cannot adequately deal with the environmental crisis. Rather it is part of the crisis and cannot itself be used as a way of dealing with the crisis. This entails, furthermore, that all of the other "divisions" of modern academia, namely, economics, sociology, political science, physics, mathematics, chemistry, the biological sciences, and so forth, cannot as separate "language games" or "universes of meaning" deal adequately with the crisis either. It is this very predilection in European intellectual history since Descartes (and its progeny) to divide and separate intellectual tasks, and, more than that, to isolate and professionalize cognitive pursuits from other dimensions of human functioning (for example, trade, commerce, kinship relations, sexuality, and so forth) that is itself reflective of our current environmental crisis. One way of putting the point is to suggest that what we thought was a Weberian process of "rationalization" (increased efficiency, goal-oriented behavior, and so forth) has turned out to be a Freudian defense-mechanism notion of "rationalization" and that, therefore, what we had anticipated to be a modern, sophisticated, efficient, and civilized world has, in fact, turned out to be yet another highly neurotic world system even more dangerous than earlier ones, since its capacity for self-destruction is global and species-wide. Nor do I think, let me hasten to add, that we can retreat into premodern religious visions or apologetically contrived modernist versions of those same visions. Such visions may provide solace or salvation or "release" from issues such as the environmental crisis, as indeed they always have, but they tend largely to be either question-begging alternatives to dealing with the environmental crisis or else, like philosophy and other isolated cognitive domains, themselves symptomatic of the crisis.

What is needed instead, in my view, is a radical reorientation of the manner in which we might construe the problem, and I would like to call attention to some fundamental considerations that should be kept in mind as we seek to determine a starting point for bringing about such a reorientation. I shall mention three such fundamental considerations which, if taken together, might provide a kind of prolegomenon to a serious discussion of the environmental crisis from the perspective of comparative studies. The considerations are hardly new, nor are they particularly controversial, but they are frequently lost sight of in contemporary discussion.

First, I should like to call attention to what Charles Hartshorne has characterized as "the fallacy of misplaced symmetry," a problem that Hartshorne finds in much of Asian thought (Śaṁkara, Nāgārjuna, and so forth) as well as in European thought (Bradley, Bergson, and so forth). In a recent paper, Hartshorne refers to the fallacy as follows:

> I find a common fallacy in these widely separated thinkers. A relation between two terms is really two relations, the one of A to B and, the one of B to A. Neither Bradley nor Nāgārjuna take this duality into account.... Past events are realities, not unrealities. This doctrine is missing both from Bradley's and from Nāgārjuna's account. Hence they clearly begged the question.... They overlooked time's arrow, the asymmetry of temporal relations. They committed the fallacy of misplace symmetry. Another way to put the matter is, both thinkers refuted only a static view of multiplicity. They spatialized time.... Relations of dependence in space are indeed symmetrical, but not those in time.

> Do effects depend on cause? Of course. Do we depend on our ancestors? Of course. Do causes depend on effects? Of course not. Do we depend on our descendants? Whatever our descendants turn out to be, we are what what we are. But it is absurd to say, whatever our ancestors were, we are what we are. Without them we would not have been, period. But our existing now is absolute fact that no occurrence or non-occurrence in the future can nullify. . . . It is the indispensable foundation of any tenable rationality. All our living implies a real past, immune to alteration, and a merely potential future, whose exact characters are in the process of being created, step by step, beginning now.[7]

Whether one would follow Hartshorne's asymmetry-of-relations argument all the way to his process metaphysics is, of course, debatable; but in this context I wish to press what I take to be a nondebatable claim regarding the environmental crisis, namely, that it is something totally new and that there is an asymmetrical relation between it and all earlier natural crises. The survival of life on earth as we now know it is seriously in jeopardy. To use a Sāṃkhya idiom from South Asia, we have reached the "curds" stage of an earlier "milk" stage in the unfolding of our natural habitat as a species, and we cannot wish away the "curds" stage by arguing for a symmetrical relationship between milk and curds. The curds can only become milk again at the time of the great dissolution (*mahāpralaya*), but, of course, when that happens, there can no longer be anything alive to enjoy the milk!

Second, I should like to call attention to what might be called the fallacy of disembedded ideas. Philosophy (and comparative philosophy) operates on the level of conceptual analysis, but we have learned from the history of Asian though (East Asian and South Asian) as well as from much of modern European thought (Marx, Weber, Freud, Wittgenstein, and so forth) that conceptual frameworks are always embedded in larger culture frames insofar as they are vital components for a "form of life." We have learned from Kuhn and Feyerabend that even our most treasured abstractions, that is to say, the conceptual apparatus of our modern science, exists or "has life" only in such environments ("paradigms," "frames," and so forth). In our comparative work we have discovered over three thousand languages on our planet and hundreds of "culture frames" and "paradigms." In our scientific work we have discovered that there is a statistical probability that perhaps as many as one million planets in our own galaxy alone are capable of supporting "intelligent life." Yet in view of all of this, we still ask ourselves if archaic notions as the "Tao," *brahman-ātman*, Allah, God, *dharma*, or whatever, all of which notions are derived from premodern, qualitative-science "frames" or "paradigms," can be disembedded, dusted off, and somehow utilized in dealing with the environment crisis. It is my inclination to think that we would do better, rather, as comparativists, to inquire into the manner in which ideas and/or concepts function in their respective "frames" and "paradigms" as a way of getting a handle on how our modern "concept clusters" might be generating and are being generated by the contemporary "frames" and "paradigms" in which we live.

Third, I should like to call attention to what might be called the fallacy of "the sovereignty of the subject." I take the expression from Foucault's *The Archaeology of Knowledge*, in which he inquires into the reasons for the difficulty of developing what he calls a "general theory of discontinuity."

> There is a reason for this. If the history of thought could remain the locus of uninterrupted continuities . . . it would provide a privileged shelter for the sovereignty of consciousness. Continuous history is the indispensable correlative of the founding function of the subject: the guarantee that everything that has eluded him may be restored to him; the certainty that time will disperse nothing without restoring it in reconstituted unity. . . .
>
> In various forms, this theme has played a constant role since the nineteenth century: to preserve, against all decenterings, the sovereignty of the subject, and the twin figures of anthropology and humanism.[8]

I am neither a post-structuralist nor a deconstructionist, but I think that Foucault's point is well taken in any properly framed discussion of the environmental crisis. I mentioned earlier that, using statistical probability, it has been determined that "intelligent life" may be present on as many as one million planets within our own galaxy. Using similar procedures in mathematical astronomy, it has been estimated that there may be as many as ten million planets within our galaxy that are able to support "life" in some sense. Our own planet is about 4.6 billion years old, and there is mounting evidence that some forms of life have been present almost from the beginning. Hominid forms have existed from about five hundred thousand years ago, but "civilized" life (in the sense of animal husbandry, agriculture, polished tools, pottery, some form of social life, and so forth) has only existed since about 6000 B.C.E (or, in other words, from what is usually called the "Neolithic" period in such areas as the ancient Near East, Mohenjo-Daro, and so forth). Cognitive and affective capacities sufficient to support "civilized" life, in other words, are quite recent from within the perspective of the "history" of life forms on our own planet, and the self-reflective images we have of ourselves as so-called "modern" hominids are a bit over two hundred years old. Remarkable strides have been made recently in such areas as brain physiology, biochemistry, mathematical astronomy, and so on, but similar strides strides have been lacking for the most part in theoretical formulations of the role and function of self-awareness in the hominid life-form. We still operate with "philosophies" and "psychologies" of self-awareness but have very little grasp of the evolutionary significance of such constructions. Putting the matter another way, we have very little understanding of the human need for or the evolutionary significance of the "meaning of meaning," and we appear to be stuck with interpretations of self-awareness that may have long outlived their usefulness. I suspect that future progress in dealing with the environmental crisis will have to address critically the problem of "the sovereignty of the subject" and to avoid the fallacy of assuming that what we think we are is in any sense an adequate, accurate, consistent, or clear measure of what we are.

The fly is still alive and well in the bottle. What I mean by alluding to Wittgenstein in this context is that we have not yet succeeded in framing our problem in a way that allows us to escape from a mind-set that is itself a part of the problem. Comparative reflection is, I think, helpful in enabling us to see this—in enabling us, as it were, to uncork the bottle.

Far be it from me to predict what will happen if the fly gets out of the bottle, but if we take seriously the sorts of considerations I have briefly been describing, namely, "the fallacy of misplaced symmetry," "the fallacy of disembedded ideas," and "the fallacy of the sovereignty of the subject," we as comparativists may well have much to contribute by way of addressing problems related to our present "environmental crisis." Let me conclude by programmatically mentioning at least a few of these contributions.

First, comparativists could be helpful in generating better metaphors for construing the problem. As comparativists, we know full well that an economic metaphor of "conceptual resources" with all of its negative connotations of exploitation and external appropriation, is not at all an apt way of thinking about conceptual frameworks. Conceptual systems or schemes are inextricably a part of a comprehensive way of life, and indeed are frequently symptomatic for helping to uncover the latent or hidden forces and/or processes that are operative in a given culture or society. In this regard, it seems to me, a biological science metaphor along the lines of comparative anatomy comes close to highlighting the problem of the environmental crisis. An arm is not a fin, and a lung is not a gill, but there are nevertheless interesting affinities between these pairs in concrete life forms over time. In a similar fashion, one conceptual framework is not identical to another, nor can simple substitutions be made between them, but nevertheless there can be illuminating comparisons about the manner in which conceptual frameworks or networks operate in larger, concrete social realities. Such a metaphor of comparative anatomy would focus attention on thinking or conception, not necessarily as a *cause* or a *prescription* for a society's or a culture's problems, but rather, as a *symptom* or a *diagnosis* regarding society's or a culture's self-understanding. Why is it, for example, that we tend to think of the environmental crisis as a "philosophical" or "ethical" problem when it is obviously so much more than that? What do our conceptual frameworks regarding "ethics" tell us about our Western historical experience, our "anatomy," as it were, as a culture or as a set of genetically related cultures? Classical Chinese thought is symptomatic of quite a different "anatomy," as is classical Indian thought, Eskimo thought, Islamic thought, and so forth. I am speaking here, of course, only in terms of metaphor, and I am not arguing for a naturalistic reductionism. I am simply suggesting that as a comparativist I find a biological science metaphor much more to the point of what is at issue in any interesting discussion of the environmental crisis. There are undoubtedly a number of other metaphors which could be helpful, and as comparativists we are nicely positioned because of our training to discuss them.

Second, comparativists could be helpful in encouraging more broadly based cross-cultural and interdisciplinary research in an area such as the environmental crisis. The comparativist in the course of training early along learns that one cannot simply do linguistics, philosophy, religion, anthropology, and so forth, nor can one isolate one's research in only one cultural context. As suggested earlier, in almost every instance, our Western scholarly specializations are little more than heuristic and historical divisions derived from the vicissitudes and power struggles of our own developing social reality. There again, "philosophy" (or any other division in the modern academy) is a a symptom of rather than a prescription for the environmental crisis. To look to "philosophy" for solutions to the environmental crisis is to acquiesce in the marginalization of the cognitive life that has been occurring in European and American culture since the seventeenth century. The time has come, perhaps, to criticize such compartmentalization as itself a major symptom of our current environmental crisis and to launch cross-cultural and interdisciplinary inquiries that bring together humanists, scientists, politicians, and the rest—inquiries that do not permit specialists to hide behind their narrowly conceived expertise.

Third, and finally, uncorking the bottle and allowing the fly to escape will surely have some intriguing political implications. If it is the case that conceptual schemes or networks make up part of the "anatomy" of every culture of society, and if it is the case that the "philosopher" must relinquish his privileged (and marginalized) isolation in order to begin to get a handle on the scope of the environmental issue, then this surely means that the conventional centers of power and wealth of our modern way of life will be seriously threatened and, as Marx has taught us, will hardly roll over and play dead. The committees for determining comfortable government grants, private fellowships, and legislative funding for higher education, all of which dispense their largess on the premises of marginalization and the maintenance of the status quo, will undoubtedly not be happy, and yet these structures of modernity are deeply implicated in the environmental crisis and cry out for criticism in any serious treatment of the issues. But by what authority would one issue the criticism or carry out a given program for political action? In the name of Reason? Civility? Justice? Fairness?

As comparativists, we know that such notions are construed differently from culture to culture and that one important factor in shaping the differences is the set of power relations that operate in a particular social reality. Hence, to begin rethinking a problem such as the environmental crisis in our modern world is at one and same time to begin reshaping the power relations in that world—in other words, the intellectual work is in an important sense a political act. To some extent, of course, this is a sort of Marxian point with the important difference, however, that unlike the Marxist, the comparativist has no platform of certainty, no "vanguard" status by means of which to pursue a political program. The comparativist is simply one more participant in an ongoing effort to attain some sort of reflexive grasp of what is happening to the human species in our own time.

To put all of this somewhat differently, the truly important task for the comparativist in the environmental debate is not to offer up non-Western, alternative "world views" for possible adoption, but, rather, precisely the opposite, namely, through comparative analysis to come to a more critical understanding of what it means to be human at a time when all of the old certainties of our Western and non-Western traditions have largely collapsed.

## Notes

1 David J. Kalupahans, "Man and Nature: Toward a Middle Path of Survival" (this volume).
2 Eliot Deutsch, "A Metaphysical Grounding for Natural Reverence: East-West" (this volume).
3 J. Baird Callicott, "Conceptual Resources or Environmental Ethics in Asian Traditions of Thought: A Propaedeutic," *Philosophy East and West* 37 (1987): 124.
4 J. Baird Callicott, "Non-anthropocentric Value Theory and Environmental Ethics," *American Philosophical Quarterly* 21 (1984): 299.
5 See Peter Berger and Thomas Luckmann, *The Social Construction of Reality: A Treatise in the Sociology of Knowledge* (Garden City, N.Y.: Anchor/Doubleday, 1967).
6 Callicott, "Conceptual Resources," p. 116.
7 Charles Hartshorne, "Śaṅkara, Nāgārjuna, and Fa Tsang," in Gerald James Larson and Eliot Deutsch, eds., *Interperting Across Boundaries: New Essays in Comparative Philosophy* (Princeton: Princeton University Press, 1988), 104–5.
8 Michel Foucault, *The Archaeology of Knowledge*, tran. A.M. Sheridan Smith (New York: Pantheon Books, 1972), p. 12.

# Part 7

# AFRICAN AND AUSTRALIAN PHILOSOPHY AND THE ENVIRONMENT

98

# WILDERNESS AND THE BANTU MIND

*G.W. Burnett and Kamuyu wa Kang'ethe**

Source: *Environmental Ethics*, 16 (1994), 145–60.

In the West, it is widely believed that, since Africans lack an emotional experience with romanticism and transcendentalism, they do not possess the philosophical prerequisites necessary to protect wilderness. However, the West's disdain for African systems of thought has precluded examination of customary African views of wilderness. Examination of ethnographic reports of Kenya's Highland Bantu reveals a complex view of phenomena that the West generally associates with wilderness. For the Bantu, wilderness is an extension of human living space, and through concerted social action rather than individual initiative, it is, or at least can be, dominated by society. Wildlife is unnatural and aliented from human society, which is natural. Because wilderness is, consequently, understood to be fearsome and hostile, it is not a place that can provide inspiration or self-actualization. Almost all forests have a special spiritual relationship with humankind, and some trees have a special relationship with God. Although traditional Bantu thought is contrary to a concept of wilderness as conserved, managed space filled with tourists and recreators, it does embrace a concept of wilderness as wildlands. The Bantu have gone to considerable length to develop an approach to wilderness that minimizes individual contact while requiring association with wilderness as a social activity. Population growth and want of vocational opportunities continue to thrust Highland Bantu into wilderness as a fundamental and traditional survival technique.

> *The game park in not an African notion.* The parks have always been unpopular with the people of Africa, and the governments of Africa are sufficiently responsive to popular feeling to avoid such disruptions where possible. It is likely that given free expression, *African opinion would be hostile to every assumption* [upon which African game parks are based].
>
> —Patrick Marnham[1]

* The authors thank Celia Nyamweru and Lisa Butler Harrington for their careful, critical review of early drafts of this paper.

## Introduction

Conventional wisdom professes that wilderness preservation, manifested most obviously by national parks, as they have evolved from America's frontier experience, is rooted in American romanticism and transcendentalism, and is so virtuous that it has been widely adopted around the world.[2] This model of wilderness conservation through park establishment intimates that for most Africans, parks are things done to, possibly *for* them, but certainly not *by* them. Furthermore, since Africans are not privy to the intellectual and philosophical traditions prerequisite to national parks, parks are things that they can neither be trusted with nor fully comprehend.[3] The undemonstrated hyphotheses that Africans lack a system of thought allowing for wilderness is likely based on Western contempt for African philosophy and religion generally. Happily, however, the growth of an independent African intelligentsia and American scholarship surrounding the integration movement have combined to reappraise the bleak assessment of African traditional philosophical and religious thought. The result has been a considerable elevation of its stature.[4]

On the basis of this reappraisal, we examine the thought of one group of people, the East African Bantu of Highland Kenya, for concepts that either support of condemn wilderness preservation in Africa. We have selected the Bantu of Highland Kenya for the simple reason that they are the African people with whom we are the most familiar. The Highland Bantu are, nevertheless, the majority of Kenya's civil service and political leadership; consequently, the major responsibility for creation and implementation of conservation policy is in their hands. How these people think, or might think, about wilderness is a subject that should concern global conservationists.

Our intent is to reconstruct Bantu thought about wilderness is a late precolonial context on the assumption that many elements of that thought still function in popular culture. For nearly a century, the Bantu have been exposed, and certainly with considerable effect, to Western thought and education. Nonetheless, a conservation ethic based on Western romanticism and transcendentalism is unappealing to Africans. Reexamination of Bantu thought about wilderness, therefore, is a likely place to begin searching for an African philosophy of conservation that will serve Africans and Africa.

## The Bantu: sources and methods

Because our data is the early ethnographic reports on the Highland Bantu, we are limited to a consideration of the Kikuyu, Kamba, and Luhya. These are large groups of people who were directly in the way of European penetration. Their numbers and location attracted both missionaries and bureaucrats, many of them both curious and well educated, interested in ethnography, and with a long residency among their clients.[5] The Highland Bantu are clearly linked by language and less obviously by history, culture and economy. They are all, to a greater or lesser extent, agriculturalists rather than pastoralists. The Kikuyu, living on the humid slopes of Mt. Kenya and nearby highlands, are agriculturalists with strong elements of pastoralism adopted from their neighbors. Spread across the plains southeast of Mt. Kenya, the Kamba are agriculturalists when and where local conditions have permitted, but living in an arid environment, they have depended more on pastoralism, hunting, and long-distance trade than other Highland Bantu. The Luhya of western Kenya are more agricultural than the Mt. Kenya people because their homelands are predictably wetter.

Our method was to examine the content of these ethnographic reports for what they reveal about the Bantu consensus of wilderness. We have centered on metaphysics, the people's world view, and to a lesser extent on their stories and traditions about wildlife and vegetation. Wilderness, either directly or indirectly, was not a popular topic of conversation between ethnologists and their informants. Furthermore, all Highland Bantu have lacked a central political authority or formal priesthood to dictate and protect orthodoxy. Because children were tutored in the traditions of their lineage, the adults reached a consensus arising from a long oral heritage; nevertheless, part of that consensus was that all were pretty well free to believe as they wished. While this portrait is not one of philosophical anarchy, it is one involving considerable freedom that eluded and deceived early observers. Informants spoke individually and only rarely about an incompletely understood consensus.

We have had, therefore, to infer much from material indirectly related to the topic; nonetheless, several themes emerge and they seem to be remarkably consistent among the Bantu groups studied. Although radically different from contemporary Western views of wilderness, these themes are not inconsistent with a wilderness ethic and are in some ways predisposed toward wilderness in ways that the West might envy. Operationally, however, these same features, many describing a frontier mentality, are decidedly hostile to wilderness under conditions in which there is a land shortage and a lack of vocational opportunities apart from the frontier.

## Bantu metaphysics

Europeans in early contact with Africans assumed the Bantu to be animists. Unfortunately, the derogatory term *animism* is not a clearly developed technical concept and, consequently, conveys little information. Although conventionally *animism* attributes a living soul to inanimate objects and natural phenomena, the term is implicitly associated with polytheism and idol worship.[6] The conventional definition inadequately describes Bantu concepts of being and the implication of polytheism and idol worship attributes inappropriate belief and behavior to the Bantu. Virtually all early missionaries and anthropologists recognized this problem, but were unable to deal with it effectively: full expression of Bantu ontology required the skill of African scholars.[7]

Although Bantu are rigorously monotheistic, it is generally understood that the Bantu see God as relatively removed from the world and as infrequently involved in human affairs.[8] Various explanations for this view have been posited: God, the Creator, having done His creating, simply removed Himself from the scene; God works His way through other forces closer at hand; because God, the Creator, is approached only on serious matters, one does not risk trifling with God. Nonetheless, it is now generally understood by African scholars that God, although remote and transcendent, is also immanent and all good. His qualities and attributes are to be emulated by humans, for the attempt raises humans above the beasts and makes them fully human.[9]

This God is not the recent invention of revisionist historians and African apologists. European travellers in East Africa, filled with preconceived notions, struggling themselves to stay alive, making ephemeral and often hostile contacts with numerous people, and lacking the language to explore esoteric theological matters, failed to perceive the one God that pervades all Bantu thought. Nevertheless, virtually every missionary who spent any substantial time with the Bantu came to recognize, and not always gracefully, the Bantu's fundamental monotheism.[10] Furthermore, God is the apogee and origin of a lively creation. For the Bantu, virtually everything in the universe has *vital energy* or *being* and interaction strengthens this being. The power to strengthen is measured in a hierarchy of five ontological categories: God, ancestors or spiritual beings, humankind, plants and animals, and non-biological things.[11]

Closely related to ontology is the issue of the Bantu's perception of time and space, subjects less well studied and understood than Bantu ontology. The Bantu view of time is controversial in that they have had a well-defined present and past but little concept of the future.[12] In this model, *sasa*, or "now," extends from present conceptions back to include the lives of all the dead who are still remembered by the living. In this period of time, the living and the dead interact regularly, the recently dead being more active in interaction than those long dead. When the last person dies who remembers a person already dead, the past, or *zamani*, begins. In the distant past, only truly heroic ancestors are remembered. Nevertheless, it contains several mythologically understood "events," at least one of which separates creation from the origin of a particular lineage or people, while another event marks the ever-changing boundary between *zamani* and *sasa*.

There is evidence that the Bantu idea of time lacks a concept of a distant future. The usual supporting evidence is that Bantu languages lack forms necessary to define the distant future.[13] This postulated view of the future is controversial, but what seems apparent is that Bantu lack a doctrine of eschatology and that they view time phenomenologically. Phenomenological time contrasts sharply with recent Western views of time as continuous and change as evolutionary.

Although concepts of space are often related to time, early ethnologists reflected little on the Bantu concept of space. Competition for space with Europeans quickly involved Bantu in controversies as they attempted to assert a concept of property, and consequently boundaries, similar to the West's.[14] It has been claimed that Africans in general perceive a plastic social space as if geography were drawn on a rubber map that can be distorted appropriately to suit changing socioecological conditions.[15] Nevertheless, the consensus of official inquiry into Kikuyu land tenure seems to be that, while Kikuyu kinship systems and tenure systems were closely linked, and space generally understood to be social, the Kikuyu did have a concept of private property and they were capable of drawing permanent boundaries in real space that are conceptually the same as boundaries in the West.[16]

## Bantu metaphysics and wilderness

An African view of space analogous to a rubber map would contrast African and Western views of property with several implications for African wilderness. The harshly demarked boundaries surrounding wilderness in the West would simply not be acceptable to people accustomed to dealing with a world on a rubber map. Upon the arrival of Europeans, Bantu, particularly Kikuyu, quickly became entwined in litigation in which it was in their best interests to assert a fundamentally Western view of space, including boundaries. Taking this acceptance of Western space at face value, further discussion of wilderness as offensive, because it bounds space, seems pointless (although bounding wilderness might still be offensive for reasons that have nothing to do with the concept of boundaries per se).

A view of time that is based on distinct events, for example, the boundary between *sasa* and *zamani*, rather than on evolution presents problems for any philosophy of conservation. Distinct events separate epochs, and epochs are without change; they are the responsibility of the Creator-Maintainer. If the environment is degrading, it is God's problem, and who are humans to interfere with His course. Understanding of time, and much more so epochs, in Bantu society is, however, speculative enough that no conclusion can be drawn relating time concepts to conservation or wilderness. The enthusiasm with which Bantu have engaged in soil erosion control and reforestation programs indicates that they do recognize change in their own time and have no scruples about engaging in interventionist practices that alter the course of change.[17]

Bantu ontological concepts, nevertheless, are highly instructive of their attitudes toward wilderness. Bantu ontology contrasts dramatically with anthropocentric and biocentric theories that have dominated Western discourse on wilderness.[18] In the anthropocentric theory, wilderness satisfies human desire, the sole locus of value. The biocentric theory emphasizes the preservation of nature for its own sake. The former theory allows humans to control wilderness absolutely for egocentric purposes while the later gives nature absolute control, but without aesthetic purpose. In either case the Western perspective results in wilderness now filled with parks, forest reserves, and game reserves. Such arbitrarily demarked wilderness falls into Bantu ontological categories four and five—plant, animal, and nonbiological things. These categorical views of wilderness are unacceptable to Bantu because wilderness is, and always has been, a portion of human habitation, a place of persistent human interaction. In fact, Bantu ontology combined with a frontier economy, renders any Western concept of wilderness tenuous at best, while opening the prospect of a truly Bantu recognition of wilderness.

## Wilderness as human space

Although Bantu understand God's omnipresence, certain places are more associated with God than others. Among these are high places and mountain tops. High elevations, as wilderness, were "sacred" by virtue of the fact that tropical crops were unadapted to cold not grow there. Even though the intoduction of such crops as cabbage, potatoes, carrots, and white maize have allowed Bantu to exploit higher, colder, and wetter environments, East Africa's highest elevations have remained well beyond any agriculture. The inclusion of many of these highest elevations in East Africa's park estate has, therefore, met with litte local opposition and has often won considerable support. The high mountain parks, for example, report fewer encounters with poaching and intrusions than do parks in the savanna.[19] Recent writers do not deny an association between the divine and high mountains and the mere title, *Facing Mt. Kenya*, by Kenya's first president, Jomo Kenyatta, represents more than a passive acceptance of the concept. There is nothing, however, to indicate that it ever occurred to any Bantu that God's association with high places called for collective or individual pilgrimage.

While some wilderness was well beyond the human province, most of it was nothing short of purely human. The Highland Bantu have several words to describe wilderness. Among the Luhya, for example, *ovulimu* is bush that has never been cultivated, but implies that it soon will be.[20] *Ovusaka* is riparian bush. *Eligyinga* is most likely equivalent to wilderness in the Western sense of the word, wild bush beyond any foreseeable human use. The word *ekiginga* is used equivalently to *eligyinga*, but implies arable land. *Ekyangalangui* is true waste land, useless to humans now and forever.

Among the Kikuyu, *githaka*, equivalent to the Luhya's *ovulimu*, is uncultivated land that will eventually be cultivated. There is no term among the Kikuyu equivalent to the Luhya's *eligyinga* and threfore no single term that is roughly equivalent to wilderness as the term is currently used in the

West. Instead, the Kikuyu distinguish between *mutitu*, uncultivated forest land, and *weru*, uncultivated open, dry bush land. Both are, like the Luhya's *ekiginga*, potentially arable and useful to humans.

Among the Akamba, *weu* has been translated as common grazing land and is understood to be the same in concept as the Kikuyu's *weru.*[21] Otieno *has, however, recently reinterpreted the term to mean "space" in the sense that is distinct from location.*[22] In this sense, *weu*, and by implication the Kikuyu *weru*, is "space" not yet imbued with the personality of "location," which is created by human activity. Space is wilderness; location is not.

Space as *eligyinga*, *weru*, or *weu* commands respect if not fear. African space is filled with many things quite capable of doing damage and the Bantu imagination further populated the wilderness with infinitely worse mythological creatures. Wagner has gone so far as to suggest that the Luhya bifurcated God (*Wele*) into two aspects, *Wele omuwanga*, or the good aspect, and *Wele gumali or evimbi*, the bad aspect.[23] *Wele gumali*, unlike *Wele omuwanga*, has definite preferences, the wild places. Wagner's interpretation is possibly extreme. Significantly, he symbolizes the ontology in terms that are literally "black" and "white." The ontology shatters the Bantu's fundamental monotheism and the widely agreed to characterization of God as *all* good and *only* good. Other observers of the Luhya have not confirmed Wagner's observations, and no concepts resembling those suggested by Wagner have been observed among other Bantu people. Nonetheless, Wagner's model clearly illustrates the Bantu's fear of wilderness.

Recorded folk tales make it clear that once humans were involved with a wilderness creature, it was capable of penetrating human location and consuming entire villages, something it did not do unless disturbed.[24] The usual way of becoming involved with a wilderness creature was to venture into the wilderness alone or in a small group. Almost inevitably, individual or small group adventures in the wilderness resulted from some particularly reprehensible social act or disobedience to authority white with a larger group in the wilderness. In a few instances, the individual is on a legitimate quest to solve a family problem that is itself unnatural of abnormal.

The folk tales teach some fairly obvious lessons. Contact with the wilderness is frightening for individuals and small groups. Individual contact with the wilderness results from anti-social, abnormal, and deviant behavior, and the consequences of the individual's behavior extend to the entire community. Fleeing to the wilderness provides no escape from social responsibility and admonishment—there is no room for Robin Hood in Bantu folklore. On a more subtle level, however, the Kikuyu folklore teaches that there is nothing so marvelous in the wilderness that humans are not preponderate. In no instance in any recorded folk tale does the human fail to prevail against the mythical forces of the wilderness. The message is clearly that society is superior to the wilderness and will, or at least can, dominate it.

Because wilderness beasts were overwhelmed by numbers, wilderness itself was best approached in large groups. Among Bantu hunters. an economically specialized minority, the wilderness was a source of meat, hides, and ivory. For the Kikuyu, the wilderness provided *riumba*, a white soft-stone substance used to decorate houses. Also, the Kikuyu used wilderness as a training ground for soldiers, and the Kamba used wilderness for retreats during male-initiation rites. For nearly all Bantu, wilderness provided honey and a source of building material. All of these uses were social and demanded at least group approval and generally group involvement.

More importantly, the wilderness provided a place for expansion by people that were frontiersmen and expanding rapidly.[25] While small areas of high elevation were beyond human economic exploitation, the words used to describe wilderness embroiled it in the settlement process and pattern. The Kikuyu around Nyeri cleared forest systematically and socially with an effectiveness of considerable anxiety even to themselves.[26] Kinship and lineage systems affected land ownership and the process by which new land was brought into production.[27] Among the Kamba, for example, the *weu* was first common grazing land and a place to hang beehives.[28] The building of a cattle post, *kienga*, in the *weu* established exclusive rights. In addition, land could be inherited from the group by demarcating land in the *weu*. This demarcation became the settlers *ng'undu*, the arable land surrounding their homesteads, *musyi*, upon which they founded their lineage.

The approach to the frontier, and therefore to the wilderness beyond it, was a social process requiring set procedures, ritual and group approval, and, just as on the American frontier, a belief that wilderness is subservient to human desires and needs. The important difference has been that Americans indulged the myth that individuals conquered the wilderness while the Bantu understood that the conversion of wilderness was social. Although the American myth seldom existed, when events closed the frontier and the wilderness was filled with parks, forests, and formally designated wilderness, no one was deprived except would-be frontier men. The closing of the American frontier converted wilderness from space where transcendent individuals strove with eternal forces, to locations where society plays with transcendentalism. The process won democratic legal approval. In contrast, the closing of the Bantu frontier was by fiat, instantly stopping an important social process. Whole peoples were denied opportunities as wilderness was converted from social space to the domain of beasts, a tourist's pleasuring ground. Bantu society has found it difficult to come to terms with such a radical and precipitous change in social organization.

## Wildlife in Bantu society

Attitudes toward wildlife, rather than attitudes toward wild space, may form a basis for wilderness philosophy; yet, the Bantu approach to wildlife is much different from the West's. A Kikuyu folk tale, reported by Kenyatta, provides a metaphor that symbolizes this difference.[29] In the tale, which is intended to explain the origin of iron smelting, animals are divided into two groups, those given by God to men and those given to women. While the men's herds grew, women began slaughtering their animals, but lacking iron, the slaughter was difficult and painful. The women's animals rebelled and wend wild. God was consequently approached about the problem. He provided iron so that slaughtered animals would die quickly and painlessly.

Kenyatta's story contrasts dramatically with the West's biblical story. In one biblical version, God creates living things, and only afterward creates mankind and gives them dominion.[30] Humans are clearly created after the rest of nature, and there is a fairly clear distinction made between wild and domesticated things god created. In an alternate version of the story, God creates humans, then creates Eden and places them in it.[31] Willfully rejecting a state of nature, the humans alienate themselves from Eden.[32] In either case, the myth presents natural things apart from humans by virtue of God having made them that way or humans having willfully alienated themselves. In the Kikuyu version of the story, in contrast, God creates a single social unity consisting of humans and animals, and it is the wildlife that elects to leave that social system. The human social system is natural, and it wildlife that is alienated and unnatural.

The Highland Bantu of East Africa are agriculturalist with a prominent pastoral component. The were apparently never hunters and most groups have little memory of a hunting tradition. By the time Europeans arrived among the Luhya, wildlife was already much depleted and hunting was insignificant.[33] The remnant of hunting in the forest margins of Mt. Elgon and the Nandi Escarpment required increasing individuality and stealth as the British imposed their game laws. The Luhya use the term *ovululu* (fierceness) to describe both the ferocious animals and the spirits of the recently dead.[34]

For most Kikuyu, hunting, or even killing wildlife except in defense of home and fields, has been forbidden, and eating of bush meat remains unquestionably repugnant.[35] Selected clans, those tracing their lineage back to groups with a hunting tradition, continued to hunt, and they became economic specialists providing others with needed wildlife products through trade.[36] In 1908, skins and ostrich feathers were commonly available in Kikuyu markets in the Nyeri area: monkey skins, several kinds of cat skins, and ostrich feather headdresses were traded for two rupees, equal to about ten pounds of honey.[37] What is noticeably absent in the comments of the time is any observation of reverence for wildlife. Folk tales and proverbs make little serious comment on wildlife except to examine the consequences of mixing wildlife and humankind.[38] The impression given by early observers is that the Kikuyu had no regard for wildlife and devoted considerable effort to building barriers to human-wildlife associations.

The Kamba, living in the driest areas of the Highlands, augmented their precarious agriculture by hunting that included the search for ivory, horn and bush meat. This hunting, however, was ritualized group process even to the extent that arrows required established clean markings.[39] Although successful hunters acquired status, hunting was not a means of self-actualization. The Kamba had a totem system that was obscure, difficult to identify, and not involved in ritual.[40] Even though the totem belonged to the clan, it was not the clan ancestor, and it passed from one generation to the next by virtue of inheritance. A mystical relationship existed between the individual and the totem. The totem was not killed except in self defense, and it was never eaten. Kindness toward the totem was expected, and in turn, it was expected that the totem might assist in times of need or desperation. Undoubtedly, many Kamba spent considerable time in the study of the clan totem, but doing so was a matter of individual election. The totem system did not require the clan member to establish a personal relationship with the totem, much less with any other species or group of animals. The clan members were thought to share personality characteristics with the totem, but again this association was inherited rather than acquired and required no active engagement with the totem.[41]

The intrusion of any of several animals into human space called for special caution and reverence among all Bantu.[42] When frogs, birds, and certain snakes, particularly pythons, appeared on the homestead, they were tolerated if not encouraged. However, the custom was not based on the characteristics of the animals, but on the likely association of their odd behavior with the presence of ancestor spirits. In the fields, they could be killed as readily as any other animals. The chameleon, in several stories reputedly responsible for the creation of death among humans (largely as a result of human insensitivity), is still treated cautiously by many Kikuyu and Luhya.[43]

The approach of the Bantu to Africa's wildlife was hostile. Africa's wildlife has been numerous, quite harmful to individuals, and seemingly capable of overwhelming society.[44] Both the Bantu and the European settler could agree in their hostility to wildlife, but the two peoples resolved their hostility through different mechanisms. While the European killed wildlife in number and without mercy, the Bantu built a structure that attempted to minimize association between humans and wildlife, and if this effort failed, the problem reverted to society except in cases of immediate self-defense. The hypocrisy of lectures on the glories of wildlife and the virtues of its conservation from a people capable of killing animals for self-actualization, sport, and pure vindictiveness, surely eludes Europeans in ways that are dreadfully obvious to Africans.

## Vegetation and Bantu society

If wildlife was an anathema to Bantu social order, vegetation is an entirely different matter. Although wildlife could generally be killed without a second thought, the clearing of forest presented innumerable problems. The pioneer would not butcher the forest without taking pains to relieve the spiritual dilemma created.[45] The usual explanation is that because trees was associated with spiritual beings, the pioneer left the forest giants where the spirits could accumulate and find safe abodes. When age and events overtook even these giants, special precautions in form of rituals were required to assure that the spirits could find new homes or were at least reconciled to the inevitable. This myth undoubtedly explained or excuses the inability of pioneers technically incapable of felling giant trees. Subsequently, the death of any tree capable of defeating human aspiration was undoubtedly worthy of special note whether among Bantu or Europeans.[46]

Certain trees maintained a special association with God. Although observers have attempted to explain why some trees had this proclivity and how they were identified, no explanation is satisfying.[47] Sacred trees were all in the same group, *Ficus*, particularly aggressive and capable of achieving elegant beauty. But why one individual would be mundane and another sacred remains mysterious, beyond the observation that social agreement and experience favored some trees over others. However they were identified, sacred trees were taken seriously. Not only was the tree sacred, but so was the land around it and everything it sheltered. The group dared to approach God about matters and events particularly troublesome to the group only at the sacred tree. Although it did happen, few outsiders were invited, or tolerated, as the group approached God at a sacred tree, which may explain why Bantu monotheism is so obscure.[48] The sacred tree was never wilderness, however. It was not a splendidly isolated, distant object, but rather a near neighbor.

There is much to suggest, however, that the Bantu viewed the forest and its proximate fate with some anxiety. The Kikuyu around Nyeri could be engaged in conversations on this topic with ease at a time and in place where the recently introduced colonial forest administration could not have been locally appreciated.[49] Furthermore, blacksmiths, arcane specialists of considerable economic importance and mystery, having special need for wood, found it necessary to reserve timber to meet their needs, and they protected these trees with potent enchantment.[50]

## Toward a Bantu philosophy of wilderness

Observation of different Bantu groups by early perceptive Westerners, or subsequently, by African and trained Western scholars varies substantially, a fact limiting contemporary comparison and generalization. In this case, available sources have restricted us to the Kikuyu, Kamba, and Luhya. Only one observer has made extensive comment on the Luhya and, while he is both prolific and careful, the nature of Western anthropology at the time of his work requires that some of his observations be approached cautiously.[51]

The Kamba, and particularly the Kikuyu, have been studied carefully by a number and variety of observers. There is a surprising consistency in these observations, and viewed comparatively, they suggest a surprising philosophical accord between the Kamba, Kikuyu, and the Luhya.

That the three groups have maintained an essentially untainted indigenous philosophy, and consequently a philosophy toward wilderness, through the colonial and post-colonial period is unlikely. However, we do believe, on the basis of our experience, that an essentially indigenous philosophy endures, even prevails, and we believe, on the basis of their actions, that many others believe so as well. For example, efforts of instruct Kenyan school children in Western wildlife and wildland values and ideas is predicated, at least in part, on an assumption of an indigenous attitude to wilderness that is unacceptable to the West.[52] This approach presumes to interject a Western perspective before the child internalizes an African perspective acquired from home and community. Efforts to instruct Africans in Western wilderness values have proceeded with little, or no, articulation of how Africans might already understand wilderness, and consequently, what ideas the interventionist seeks to change. It would be far easier and produce fewer conflicts if the Bantu concept of wilderness could be articulated and developed as an indigenous philosophy of wilderness.

The Bantu approach to wilderness can be summarized as: (1) The wilderness is not a source of spiritual self-realization or inspiration for individuals or groups. (2) With rare exceptions, wilderness is conceived of as essentially an extension of human living space or location. (3) Wilderness will be, or at least can be, dominated by humans through concerted social action. (4) Wildlife is unnatural, a portion of God's creation that is alienated from human society, which is natural. (5) Wildlife is only an objective manifestation of a wider variety of unnatural or supernatural beings that inhabit wilderness and compete with humankind for control of wilderness. (6) Consequently, individual experience within wilderness-like conditions is likely to be harmful to the individual and the group that the individual belongs to.

Taken at face value, such a set of ideas place humans and wilderness in hostile and direct competition and accord poorly with wilderness as a place to be conserved by packing it with designated wildlands, parks, and game reserves. On closer examination, nevertheless, the ideas accord very well with wilderness as wildlands. The West, after all, argues that wilderness must have utility and that utility justifies wilderness. Hurly-burly produces cash flow and the profits from that flow pays for wilderness. The Bantu have no such needs, and, in fact, it is to be wondered if Western attempts to justify wilderness by filling it with Western style activity, thereby gaining from it Western-style profits, defiles wilderness in the Bantu mind. For the Bantu, wilderness simply *is*. In time, humankind can dominate it if humankind so wills and the need arises. In the meantime, it is to be approached cautiously, if at all. There is little to be gained by driving through it, walking through it, photographing it, or playing in it. It can remain authentic, vicious, and wild so long as its viciousness stays out of human location. Given this preception and the social nature of wilderness, parks the use zoning and permit at least some traditional uses such as the multiple-use land of the U.S. Forest Service, might be more popular in Africa than the bastion model of the national park—provided that Africans can also solve some of their demographic and economic problems.

Had the West never intruded itself into Bantuland, the problem of wilderness conservation would have eventually presented itself. Imagine a council of elders gathering to consider the question, "Shall we consume or conserve our now much depleted *weru*?" How the council might have acted pales compared with the fact that the question never had the chance to arise. Western intrusion stopped slapdash an important social process that allowed excess population to convert wilderness to human space and to acquire status, wealth, and security for their efforts. The sudden halting of this process would have been reason enough to resent the formalization of wilderness, but European intrusion compounded the matter by allowing rapid population increases and no economic substitute for the frontier except increasingly marginal subsistence farming and urban poverty. Western wilderness symbolizes the disasters engendered by Westernization while the Western solution, to raise the Gross National Product by exporting landscape views,[53] is immaterial to those who have no alternative but to survive on the land where they stand.

## Notes

1 Patrick Marnham, *Dispatches from Africa* (London: Abacus, 1981), p. 8; emphasis added.

2 Throughout this paper, wilderness is equated with formally designated national parks and equivalent reserves as identified by the International Union for the Conservation of Nature and Natural Resources, *IUCN Directory of Afrotropical Areas* (Gland, Switzerland: IUCN, 1987). The model is harsh, but there is little "wilderness" outside formal national parks in Africa. While other conceptualizations of wilderness are available, this paper is based on the work of Roderick Nash, "The American Invention of National Parks," *American Quarterly* 22, no. 3 (Fall 1970): 726–35; Alfred Runte, *National Parks: The American Experience* (Lincoln: University of Nebraska Press, 1979), pp. 1–47; Roderick Nash, *Wilderness and the American Mind* (New Haven: Yale University press, 1982), pp. 44–66, 84–95; and Richard Grove, "Early Themes in African Conservation: The Cape in the Nineteenth Century," in David Anderson and Richard Grove, eds., *Conservation in Africa: People Policies and Practice* (Cambridge: Cambridge University Press, 1987), pp. 21–39. The wisdom of exporting the American national park is not accepted by all observers. See, for example, David Harmon, "Cultural Diversity, Human Subsistence, and the National Park Ideal," *Environmental Ethics* 9 (1987): 147–58, and Ramachandra Guha, "Radical American Environmentalism and Wilderness Preservation: A Third World Critique," *Environmental Ethics* 11 (1989): 71–83. The IUCN and World Wildlife Fund have long sought methods of

softening the fortress aspect of the national park. A recent collection of papers edited by David Western and Mary C. Pearle, *Conservation for the Twenty-First Century* (New York: Oxford University Press, 1989), suggests that Third World parks must undergo radical evolution or they will not survive the next century.

3 G. W. Burnett and Richard Conover, "The Efficacy of Africa's National Parks: An Evaluation of Julius Nyerere's Arusha Manifesto of 1961," *Society and Natural Resources* 2 (1989): 251–59; R. H. V. Bell, "Conservation with a Human Face: Conflict and Reconciliation in African Land Use Planning," in Anderson and Grove, *Conservation in Africa*, p. 81.

4 See, for example, Placide Temples, *Bantu Philosophy* (Paris: Presence Africaine, 1959); Kwame Gyekye, *An Essay on African Philosophical Thought: The African Conceptual Scheme* (New York: Cambridge University Press, 1987); and Paulin J. Hountondji, *African Philosophy: Myth or Reality* (Bloomington: Indiana University Press, 1983).

5 The Origin and classification of the Highland Bantu are summarized in Robert M. Maxon, *East Africa; An Introductory History* (Nairobi: Heinemann, 1989), pp. 63–73, 93–100. Our initial study group was the same as that listed by George P. Murdock, *Africa: Its Peoples and Their Cultural History* (New York: McGraw Hill, 1959) pp. 342–47, except that we did not include the Pare, associated Taveta, and the Shambala, since they are predominantly in Tanganyika, or the Teita, who are generally associated with the Coastal Bantu. Available sources, for all practical purposes, limit our discussions to the Kikuvu, Kamba, and Luyha.

6 William Lithe, *The Oxford Universal Dictionary on Historical Principles*, 3d ed. (Oxford: Clarendon Press, 1964), p. 68.

7 Paul Bohannan and Philip Curtin, *Africa and Africans*, 3d ed. (Prospect Heights, Illinois: Waveland Press, 1988), pp. 191–207, ascribe monotheism to all African traditional religions, Bantu and otherwise (p. 191), and examine Western myths about African religion. The first African to examine the God concept among any African people was likely Jomo Kenyatta, *Facing Mt. Kenya* (1938; reprint ed., New York: Vintage Books, 1965), pp. 222–58, and the topic is thoroughly developed by J. S. Mbiti, *Concepts of God in Africa* (New York: Praeger, 1969), although a more manageable version of this essay is found in J. S. Mbiti, *African Religions and Philosophy* (Garden City, N.Y.: Anchor Books, 1970), pp. 37–49.

8 Bohannan and Curtin, *Africa and Africans*, p. 195; P. P. Cayzac, "La Religion des Kikuyu (Afrique Orientale)," *Anthropos* 5 (1910): 310.

9 Mbiti, *African Religions and Philosophy*, p. 33; E. Bolaji Iiowu, *African Traditional Religion* (New York: Orbis Books, 1973), pp. 140–64.

10 For example, W. Scoresby Routledge and Katherine Routledge, *With a Prehistoric People: The Akikuyu of British East Africa* (London: Edward Arnold, 1910), pp. 225–28, identify the Kikuyu as monotheists without question while the Roman Catholic priest C. Cagnolo, *The Akikuyu: Their Customs, Traditions and Folklore* (Nyeri, Kenya: The Mission Printing School, 1933), pp. 26–27, is positively enthusiastic about the characteristic. Cayzac, "La Religion des Kikuyu," p. 309, however, grants the Kikuyu two gods, "celui des Kikuyu" and "celui des Kamba et des Massai." The observation must be viewed with suspicion since Cayzac thinks of the Kamba and Massai as enemies of the Kikuyu. He also names none of the gods except as "black" and "white." Massai and Kikuyu generally use the same name for God, N'gai. Gerhard Lindblom, *The Akamba in British East Africa: An Ethnological Monograph*, 2d ed. (1920; reprint ed., New York: Negro University Press, 1969), p. 249, emphatically asserts a single God among the Kamba. Working with the Luhya, however, Gunter Wagner, *The*

*Bantu of Western Kenya with a Special Reference to the Vugusu and Logoli*, vol. 1 (1949; reprint, ed., London: Oxford University Press for the International African Institute, 1970), p. 167–77, admits the Luhya's monotheism but only grudgingly and with considerable qualification (for example, note 23). Charles Miller, *The Lunatic Express* (New York: Ballentine Books, 1971), pt. 1, pp. 13–272, ably portrays the intellectual and psychological capacities of the many early "explores" to intrude on the Highland Bantu.

11 Mbiti, *African Religions and Philosophy*, p. 20.

12 Ibid., pp. 19–36; Cagnolo, *The Akikuyu,* pp. 191–201, also discusses time among Kikuyu specifically though less philosophically.

13 Mbiti, *African Religion and Philosophy*, pp. 22–23.

14 The struggle over space has been particularly keen for the Kikuyu and this is the central theme for Kenyatta, *Facing Mt. Kenya*. The plethora of official inquires that the Kikuyu-European land conflicts generated are briefly reviewed and cited in Godfrey Muriuki, *A History of the Kikuyu, 1500–1900* (Nairobi: Oxford University Press, 1975), pp. 5–13.

15 Bohannan and Curtin, *Africa and Africans*, pp. 129–38.

16 See, for example, Kenyatta, *Facing Mt. Kenya*, pp. 26–27.

17 For example, L. A. Lewis and L. Berry, *African Environments and Resources* (Boston: Unwin Hyman, 1988), pp. 263–72. Asenath Otieno, "Soil Erosion in Ukambani: A Geographic and Historical Perspective" (M.A. thesis, Ohio University, 1984), in her general review of soil erosion among the Kamba identifies no philosophical resistance to environmental intervention.

18 Jay Hansford C. Vest, "The Concept of Wilderness: A Proprietary Right over the Land," *Western Wildlands* 17, no. 2 (1991): 2–6.

19 IUCN, *Directory of Afrotropical Areas*, pp. 249–314.

20 Wagner, *The Bantu of Western Kenya*, vol. 2, p. 76.

21 John Middleton and Greet Kershaw, *The Central Tribes of the North-Eastern Bantu, pt. 5 of The Ethnographic Survey of Africa, ed. Daryll Forde (London: International African Institute, 1965)*, pp. 77–78.

22 Otieno, "Soil Erosion in Ukambani," pp. 32–35. Her revisions are based on conceptualizations by Y. F. Tuan, *Space and Place* (Minneapolis: University of Minnesota, 1981).

23 Wagner, *The Bantu of Western Kenya*, vol. 1, pp. 175–77.

24 Our analysis of folk tales is based on the Kikuyu for whom a number of stories, including variants, were recorded soon after initial European penetration. The best collection is found in Routledge and Routledge, *With a Prehistoric People*, pp. 290–328. Other sources are W. E. H. Barrett, "A'Kikuyu Fairy Tales (Rogano): The Three Warriors and the Masai Woman's Head," *Man* 12 (1912): 112–14; W. E. H. Barrett, "A'Kikuyu Fairy Tales (Rogano): The Drinking Place of the Irimo (Evil Spirits)," *Man* 13 (1913): 4–6; W. E. H. Barrett, "A'Kikuyu Fairy Tales (Rogano): The Adventures of Kamu and Njeroge," *Man* 13 (1913): 73–74: W. E. H. Barrett, "A'Kikuyu Fairy Tales (Rogano): The Warrior and the Irimo," *Man* 13 (1913): 74–75. A number of tales from the Kamba are summarized throughout Lindblom, *The Akamba in British East Africa.*

25 Charles Dundas, "The Organization and Laws of Some Bantu Tribes in East Africa," *Journal of the Royal Anthropological Institute of Great Britain and Ireland* 45 (1915): 240–41. The population growth and expansion of Bantu people is one of the most remarkable demographic events in history (Murdock, *Africa*, p. 271).

26 Routledge and Routledge, *With a Prehistoric People*, pp. 5–8.

27 Kenyatta, *Facing Mt. Kenya*, pp. 21–51.

28 Middleton and Kershaw, *The Central Tribes of the North-Eastern Bantu*, pp. 77–78. Charles Dundas, "Native Laws of the Bantu Tribes of East Africa," *Journal of the Royal Anthropological Institute of Great Britain and Ireland* 51 (1921): 273; Lindblom, *The Akamba in British East Africa*, pp. 164–65.
29 Kenyatta, *Facing Mt. Kenya*, pp. 68–69.
30 Genesis 1:21–31.
31 Genesis 2:4b–24.
32 Genesis 3.
33 Wagner, *The Bantu of Western Kenya*, vol. 2, pp. 53–55.
34 Wagner, *The Bantu of Western Kenya*, vol. 1, p. 165.
35 Routledge and Routledge, *With a Prehistoric People*, pp. 49–50.
36 Muriuki, pp. 37–109, examines the migration of Bantu people into central Kenya and describes the process by which groups and consequently traditions both merged and specialized. There being no basis for it, trade between agriculture peoples was limited; however, between agriculturalists, pastoralists and hunters, trade was lively. Kenyatta, *Facing Mt. Kenya*, pp. 66–67.
37 Routledge and Routledge, *With a Prehistoric People*, pp. 106–07.
38 Sources for folk tales are given in note 24. Cagnolo, *The Akikuyu*, pp. 214–25, records Kikuyu proverbs.
39 Otieno, "Soil Erosion in Ukambani," p. 41; Lindblom, *The Akamba in British East Africa*, pp. 134–35.
40 Lindblom, *The Akamba in British East Africa*, pp. 113–41; Middleton and Kershaw, *The Central Tribes of the North-Eastern Bantu*, p. 7.
41 Lindblom, *The Akamba in British East Africa*, pp. 117–21.
42 Cagnolo, *The Akikuyu*, p. 22, considers this semitotemism but other writers do not agree. Wagner, *The Bantu of Western Kenya*, vol. 1, pp. 198–200; Lindblom, *The Akamba in British East Africa*, pp. 212–16.
43 The chameleon's mythological association with death is examined by Wagner, *The Bantu of Western Kenya*, vol. 1, p. 170.
44 A. D. Graham, *The Gardeners of Eden* (London: Allen and Unwin, 1973).
45 Routledge and Routledge, *With a Prehistoric People*, pp. 5–8, for example, explore the process of forest clearing.
46 Keith Thomas, *Man and the Natural World: Changing Attitudes in England, 1500–1800* (Harmondsworth, England: Penguin, 1983), pp. 212–23.
47 Lindblom, *The Akamba in British East Africa*, pp. 219–20, 231, 240–43; Routledge and Routledge, *With a Prehistoric People*, pp. 38–39, 226, 229–35; Kenyatta, *Facing Mt. Kenya*, for example, pp. 134–36; Mbiti, *African Religions and Philosophy*, for example, p. 163; Cayzac, "La Religion des Kikuyu," p. 39.
48 Routledge and Routledge, *With a Prehistoric People*, pp. 229–35.
49 Ibid., pp. 5–8.
50 C. W. Hobley, *Bantu Beliefs and Magic* (London: Frank Cass and Company Limited, 1922; New York: Barnes and Noble, 1967), pp. 170–71.
51 Wagner, *The Bantu of Western Kenya*, vols. 1–2.
52 Sandra Price, "Conservation Education: The wildlife Clubs of Kenya," *Parks* 3, no. 3 (1978): 1–2.
53 Roderick Nash, *Wilderness and the American Mind*, pp. 343–46.

99

# MAPPING THE MYTHOLOGICAL LANDSCAPE

## An aboriginal way of being-in-the-world

*Paul Faulstich*

Source: *Ethics, Place, and Environment,* 1:2 (1998), 97–221.

### Abstract

Warlpiri Aborigines utilize graphic and cognitive systems to represent their connections to landscape. The Dreaming is the primary mechanism through which Warlpiri organize and understand the significance of places. Each Dreaming myth has an accompanying graphic map, which references incidents and places associated with Ancestors. The maps recount sites along Dreaming tracks, and provide assessments of resources. Warlpiri create these coded images to coordinate physiographic and mythical components of the landscape. They structure knowledge about the world and facilitate the recollection of activities in space. Dreaming maps signify, among other things, aspects of cultural ecology around which society is organized.

> The mind is a part of the nature of things; the world is a divine dream.
>
> Ralph Waldo Emerson, *Nature*.

### Placeness

In Australian Aboriginal studies, virtually no ethnographic work passes without substantial mention of the people's relationship with the natural world. The reason for this is simple: in the company of Aboriginals one cannot escape their consuming place-orientedness.

Warlpiri Aborigines of the Australian Central Desert utilize graphic and cognitive systems to symbolically represent their connections to the landscape. The natural environment sustains the Warlpiri worldview, and this central Australian landscape exists not only as physical terrain, but as a geography of the mind.

The Dreaming is the most powerful mechanism through which Warlpiri organize and understand the significance of places.[1] It has been, perhaps, the most avidly discussed aspect of Australian Aboriginal societies. Another look at its premises will illuminate the ecological framework by which Warlpiri are situated in the landscape.

The Dreaming has various levels of meaning: it is the mythological realm of totemic Ancestors; it is the embodiment of metaphysical potency in the land; it is the 'Law' to which humans must conform; and it is the spiritual identity of the individual. Knowledge of the Dreaming is an essential element in the attachment of Warlpiri to the landscape. But ultimately, the Dreaming is a *process*, a dynamic unfolding of the universe which provides the dominant reference points for human identity, intellect, and action in relation to the land (see Napaljarri and Cataldi, 1994).

Warlpiri Aborigines are a formerly foraging people whose territory lies in the central-western part of Australia's Northern Territory. Inhabited by Aboriginals for millennia, this desert landscape is rich in diversity and expanse. Knowledge of its environments, mapped from generation to generation through graphic and oral genres, includes copious symbolic lore, as well as intimate bioregional details.

## Methodology and the nature of the inquiry

All peoples have culturally specific ties to the land. In anthropology, the role of culture in providing order and meaning to people's social and physical universe, which includes the land, has been of central interest. At the heart of this essay lies a fundamental question concerning the nature of these cultural ties among Warlpiri Aboriginals: How and what does the land mean?

My interest in this question is, on the surface, primarily aesthetic. But its implications run much deeper. In an immediate and direct way, cultural perceptions of the environment are relevant to many of the problems facing indigenous peoples. In Australia's Northern Territory, for example, the onslaught of mining interests over the past three decades has been followed by numerous Aboriginal land claims field in Territory suits. Contemporary relationships of peoples with the land are expressed not only through social organization and subsistence, but also through ritual and myth, emotions, and graphic arts. As I intend to show, an exploration of these aspects of culture yields a richer understanding of a people's ties to the land.

The sense of place has been on the periphery of anthropological interest, but it has not occupied a central theoretical position; anthropological appraisals of emotional and perceptual responses to environmental stimuli have been minimal. Generally, social scientists have viewed the natural environment in terms of its biological effects on adaptation, or its corresponding socio-political constructs. I situate my considerations of Warlpiri actions within a different view of environment; one which emphasizes symbolic constructs of it. I am interested in how cultural meaning is constructed out of the phenomenal elements of the land.

The present study is framed within a contact situation in which contemporary Warlpiri are existing within the reality of post-colonial, white Australia. Field research for this project was carried out in several stages. I first arrived in the Northern Territory in June 1980 to explore the bonds between Warlpiri Aborigines and the environment. In particular, I was interested in how Aboriginal artists experience the landscape and express their perceptions of it. During my ten months of research, I lived in Yuendumu and surrounding Warlpiri communities (principally Nyirripi). In 1988 I returned to the Warlpiri outstation of Nyirripi for a year of intensive field research. A central goal of this project was to begin to understand how Warlpiri symbolic systems augment and express human/nature relationships. Over the past decade I have revisited Aboriginal lands in the Northern Territory on several occasions, and each time have had the opportunity to rethink aspects of cultural geography.

The Nyirripi settlement began as a small, breakaway community (known as an outstation) in 1975. Its establishment reflected the desire of many Pintupi and Warlpiri to renew their longstanding social and symbolic attachments to the land. Outstations are, in effect, homeland places, and the motives for establishing them are to escape the fighting and drinking of larger settlements, to have access to richer foraging areas, and most importantly, to care for the country; to 'look after, 'im properly' as Warlpiri say. The outstation movement also began, in part, as a way for Aboriginals to gain more political autonomy. Outstation living reduces Warlpiri participation in while Australia to a minimum; as an assertion of independence, the outstation movement has enabled the reemergence of Aboriginal authority in spiritual and political matters.

Since its establishment, Nyirripi has grown rapidly, and now has a fluctuating population of about 200 people. Cinderblock houses are interspersed with makeshift shelters (humpies), and widespread trash and gutted cars contrast with the expanse of red earth. Most Nyirripi residents rely on regular social service payments, which are distributed fortnightly to pensioners, widows, and the unemployed. This includes virtually every adult in the community, although a few are employed on a part-time basis as shop and school workers. Warlpiri refer to social service payments as 'sit-down money' since they are required to secure their check from the Nyirripi shop, and thus feel obliged to stay put in the community. Mobility is highly valued by Warlpiri, and 'sit-down money' is seen as putting a restriction on their preferred lifestyle. But despite bureaucratic impositions, Warlpiri regard Nyirripi as distinctly *their* community. As Banjo Jungarrayi, a long-time Nyirripi resident, put it; 'This good place here. No whitefella push 'im (us) around, only *yapa* (Aboriginals)' (for a detailed discussion of the Nyirripi community see Faulstich, 1990).

My research at Nyirripi basically conformed to the pattern of daily life; typically I spent the mornings under the communal shade shelter, and the afternoons in camp talking with old men under the shade of bloodwood trees.[2] It is with these men that I spent the great majority of my time. On a casual basis, we engaged in the daily discourse of gossip, disputes, and Dreaming talk. On a more formal basis, I solicited specific types of information, and engaged in open-ended discussion with consultants. Over the months of field research, I developed stronger relationships with some of the men than with others: I came to rely upon these people as my primary mentors.

Warlpiri themselves do not ask many questions. Instead, their tendency is to learn from observation. In attempting to conform with their preferred behavior, most of my research was conducted through participant-observation. At times I taped conversations of Dreaming stories, but generally I simply listened ant took notes.

At the request of many Warlpiri, and with the intention of collecting data on graphic art and Dreaming affiliations, I supplied acrylic paint and canvas to several individuals. With interests in specific places and myths, I commissioned (from people with the appropriate affiliations) paintings of specific Dreamings. Generally, however, I supplied the materials and let the painters decide what Dreaming to depict and what colors to use. Through this process I was able to obtain information about the land, myths, and Dreaming ties. This proved to be very fruitful for collecting data, and Warlpiri appreciated access to artist's materials. I purchased some of the paintings, but most were left with the artists for them to sell. The people whose works are illustrated herein gave permission to have their works reproduced in the context of this essay.

Between 1962 and 1973 Munn published a series of acclaimed works concerning Warlpiri graphic symbols. In addressing the function of Warlpiri visual representational systems, she has looked at the broader issue of the relationship between symbols and the objects that they represent. Munn's celebrated 1973 book, *Walbiri Iconography*, provides an interpretation of Warlpiri graphic designs. Munn (1973) maintains that Warlpiri art is iconically based and can therefore be considered non-arbitrary since a design has an iconic relation with what it represents.

Land-based models of Aboriginal authority and social structure are common (see Radcliffe-Brown, 1930; Strehlow, 1965). Such approaches have matched social units with land-use patterns in recognition that a spatial component underlies social organization among hunting and gathering peoples. In the 1970s and 1980s, inquiries concerning non-Australian hunter and gatherer territorial organization have offered flexible ecological models of land use (see, for example, Lee, 1976). These models, however, have viewed the environment as a resource base, and have made few attempts to situate 'landscape' within a broader context of personal interpretations, emotions, or ideals. More recently, the environmentally-based ethnographic writings of Deborah Bird Rose and others have shown the importance of human ecology in grappling with issues of justice, sustainability, and intercultural understanding (see Rose, 1992, 1996; for environmental ethnographies see Basso, 1996; Feld, 1982; Nelson, 1983; for an environmentally-based chronicle of an anthropologist's experience among the Warlpiri, see Jackson, 1995).

An excellent ethnography out of central Australia is Fred Myers' (1986) *Pintupi Country, Pintupi Self*. Pintupi are neighbors of the Warlpiri, their territory lying to the immediate south of Warlpiri country. The two peoples share many cultural traits, including important myths and Dreaming tracks. They occasionally participate in joint social and ceremonial events, and travel between their two territories is common. Myers (1986) stresses Pintupi social attachments to the land and organizes these relationships through a framework that analyzes the political basis of landownership. There is, also, a developing body of literature discussing the traditional ecological knowledge of indigenous peoples; Peter Latz (1995) for example, provides an ethnobotany of central Australia, and situates his discussion in a broad socio-ecological framework. Within this ethnoecological context, Aboriginal fire management is being further understood, and is gaining greater respect (see Rose, 1995; Williams and Baines, 1993). Ethnoecology documents the vast environmental knowledge of many indigenous peoples, and recent work in this area suggests the value of this knowledge in addressing contemporary socio-ecological problems and sheds light on diverse ontologies of knowledge.

## Landscape and place

Landscape provides the taken-for-granted backdrop for human activity and behavior. In everyday experience, landscape is the most pervasive but least consciously felt aspect of regional character. Landscape is fluid—it flows around us and encompasses us. It is not, however, external to the individual; landscape is an assembly of sensory information (including natural and artificial features) which generates a seen and felt experience of the world. It is a symbolic constitution of the environment within which humans exist.

Landscape is an active medium of communication, embodying not only values and ideals, but directly influencing perception and behavior. As a matrix of meaning, a landscape comes to embody the sensibilities of those who inhabit it. The social significance of landscape can only be discovered through consideration of the cultural processes to which it is linked.

Landscape is part of the iconography of Warlpiri culture. It is, for Warlpiri, an 'image' of the invisible; a physical link to the creative forces contained within it. Not only are there practical bonds of subsistence between Warlpiri and landscape, but there are also potent mythical, social, and emotional bonds. Landscape gives and receives life; the country and the life-forms it sustains must take care of each other. It is through these bonds that Warlpiri develop a sense of place; affinities with particular locales. Warlpiri have an acute sense of bioregionalism in which people are viewed as interconnected with the land and its non-human inhabitants.

Part of the Warlpiri quest for meaning involves the ordering of landscape into places. Whether or not a place is seen to be a discrete unit of space, Warlpiri understand them to be inextricably linked to the oeuvre of other Dreaming places. Places are centers of cultural and personal meaning; they exist as foci of emotional attachment.

Places are distinguished from landscapes not in their scale but in their qualitative experiential dimension. Whereas *landscape* is a part of any immediate encounter with the world, *place* is further constructed through symbols and affections (Relph, 1985). Place embodies the experiences and aspirations of people, and the experience of place is a product and expression of the person encountering the place. Places of nature present themselves to us existentially, and are a source of inspiration and insight. A 'sense of place' begins with the interactions between people and world; it extends beyond location to encompass history and purpose.

While Warlpiri do not express the precise conceptual categories of landscape and place as explicated above, they do have words which translate as 'country' (*nguru*) and 'site' (*ngurra*). Admittedly, linguistic translations are not always conceptually accurate, but translating the geography or *Jukurrpa* (the Dreaming) into familiar English concepts can help us to achieve some understanding of conceptual systems vastly different from our own. Such conceptual alignment can help us to better comprehend the various and variable perspectives from which people understand their lived environment.

The Warlpiri 'sense' of place is twofold: on one hand Warlpiri feel what it is like to be at a place, and on the other hand they intellectually grasp the meaning of the place. For Warlpiri a sense of place unfolds through the religious, moral, and aesthetic discernment of specific locations. Through the synergistic sense of place, the boundaries between person and 'other' become blurred. Warlpiri become saturated with placeness, and become, they maintain, a part of the landscape.

To Warlpiri, places have personality; they possess the distinctive qualities of the mythological Ancestors who created them. The personality of a place is composite of physical features and symbolic meanings. For Warlpiri, the significance of place is evident in daily discourse; no matter what topic of conversation, it inevitably returns to 'place'.

Places are fundamental expressions of human involvement with the world. They provide foundations for existence, imparting not only a geographic context to activity, but providing security and identity as well. Through natural places humans gain insight into their existence, for there is self-discovery in place.

## Warlpiri cosmology

The landscape of Walpiri country is a manifestation of the natural and mythological histories explicated in the Dreaming. Like other Australian

Aboriginal peoples, Warlpiri Aborigines tell of a realm in which the Earth and animals do not exist in their present forms. In this realm, mythological Ancestors emerge from a featureless earth, transform it, and create the landscape. The clouds and hills, billabongs, grasses, and trees are created during this period, as are animals, and kinship patterns, taboos, and other tribal laws. When the Ancestors complete their creative wanderings they change into Spirit Beings, and they continue to dwell in special places within the land.

The Dreaming (an English gloss of the Warlpiri concept of *Jukurrpa*) is a time out of time that contains the forces of creation; it is the inescapable and power-filled ground of existence (for an important account of the origins of the term 'Dreamtime' see Wolfe, 1991). The Dreaming persists in a realm separate from, yet accessible to, contemporary existence.

Existing simultaneously in the past, present, and future, the Dreaming is a force which overlaps and envelops contemporary Aboriginal existence. Although English-speaking Warlpiri generally refer to Dreaming events as past occurrences, this is only a convention employed to emphasize the distinction between the immediate present and the intangible Dreaming.

The grammatical structure of many central Australian languages enables the teller of a myth to present it as an ongoing reality, not just an inaccessible event from the past (Yengoyan, 1987). The imperfective tense combines time and thought into a single coherence. The Warlpiri universe, as explicated in myth and expressed in language, is in a continual state of emergence.

In semantic contrast to *Jukurrpa* (Dreaming) is *yijardu*, which means 'real' or 'here'. *Yijardu* refers to the palpable world; the ongoing present and events that occur within the sensory field. The distinction between *Jukurrpa* and *yijardu* is not one of simple opposition however, since *Jukurrpa* is the foundation of all being.

Representing neither the simple past nor the distinct present, the Dreaming is difficult to discuss in English. The distancing involved in its translation is compounded when we acknowledge that when one describes the Dreaming one prescribes it, thereby limiting it. In the Warlpiri view, just because an object was created in the Dreaming and continues to exist in the constituted world does not mean that it is not still being created. The Dreaming entails no single finite process.

The Dreaming is dynamic and ongoing, providing the spiritual foundation for Warlpiri existence. It contains the structure of the world and unites humans with animals, the inanimate environment, and mythological personages. All things of the world are instituted in the Dreaming and are related in its self-created and enduring system.

Within this cosmological realm, totemic Ancestors (themselves Dreamings) change the earth into its present form. These Beings emerge from a vacuous earth and travel about. As they wander, they not only perform acts of unusual and often superhuman prowess, but they also live very much like Aboriginal people (*Yapa*) live today; they camp, gather and hunt, engage in rituals, defecate, and so forth. Although these Beings are ancestral, they continue to act upon the contemporary world of nature.

## Dreaming the country

Dreamings themselves are noumena; they are inaccessible to experience except through the phenomenal realm. Ultimately, it is only the phenomenal realm, as a remnant of the Dreaming, that remains accessible to human perception. As Banjo Jungarrayi put it; 'You can't see 'im [the Dreaming], you can't feel 'im. Nothing'.

Landscape is one way that Warlpiri give rational objectification to the Dreaming. Condensed within the landscape are two aspects of a single reality; the potency of the Dreaming and the tangibility of the world. The landscape, as a transformation of Dreaming potencies, is free from specific historical location and the constraints of temporality (Munn, 1970).

The Dreaming gives meaning to the land. Virtually every prominent physiographic feature, and many seemingly undistinguished ones (to untutored eyes), are imbued with mythological significance. Contained within the visible terrain is a codified landscape rich in mythological lore, ritual traditions, and personal affiliations. There is no geography without meaning.

Warlpiri maintain two inseparable views of geography; one is of the physical landscape and the other is the mythological landscape. Through mythological interpretation of the land, geography is transformed into cosmology. And through the Dreaming, Warlpiri are inextricably woven into the fabric of wild nature.

Warlpiri have produced a pattern of land tenure in which there is a 'heartland' or 'country' from which Dreaming tracks run (for details on patterns of land use of this area, see Peterson *et al.*, 1978). Dreaming tracks are the paths created by Ancestral Beings during their travels. These tracks are often several hundred kilometers long, with dozens of named sites along the route. The complex of Dreaming tracks creates a pattem of mythological trails that cross and re-cross each other.

Warlpiri country—in fact the entire Western Desert of Australia—is criss-crossed with mythological tracks, each with an accompanying mythic narrative, song-cycle, dance enactment, and ritual caretakers (Liberman, 1980). Each of these Dreaming tracks consists of a series of sacred sites and the paths between these sites. The myths associated with these tracks recount the actions of the Ancestors; their subsistence activities, their fights, their love-making, their ceremonies.

*Sacred* places are always defined by mythological events. The totemic site of *Puwarripuwarri*, for example, is dug as a roasting pit by the giant woman *Warrpalypardu*. In the Dreaming, *Warrpalypardu* crushes people with large rocks, and cooks them in these pits. The waterholes exist because of the actions of *Warrpalypardu*, and their existence is proof to Warlpiri of her Dreaming activities.

The landscape is filled with signs that are continually interpreted by Warlpiri as having resulted form the activities of Ancestral Beings. Sacred sites are scattered across Warlpiri country, connected by Dreaming tracks, which are essentially lines of essences containing the potency of the Dreaming.

The places where Dreaming tracks intersect are not necessarily more significant than other named places. They do, however, provide access to sites and stories related to other Dreamings, thus enlarging the world of mythical-conceptual space. The entire complex of related Dreaming sites, then, is understood as one 'story'; an intertwined series of natural histories explicating the origins of the world. Individual Warlpiri locate their identities in as many geographic contexts as possible, thus extending the self outward into the tangible world.

Mythological concepts and images permeate the landscape. As such, topography is infused with spirit and brought alive through a series of places which articulate physically that which is invisible, i.e. the potency of the Dreaming. Warlpiri see a causal relationship between totemic Ancestors and features of the land. My experiences with Warlpiri suggest that their myths and rituals can best be understood in terms of places, for it is the landscape which provides the most obvious and enduring evidence of Dreaming occurrences. But for Warlpiri, land is more than simple evidence, it is the actual transfiguration of Ancestral Being. The Land *is* the Dreaming.

*Marlujarra* Dreaming is one of the most important Warlpiri. Based on the creative adventures of Two Kangaroos (*Marlujarra*), the Dreaming story describes how these Ancestors travel across Warlpiri country and form geographical features along the way. *Marlujarra* Dreaming begins where the Ancestors emerge from the ground at *Marlunguru* ('Kangaroo Country'), about 300 km north of the Aboriginal community of Nyirripi. The land is mapped verbally through the telling of myth. What follows is a summary of Tiger Japaljarri's telling of *Marlujarra* Dreaming as it relates to the Kangaroos' travels near Nyirripi (a full version of the Marlujarra travels through Warlpiri country is give by Meggitt, 1966; Meggitt's version of the myth lists 53 named sites between Marlunguru and Nyirripi):

> The Kangaroo Ancestors (one kangaroo and one euro, a smaller relative of the kangaroo) become frightened by the approach of a dingo. They bound away, and one of them (who at this point has a spear for a tail), drags his tail along the ground and gouges out a channel to form Waite Creek. They run off to *Walguru*, where they camp. The next morning they proceed to *Jiluru* to 'paint up' (decorate) each other and perform ceremonies telling of their narrow escape from the dingo.

> The Kangaroo Ancestors travel on, camping next at *Purpurulu*, where they climb a hill to survey the country. They move on to *Wadilwanu*, and having gone without food for several days they become quite hungry. Using hairstring that they made for ceremonial use, they fashion waistbelts, which they cinch tight to relieve their hunger pangs. They camp at *Nyirripi*. A huge Snake at *Kunajarrayi* smells the Kangaroos and travels to *Nyirripi* in the hope of catching and eating them. The fearful Kangaroos flee in different directions, losing the Snake. Later, the Kangaroo Ancestors meet up at *Nyinyiripalangu*, where they camp.
>
> They travel on, and at *Kilipali* waterhole they see Emus coming to drink. At *Kuritjinyungu* they gather and eat honey; what they cannot finish they coat themselves with, as protection from the heat. Refreshed, they make two large string crosses, decorate themselves, and perform increase ceremonies. One of the Kangaroo Ancestors dances with the cross vertical behind his back, simulating the swaying branches of the eucalyptus trees that grow beside the waterhole.
>
> The next evening they camp at *Jurru* where one of the Kangaroos makes a new handle for his spearthrower. The following night is spent at *Jiliparnta* (the site of the *Nyirripi* community). A soakage is formed where they lay down to sleep and eucalyptus trees sprout, embodying some of the Ancestral power. At *Panggunapanda* they grind acacia seeds into meal and cook their cakes in the ashes of a fire. As they approach *Karrku*, the place of red ochre, they hear dingoes howling. The dingoes are performing ceremonies at the hill of *Wirnijarru*. The frightened Kangaroo Ancestors flee, being chased by the dingoes.

Having predictable sources of water, the Nyirripi area has long been an important camping place for Warlpiri. Each of the named places along this section of the *Marlujarra* track have natural features—from ochre colored stones to tall eucalypts—that are transformed objects of the Dreaming.

Each myth has an accompanying graphic map (discussed momentarily) and a song, which refer to incidents and places associated with the Ancestors. To Warlpiri, myth, design, and song reinforce each other and share in the virtue of the Dreaming. The songs and maps recount the sites along Dreaming tracks, and give accounts of the ritual and economic resources in the country.

Songs follow the travels of Ancestral Beings across the desert, celebrating places associated with their Dreamings. They are lyrical maps of the land that lay cartographic claim to particular tracks of country. The recital of place names is fundamental to Warlpiri myths and songs. Some songs, for example, consist solely of place names that are chanted virtually in list form. Likewise, it is not uncommon to hear a Warlpiri man detailing the names of sites within his country; with a finger snap and a gesture toward some unseen place, Warlpiri men will spontaneously recite a series of mythologically connected place names.

According to Warlpiri, place names are nearly always descriptive, and they are prophetic. The extremely sacred site of *Wantunguru*, for example, derives its name from *watunguma*, referring to the winged phase of an edible termite, and *nguru*, meaning 'country'. *Wantunguru*, then, is the main place associated with the Flying Termite Dreaming, from which it gets its name. Place names also commonly refer to geological features, such as *Karrku*—the place of red ochre.

The initial naming of Dreaming sites is attributed to Ancestors, and place names verify the existence of totemic Ancestors. Once the Ancestors give names to places and other phenomena, those things begin to truly exist. Through their semantics, place names map the terrain.

## Being-in-the-world

Places provide reference points within which Warlpiri locate experience and orient position. David Lewis (1976) has shown how, for Pintupi Aborigines, direction is reckoned through the fixed location of sacred sites, not through the transitory location of people. Directions are not given from where one happens to be at a given moment, but from significant places with geographically fixed positions.

Warlpiri elders sometimes give directions similarly. Navigation and location are determined through mythological reference points that locate people and places in relation to Dreaming events. At *Nyirripi* I asked Mosquito Jungarrayi to indicate the direction to *Pikilyi*, a freshwater spring and an important Two Kangaroo Dreaming site. He indicated that *Pikilyi* is west of *Walguru* (*walgurungka karlarra*; at Walguru west). *Walguru*, associated with the same Two Kangaroo Dreaming, is 40 km north from *Nyirripi*, and *Pikilyi* is 30 km east from that. The direction to *Pikilyi* was reckoned not from where we were at the time (*Nyirripi*), but from *Walguru*, an important totemic site. It is, then, *places*, not people, from which direction and spatial orientation are often determined.

In preparation for initiation ceremonies, men face their father's country—however distant—as they pierce their arm veins to draw blood essential for certain ritual adornment. Facing one's paternal country is an expression of continuity between the land and the self. Spatial coordinates function in this way to mediate between a person and the Dreaming.

The Warlpiri way of positioning oneself in reference to the world of places indicates the symbolic importance of geography. Places are important to Warlpiri not only because they are good to visit, but because they are good to think (for discussions on how *animals* are 'good to think' see Levi-Strauss, 1964, 1966). By learning about the country people become, Warlpiri say, 'stronger'. Through knowledge of the land, they develop into mature individuals, distinguished by their affiliation with organic nature and its Dreaming powers.

As a way of coordinating physiographic and mythical components of landscape, Warlpiri create coded images (cognitive or mental maps) which function as mnemonic devices. Mental maps structure and contain knowledge about the world, and facilitate the recollection of activities in space. Symbolically constituted, socially transmitted, and individually applied, mental maps are reconstructions of reality. Mapping, however, is far from a neutral mode of representation; it is a performance entangled in epistemological assumptions about the lay of the land.

Landscape is a continuous entity, punctuated by sacred sites where particular mythological actions occur. Dreaming maps may be extensive by linking geographically distant places, or they may be intensive by explicating the details of a single place.

Mental maps are geographical profiles created through perception and intellect. They involve the integration of images, information, and attitudes about the environment. Being symbolically structured, Warlpiri graphic maps do not necessarily correspond in spatial terms with maps prepared by Western cartographers. They express *Warlpiri* ordering of landscape and place, which is decidedly different from European spatial organization. Imagery coding of the landscape can exist completely within the mind, or it can be tangibly presented, as in Jampijinpa's sketch of his 'country'. (Women also hold 'secret' and 'dangerous' knowledge which is not necessarily less than that of men's; consider, for example, the work of Bell (1983) who demonstrates the extensive Dreaming knowledge held by Warlpiri women.)

## Mapping the dreaming

Warlpiri graphic arts are concerned with mapping the mythological landscape. The paintings being done by Warlpiri today belong to a class of Aboriginal art that has come to be known as the 'Western Desert Style'. The canvas paintings, executed in acrylics, are enmeshed in the larger system of Warlpiri social, political, and ecological values. Derived from traditional designs, these paintings and expressive of Warlpiri emotion, purpose, and place within the landscape, but they are decidedly modern in their medium and in their role in interfacing between Aboriginal and non-Aboriginal interests.

Contemporary acrylic paintings share an uncanny resemblance to prehistoric rock art, the oldest surviving artistic expression in central Australia. The present-day paintings function as Dreaming maps of important places and events; they map the travels of totemic Ancestors, and depict the sacred places these Ancestors create.

The visual style of Warlpiri art mimics the narrative style of Warlpiri myth. In myth, Dreamings are recounted as the Ancestors travel through the country. Warlpiri paintings depict these travels and the sites associated with them. The narrative style is evident in the interconnected circles and lines that are so prevalent in Warlpiri paintings. The circle/line composition is widely used to illustrate the journeys of Ancestral Beings and the place that they create; the sites represented by circles, the paths that connect the sites being represented by lines. This site/path structure is a visual device that graphically maps the Dreaming.

The site/path combination iconically illustrates the movement of Ancestral Beings across the land. It provides an essential structure that links Dreaming events to geographical places and life experiences. In Warlpiri art, circles and connectors are visual models of things in the world.

The line motif reflects Warlpiri mobility and the emphasis on movement across the country. It illustrates travels through the landscape in determined and knowledgeable ways, linking sites in distinct paths. Lines, and other less direct connectors, are employed by Warlpiri to depict the tracks of Ancestral Beings. In this sense they are iconic.

Conversely, circles are used to depict places. Initially this may seem problematic if we are to think of Warlpiri maps as essentially iconic, for most places do not really *look* like circles. However, major campsites and sacred places are usually associated with water supplies, caves, or trees. These objects are basically circular in form at the point where they intersect the earth, and it may be through this organic connection that sites came to be represented as circular forms. Though the circle may actually *depict* a waterhole, it *denotes* broader social and mythological aspects of the site (for a further discussion of the iconicity of Western Desert Art, see Morphy, 1980, and Anderson and Dussert, 1988).

Compositionally, there tends to be a difference between Dreaming maps produced by men and those produced by women. The site/path structure is employed more in men's paintings than women's. Men's paintings, in other words, are largely concerned with Ancestral travels across the land, whereas women's paintings tend to depict the rhythms of family life and localized activities.

In general, Warlpiri equate circular forms with women and elongated forms with men. The sexual content of this graphic symbolism is easily recognizable, but its social ecology is more elusive. Circular forms are associated with camps and connote domestic activities frequently analogous to women. Elongated forms, on the other hand, are associated with journeys and connote more the activities of men. This is not to say that these classes of activities are the sole domain of women and men, respectively, but rather that Warlpiri generalize along these lines when it comes to mapping gender associated pursuits.

The symmetry of Warlpiri art—sometimes elusive, but virtually always present—assists in the ordering of experience and space. Through symmetrical compositions, Warlpiri impose a structure on phenomena that may otherwise lack this quality. The uneven distribution of sacred sites, for example, is given compositional balance, as demonstrated in Naparrula's painting. Warlpiri ceremonial grounds are mapped out with a similar aesthetic, providing in miniature a symmetrical reconstruction of the Dreaming landscape. By restructuring disparate aspects of geography, Warlpiri create a semblance of unity. Symmetry, along with repetition, assists Warlpiri in comprehending the universe as a synthesized whole.

Every Warlpiri graphic design represents both an identifiable locality and its mythological association, but the knowledge to interpret the design is gained only through synthetic understanding of the Dreaming and the land. But Dreaming maps are not just about far-removed myths. They signify, among other things, aspects of cultural ecology around which society is organized. They are highly charged images expressive of kinship rules, rights to resources, ecological insights, the distribution of sacred knowledge, and other elements of social and environmental organization.

Warlpiri paintings diagram religious, economic, social, and ecological linkages between individuals and groups. They articulate rights over resources, and have found their way into the claim process as cartographic evidence of ownership. Acrylic paintings have emerged from the intersection of a system of mapping (which is profoundly deliberate, unspontaneous, and hierarchical) and postcolonial influences that have been overwhelmingly disruptive but occasionally revitalizing (see Bardon, 1986). Warlpiri iconography employs a visual complex which, like language, must be learned. To learn the meanings of the designs to which one belongs is part of the process of developing an identity; of becoming a distinct Warlpiri individual. Warlpiri assert that they paint to maintain the Dreaming; to 'keep it alive' as they say, and to pass Dreaming stories on to their children.

Painting recalls the Ancestral landscape, and Warlpiri art visually maps the interdependent relationship between humans and natural systems. The paintings are graphic projections of cognitive maps, laying out the symbolic structure of the land. In 'reading' paintings, Warlpiri interpret that land and their place within it. For Warlpiri, Dreaming maps distinctively express the connections between person and place. Although maps and knowledge are not one and the same, both imply an understanding of the world.

Warlpiri paintings have habitats, which is to say that they belong with particular places. They are rooted in specific geographical locales, and they metaphorically relate to the Ancestors who created those places. Graphic designs illustrate the way in which Warlpiri view themselves within the context of the world and its origins. Through these maps, the mythological potency of landscape is made accessible.

For Warlpiri, mapping manifests cultural ideas of the nature of reality. The task of the Warlpiri 'cartographer', then, is to express a sense of ultimate reality through significant form. As expressions of cosmology, Warlpiri paintings graphically illustrate how they perceive the land, with all of its physical and spiritual properties.

The power of an image lies in its ability to provoke an emotional response. For Warlpiri, graphic designs are meaningful insofar as they depict the Dreaming. But these maps not only represent and help maintain the Dreaming, they also give it clarity and coherence. As an iconographic expression, Warlpiri graphic mapping defines the order of the world as Warlpiri conceive it to exist.

Symbols attempt to disclose structures of cultural reality, and to this end Warlpiri graphic arts connect mythic symbols (the Dreaming) with physical symbols (the land). Making the invisible visible, translating cosmological essences into apprehensible appearances, the central concern of Warlpiri mapping is to expound upon reality. Graphic art is one expression of the ways in which Warlpiri perceive the world. It is a manifestation of the forces connecting humans with the landscape. Warlpiri maps give significant form to esoteric relationships with the world.

## An economy of Dreaming

The ability to recognize mythical resources requires knowledge of Dreaming events. Charlie Jampijinpa, pointing out a nondescript bloodwood tree marking the sacred site of *Wannatunji*, remarked 'that one not timber, that one Dreaming. Dreaming that one'. The message, of course, was 'that tree is more than just wood, it is pregnant with the essence of the Dreaming. In fact, that tree is Dreaming'.

To Warlpiri, goodness is resources in abundance. These resources can have either economic or mythological bases. When I asked Paddy Japananka what was good about his country, he replied 'plenty *ngapa* [water], plenty *kuyu* [meat]'. And George Jamijinpa, expounding the virtues of his country, said: 'Oh, too much *marlu* [kangaroo], too much goanna. Too much.' 'Natural resources' (a Western concept) are, logically enough, important to Warlpiri as a subsistence base.

Mythical resources, however, are not an entirely different order. The spiritual aspects of the Dreaming and its socio-economic implications are not completely seperate for Warlpiri. Aside from direct economic value of the Dreaming, access to mythological resources and sacred knowledge provides political stature as well as spiritual security. For Warlpiri, knowledge commands respect.

During a visit to the very sacred site of Kunajarrayi, a Banjo Jungarrayi asked me to pay him for the Dreaming knowledge he imparted. When I explained to him that I wasn't prepared to give him what he asked—$10 000!—he said, 'Well, okay, three dollar fifty ($3.50)'. The actual dollar of my payment, it would seem, was relatively unimportant to him. What Banjo was asking for was acknowledgment of his erudition. Among Warlpiri, reciprocity is central to the exchange of sacred knowledge. Younger men present gifts to the elders accompanying them to sacred sites. Indigenous Australians also routinely ask for monetary compensation when government and other agencies request information about their sites, and it is within this framework that this senior man initially asked (*almost* jokingly) that I reciprocate with such a huge sum.[3]

At this point it will be useful to look at an example of a Dreaming place that has direct economic as well as mythological import. *Karrku*, one of the most striking landforms in the *Nyirripi* region, appears as though it was pushed up from a flat earth. According to Paddy Jupurrula, the *Karrku* tableland was formed in the Dreaming by the combined actions of several Ancestors:[4]

> A Jupurrula Man (a 'Devil') steals red ochre from a quarry in Alyawarra country to the east of Nyirripi. The ochre is the congealed blood of a butchered man. Traveling west, carrying the ochre on his head, he passed several named sites, finally reaching *Warrinjirrinijirri*, where he camps. Traveling, the next day, through *Yalapalong*, Jupurrula hunts two goanna (*kurrkardi*) and a snake (a 'quiet one', i.e. non-poisonous), and camps at *Yariyarriri*. Waking in the early morning, he walks to the site of *Karrku*, where he builds a shelter ('humpy'). Placing the ochre on the top of the shelter, he leaves on an extended hunting trip. While away he steals more ochre, and kills a Snake (*Warna*) and an Eaglehawk (*Warlawurru*). He brings them back to a nearby windbreak (*Yunta*, which subsequently transforms into a hill of the same name) to cook and eat them. He camps at *Yunta*, returning to his shelter when he sees Rain Dreaming (*Ngapa jukurrpa*) coming from the west. (A Rain Dreaming site is at the base of *Karrku*.) Jupurrula now has a large shelter inside *Karrku*, which at this point is a rock outcropping of modest size. Placing his ochre on top of the shelter once again, he crawls inside and goes to sleep. Two women, a Napaljarri and a Nungarrayi, come from the west to drink water from a catchment on the top of the hill. Smelling the women approaching, the hill swells and rises. Napaljarri and Nungarrayi throw a rope over the hill and pull it back down. They dislodge the ochre on the top, which tumbles to the base of the cliff.

Dominating the southern horizen from *Nyirripi, Karrku* commands attention. Important as a source of highly valued red ochre, it is a significant feature of the *Nyirripi* landscape. Ochre from *karrku* is coveted by Warlpiri, who use it for body and ground paintings, as well as to decorate shields, boomerangs, and other objects.

*Karrku* ochre is valued for its rich silvery sheen, which results from an admixture of mica. Ordinarily, red ochre is called *yurlpa* in Warlpiri. This particular ochre, however, is referred to as *karrku*, echoing the name of the site in which it is found. This association between place name and object name indicates the special consideration given to this deposit. The ochre (*karrku*), even after being removed and processed, retains its relationship to its source (*karrku*). In the Warlpiri language, semantic relations exist such that a single word can designate both an entity and its source. This association between an entity and a place is basic to the integrative world view of Warlpiri: *karrku* the mineral implies *karrku* the place, and vice versa.

## Dreaming relations

*Nguru* means, basically, 'country' or 'home country'. It connotes a range of social and spatial referents linking individuals to Dreaming Ancestors, sacred sites, tracks of country, and local groups. The term carries profound social and emotional meaning for Warlpiri, for whom person and place are intertwined. People, like the Ancestors themselves, are not conceptualized as independent of locality.

*Ngurra*, a concept related to *nguru*, refers both to one's 'camp' and to one's 'home place'. Warlpiri use *ngurra* to designate not only the place one sleeps, but with whom one is living (see Myers, 1986). Relating to the Dreaming as well as to social groups and places, *ngurra* embodies spirituality as well as sociality. 'It is my home,' said one Warlpiri, 'the place, the Dreaming' (*Ngajunyangu Ngurra, ngurra, Jukurrpa*). *Ngurrara* has the referent of 'home' or 'fatherland', and *ngurrarntija* identifies a local person, a native to an area. Warlpiri refer to individuals who share their Dreaming as 'countrymen'.

Individuals, as well as social aggregates, identify intimately with specific tracks of land and series of special places, for it was their Ancestors who were responsible for creating their country. Most myths are concerned with identifying the rights and obligations of particular social groups to specific areas of land, and in this way geographical space is socially segmented. Mythological occurrences often symbolize relationships of alliance or political disjunction between territorial groups.

Myths, however, are not explicitly normative, but are primarily explanatory; the things of the world are the way they are because of the actions of Ancestral Beings. These Beings are both virtuous and sinister, beautiful and offensive: in addition to Rain (*Ngapa*) and Fire (*Warlu*) Dreamings, there are Excrement (*Kuna*) and Vomit (*Yurlkulyu*) Dreamings. Ancestral Beings not only establish the 'Law', but they also break the law that they enact. The same totemic Ancestor might appear as both morally commendable and morally reprehensible on different occasions in the same myth.

The very mention of certain place names can spark emotional responses in Warlpiri. I have witnessed old men weep at the mention of certain places, overcome with sentiment. I have observed people carefully avoiding the mention of particular places because they are related to a deceased person (for a fuller discussion of this custom of a speaking taboo, see Faulstich, 1995). And I have listened as people spoke of places in the first person, or as if they were a relative. One of my Warlpiri companions refers to a particular place as his 'daddy'.

## Mapping and being

Places are allegorical, and through them mythological histories are made tangible and visible. For Warlpiri, identifying and mapping places entails knowledge of the natural order, an order established in the Dreaming and manifest in the land. Warlpiri do not evaluate the land through a canon of physical beauty as many Westerners do. Instead, the landscape is read for the Ancestral reality which it presents. Mediating between the Ancestral realm and the human realm, the landscape is an index of sacred as well as secular events.

Places of the land are symbolic for Warlpiri in that they have the power to suggest a larger assemblage, transcending their immediate physical forms. They encompass a penumbra of meanings, and exist partly as storehouses of information relating analogically and metaphorically to the Dreaming. Mythological constructions are projected onto the land, and the physical endurance of sacred places reinforces the significance and continuity of the Dreaming.

Landscape is layered with meaning, and it communicates to those who can read its message. Warlpiri interpret the landscape on various levels, dependent on the viewer's sophistication and knowledge. For Warlpiri, places provide the basic reference points for locating experience, and understanding one's position in the world. The synthesis of located experience—including perceptions, affections, and myths—give individuals a sense of place and a sense of being. The Warlpiri landscape, then, is not spatially and temporally external; it is ontologically significant.

Through the Dreaming and its associated myths and graphic designs, Warlpiri existence is situated within the landscape. As a model for identity, the Dreaming defines, predicts, interprets, and integrates. It unifies disparate events and experience by mapping—and thereby making meaningful—the totality of Warlpiri experience.

## Acknowledgements

I would like to thank my Warlpiri 'countrymen' for imparting some of their knowledge, and I am grateful to the anonymous reviewers whose critical readings and thoughtful suggestions have made this a stronger essay. This paper builds on earlier work, especially Faulstich (1993, 1994).

## Notes

1 As Strehlow (1965, p. 694) has noted, 'Dreaming' is not an adequate translation of the Aboriginal words for their mythological creative era. More precise translations

would include terms like 'eternal' or 'eternal present'. Literate Aborigines sometimes translate this notion as 'power'. Anthropologist Francois Dussart has suggested the term 'Ancestral Present'. In keeping with the anthropological norm, I continue to use the familiar phrase 'Dreaming'. I do so for the sake of clarity, realizing that with such a decision I am sacrificing a certain degree of accuracy.

2 'Old men' (*purlka*) is a respectful term, one which confers status and acknowledges wisdom.

3 As a young man, Banjo Jungarrayi was forcibly rounded up at *Kunajarrayi*, where he had grown up 'naked, like a wild man'. He was transported to a cattle station where he worked briefly. During World War II, Jungarrayi was employed by the army as an unskilled laborer, and he was relocated to the government settlement at Yuendumu in the mid-1940s. Jungarrayi now resides at Nyirripi, the closest community to his 'country', but he wants to establish an outstation at *Kunajarrayi*. Jungarrayi expects a bore to be sunk, houses to be built, and the road to Kunajarrayi to be graded. Jungarrayi assumes that since he had worked for *Kardiya* (whites) they should continue to 'look after' him. In substantiating his right to outstation services, Jungarrayi also points out that he is a traditional owner of Kunajarrayi and has instructed many people on its significance. His knowledge, he asserts, entitles him to the services.

4 This is an abbreviated, annotated, and paraphrased account of the telling of this Dreaming story.

## References

Anderson, Christopher and Dussart, Francois (1988) Dreamings in acrylics: Western Desert Art, in: Sutton, Peter (ed.) *Dreamings: The Art of Aboriginal Australia*, New York: The Asia Society Galleries, 89–142.

Bardon, Geoff (1986) The origin of the painting movement, in: *Dot and Circle: A Retrospective Survey of the Aboriginal Acrylic Paintings of Central Australia*, Melbourne: Royal Institute of Technology, Communication Services Unit.

Basso, Keith (1996) *Wisdom Sits in Places*, Albuquerque, NM: University of New Mexico Press.

Bell, Diane (1983) *Daughters of the Dreaming*, Melbourne and Sydney: McPhee Gribble and George Allen Unwin.

Faulstich, Paul (1990) Landscape perception and visual metaphor in Warlpiri world view, PhD thesis, University of Hawaii.

Faulstich, Paul (1993) You read 'im this country: landscape, self and art in an Aboriginal community, in: Rose, Roger and Dark, Philip (eds) *Artistic Heritage in a Changing Pacific*, Honolulu: University of Hawaii Press; Bathurst: Crawford House Press, 149–161.

Faulstich, Paul (1994) Dreaming place: land and myth at Nyirripi, *Environments*, 22(2), 3–12.

Faulstich, Paul (1995) A man in the landscape: reflections on the person as a place, in: Oelschaeger, Max (ed.) *The Company of Others: Essays in Celebration of Paul Shepard*, Durango: Kivaki Press, 65–74.

Feld, Steven (1982) *Sound and Sentiment: Birds, Weeping, Poetics, and Song in Kaluli Expression*, Philadelphia, PA: University of Pennsylvania Press.

Jackson, Michael (1995) *At Home in the World*, Durham, NC: Duke University Press.

Latz, Peter (1995) *Bushfires and Bushtucker: Aboriginal Plant Use in Central Australia*, Alice Springs: Institute of Aboriginal Development Press.

Lee, Richard (1976) !Kung spatial organization: an ecological and historical perspective, in: Lee, R. and De Vore, I. (eds) *Kalahari Hunter–Gathers*, Cambridge, MA: Harvard University Press.

Levi-Strauss, Claude (1964) *Totemism*, London: Merlin.

Levi-Strauss, Claude (1966) *The Savage Mind*, London: Weidenfeld & Nicolson.

Lewis, David (1976) Observations on route finding and spatial orientation among the Aboriginal peoples of the Western Desert Region of Central Australia, *Oceania*, 46, 249–282.

Liberman, Kenneth (1980) Two world-creative carpet snakes from the Australian desert, *Folklore and Mythology Studies*, 4, 6–15.

Meggitt, M.J. (1966) Gadjari among the Warlpiri Aborigines of Central Australia, *Oceania Monograph* No. 14, Sydney: University of Sydney Press, 98–104.

Morphy, Howard (1980) What circles look like, *Canberra Anthropology*, 3, 17–36.

Munn, Nancy (1970) The transformation of subjects into objects in Walbiri and Pitjantjara Myth, in: Berndt, R. (ed.) *Australian Aboriginal Anthropology*, Nedlands: University of Western Australian Press, 141–163.

Munn, Nancy (1973) *Warlpiri Iconography: Graphic Representation and Cultural Symbolism in a Central Australian Society*. Ithaca, NY: Cornell University Press.

Myers, Fred (1986) *Pintupi Country, Pintupi Self*, Washington and Canberra: Smithsonian Institution Press and Australian Institute of Aboriginal Studies.

Napaljarri, Peggy Rockman and Cataldi, Lee (trans.) (1994) *Warlpiri Dreamings and Histories*, San Francisco, CA: HarperCollins.

Nelson, Richard K. (1983) *Make Prayers to the Raven: A Koyukon View of the Northern Forest*, Chicago, IL: University of Chicago Press.

Peterson, Nicolas, McConvell, Patrick, Wild, Steven and Hagen, Rod (1978) *A claim to Areas of Traditional Land by the Warlpiri and Kartangarurru-Kurintji*, Alice Springs: Central Land Council.

Radcliffe-Brown, A.R. (1930) Former numbers and distribution of the Australian Aborigines, *Official Year Book of the Commonwealth of Australia* 23, 687–696.

Relph, Edward (1985) Geographical experiences and being-in-the-world: the phenomenological origins of geography, in: Deamon, David and Mugerauer, Robert (eds) *Dwelling, Place, and Environment: Towards a Phenomenology of Person and World*, Boston, MA: Martinus Nijhoff, 15–32.

Rose, Deborah Bird (1992) *Dingo Makes us Human: Life and land in an Aboriginal Australian Culture*, Cambridge: Cambridge University press.

Rose, Deborah Bird (ed.) (1995) *Country in Flames: Proceedings of the 1994 Symposium on Biodiversity and Fire in North Australia.* Canberra and Darwin: Department of Environment, Sport and Territories, Biodiversity Unit and the Australian National University, North Australian Research Unit.

Rose, Deborah Bird (1996) *Nourishing Terrains: Australian Aboriginal Views of Landscape and Wilderness*, Canberra: Australian Heritage Commission.

Strehlow, T.G.H. (1965) Culture, social structure and environment in Aboriginal Central Australia, in: Berndt, R.M. and Berndt, C.H. (eds) *Aboriginal Man in Australia*, Sydney: Angus & Robertson, 212–245.

Williams, Nancy M. and Baines, Graham (eds) (1993) *Traditional Ecological Knowledge: Wisdom for Sustainable Development*, Canberra: Center for Resource and Environmental Studies, Australian National University.

Wolfe, Patrick (1991) On being woken up: the Dreamtime in anthropology and in Australian settler culture, *Contemporary Studies in Society and History*, 33, 197–224.

Yengoyan, Aram A. (1987) Economy, society and myth in Aboriginal Australia, in: Edwards, W.H. (ed.) *Traditional Aboriginal Society: A Reader*, Melbourne: Macmillan Company of Australia, 203–224.

# Part 8

# PHILOSOPHY AND ECOLOGY

# 100

# THE METAPHYSICAL IMPLICATIONS OF ECOLOGY

*J. Baird Callicott*

Source: *Environmental Ethics*, 8 (1986), 301–16.

Although ecology is neither a universal nor foundational science, it has metaphysical implications because it profoundly alters traditional Western concepts of terrestrial nature and human being. I briefly sketch the received metaphysical foundations of the modern world view, set out a historical outline of an emerging ecological world view, and identify its principal metaphysical implications. Among these the most salient are a field ontology, the ontological subordination of matter to energy, internal relations, and systemic (as opposed to oceanic) holism. I treat moral psychology as a special case of the metaphysical implications of ecology. Ecology undermines the concept of a separable ego or social atom and thus renders obsolete any ethics which involves the concepts of "self" and "other" as primitive terms.

## I

The subject of this paper is the metaphysical implications of ecology. From an orthodox philosophical point of view, not only is value segregated from fact, but philosophy is substantively informed only by the universal and foundational sciences.[1] The idea, therefore, that ecology—a scientific newcomer and a science remote from the more fundamental natural sciences—might have *metaphysical implications* may appear, on the face of it, ridiculous. So here at the beginning let me enter a couple of apologetic caveats.

Although it is not a foundational science like physics or a universal science like astronomy, ecology has profoundly altered our understanding of the proximate terrestrial environment in which we live, move, and have our being. And by "implications" I do not mean to suggest that there are logical relationships between ecological premises and metaphysical conclusions such that if the former are true the latter must also be true. *Imply*, *implicate*, and *implication* have a wider meaning evolved from the Latin root, *implicare*—to infold, involve, or engage—which I wish to evoke. Ecology has made plain to us the fact that we are enfolded, involved, and engaged within the living, terrestrial, environment—i.e., implicated in and implied by it. (This proposition is itself among the metaphysical implications of ecology.) Therefore, ecology also necessarily profoundly alters our understanding of ourselves, severally, and human nature, collectively. From this altered representation of the environment, people (personally and collectively), and the relationships in it and between it and ourselves, we may *abstract* certain general conceptual notions. These abstractive distillates are the metaphysical implications of ecology to which I draw attention in this discussion.

Ecology and contemporary physics, interestingly, complement one another conceptually and converge toward the same metaphysical notions. Hence, the sciences at the apex of the hierarchy of the natural sciences and those at the base, the "New Ecology" and the "New Physics," respectively, draw mutually consistent and mutually supporting abstract pictures of nature in its most elementary and universal and in its most complex and local manifestations.[2] A consolidated metaphysical consensus, thus, appears to be presently emerging from twentieth-century science which may at last supplant the metaphysical consensus distilled from the scientific paradigm of the seventeenth century.

## II

To bring dramatically to light the metaphysical implications of ecology, let me begin with a foil: the metaphysical ideas, just mentioned, implicated in modern classical science including pre-ecological natural history. Modern classical science adopted and adapted an ontology first set out in Western thought by Leucippus and Democritus in the fifth century B.C.—atomic materialism.[3]

The classical atom is essentially a mathematical entity and its so-called primary qualities may be precisely and quantitatively expressed as aspects or "modes" of geometrical space. An atom's solid mass was thus understood, mathematically, as a positive or "full" portion of negative or "empty" Euclidean space, shape its plane limits in a three-dimensional continuum, size its cubic volume, and motion its linear translation from one location (point) to another.[4]

The void and the simple bodies (or atoms) it contains were conceived by Democritus to be uncreated and indestructible. The theistic moderns conceived space, time, and the atoms to have been uniquely created by God as the permanent theater and immutable constituents of the universe.[5]

Composite bodies, the macroscopic things composed of atoms, however, routinely come into being and pass away. The "generation" and "corruption" (or "coming into being" and "passing away") of composite bodies were understood as the temporary association and dissociation of the atoms in the course of their ceaseless jostling and shuffling.[6]

Atomism, thus, is reductive. A composite body is ontologically reducible to its simple constituents. And the career of a composite body—its generation, growth, corruption, and disintegration—is reducible to the local motions of its several constituents.

And atomism, thus, is mechanical. All causal relations are reducible to the motion or translation from point to point of simple bodies or the composite bodies made up of them. The mysterious causal efficacy of fire, disease, light, or anything else is explicable, in the last analysis, as the motion, bump, and grind of the implacable particles. Putative causal relations which could not be so conceived—those postulated in astrology, magic, witchcraft, priestcraft, Newton's gravitational theory, the Faraday-Maxwell representation of magnetism, etc., etc.—were either dismissed as superstitions and their existence denied or regarded as physical problems awaiting a mechanical solution.[7] Only a mechanical solution could be satisfactory, since only a mechanical solution implicated exclusively the fundamental ontology of atomic materialism.[8]

This material, reductive, particulate, aggregative, mechanical, geometric, and quantitative paradigm in physics governed thought in other areas of philosophical interest, for example in moral psychology and biology.

Although Democritus, Lucretius, and Hobbes were thoroughgoing materialists and attempted to treat mind in exactly the same mechanical terms as, say, fire, light, and heat, dualism as espoused by Pythagoras, Plato, and Descartes became more characteristic of the dominant psychology of modern classical science.[9]

Mind, nevertheless, was derivatively and analogously conceived by the dualists in atomistic terms—as a psychic monad. Each mind, in other words, was a discrete psychic substance insulated within an alien (to its own nature) material cladding.[10] The mind was passively bedazzled and deluded by the bodily senses which were mechanically excited by the local "external" world. But minds were not otherwise informed by interaction with matter. That is, the rational structure of the human mind together with its passions and volitions was regarded as an independent given. By carefully sifting and sorting the raw, confused data afforded by sensation, disciplined rational minds could figure out the mechanical laws of the foreign material world and apply that knowledge to practical problem solving.

Given a monadic moral psychology, there are two fundamental options for ethics. For example, as represented most clearly by Hobbes, ethics might consist in finding the most felicitous rules to harmonize the inertia-like appetites of individual egos (or social atoms).[11] Or, a conceptual talisman to overcome the appetitive egoism of the discrete psychic monads might be posited. The concept of reason functions as such a transcendental principle in Kant's ethic.[12]

In biology, an even more subtle "conceptual atomism" prevailed. To explain the existence of natural kinds or species had been a major burden of Plato's theory of forms.[13] For each species or natural kind there was a corresponding eternal form or idea. Individuals acquired their "essences," their specific, discrete natures, by "participation" in the forms. Thus, lions were lions and differed from panthers because lions participated in the form lion, and panthers in the form panther. And so for horses, cows, and all other living specimens, each acquired the specific characteristics they to one degree or another possessed through their association as token to type with specific forms.

Although Aristotle (whose relationship to subsequent Western biology is comparable to Pythagoras' relationship to subsequent Western mathematics) rejected Plato's theory of independently existing forms, he retained the more insidious Platonic doctrine of essences. According to Aristotle, a thing's essence was its definition, given in terms of a classificatory hierarchy.[14] The universals of this hierarchy (later modified and refined by Carl Linnaeus)—species, genes, family, order, class, phylum, and kingdom—were not real or actual; only individual organisms fully existed. Nevertheless, for Aristotle, a species acquires its peculiar characteristics, not through interaction with other species, but through the place it occupies in a logically determined classificatory schematism.

Aristotle's teleological conception of nature introduced into biology a hierarchy of another sort. Some species were "lower," others "higher" on the scale of ends. Lower organisms existed for the sake of higher ones.[15] This habit of calling evolutionarily more venerable beings "lower organisms" persists today as an Aristotelian residue in modern biology much as the habit of referring to certain numbers as "square" or "cubic" persists as a Pythagorean residue in modern mathematics. The former, however, seems, somehow, more than a quaint and harmless terminological legacy of classical antiquity; it seems to impute a distinct pecking order to nature.

In sum, then, the endemic Western picture of living nature prior to its transformation by ecology might be characterized (or caricatured) as follows. The terrestrial natural environment consists of a collection of bodies composed of molecular aggregates of atoms. A living natural body is in principle a very elaborate machine. That is, its generation, gestation, development, decay, and death can be exhaustively explained reductively and mechanically. Some of these natural machines are mysteriously inhabited by a conscious monad, a "ghost-in-the-machine." Living natural bodies come in a wide variety of types or species, which are determined by a logico-conceptual order, and have, otherwise, no essential connection to one another. They are, as it were, loosed upon the landscape, each outfitted with its (literally God-given) Platonic or Aristotelian essence, to interact catch-as-catch-can.

Anthony Quinton has recently characterized the modern classical world view similarly, but even more graphically. According to Quinton:

> In that conception (the Newtonian) the world consists of an array of precisely demarcated individual things or substances, which preserve their identity through time, occupy definite positions in

space, have their own essential natures independently of their relations to anything else, and fall into clearly distinct natural kinds. Such a world resembles a warehouse of automobile parts. Each item is standard in character, independent of all other items, in its own place, and ordinarily unchanging in its intrinsic nature.[16]

## III

Ecology was given its name in 1866 by Ernst Haeckel, but the concept of an "economy of nature" had been current in natural history since Linnaeus had devoted a treatise to it a century earlier.[17] Although the idea of an orderly economy of nature was an improvement over the Hobbesian picture of nature as a chaotic free-for-all, Linnaeus and his exponents explicitly represented it in mechanical terms. Living nature is, as it were, a mechanical Leviathan, a vast machine which is itself composed of machines. "Like a planet in its orbit or a gear in its box, each species exists to perform some function in the grand apparatus."[18] The grand apparatus and its functions, to which each species is fitted, were, like the component species, believed to be designed by God. So, all natural relations and interactions remain external.

The subsequent Arcadian and Romantic intellectual countercurrents to eighteenth-century rationalism and mechanism, however, gave the proto-ecological notion of a natural economy a more integrative and holistic cast. Ecology as it eventually emerged as a distinct subdiscipline of natural history was shaped by a complex of governing metaphors derived from these minority traditions. Natural relations among species were portrayed, for example, by Gilbert White in the late eighteenth century, as a "harmony" and as a felicitous "balance"—balance both in the physical sense of a dynamic equilibrium and in the distinctly aesthetic sense of a tension and resolution of opposites, as in beautiful painting, poetry, and music.[19]

In contrast to the Linnaean designed, reductive, mechanical Leviathan, John Burroughs in the late nineteenth century posited an evolving and animated organic Leviathan, an idea later given theoretical definition and articulation by William Morton Wheeler.[20] Wheeler's exemplar was a beehive, a superorganism composed of multicelled organisms—the bees—which were in turn composed of single-cell organic units. Populations of bees, other insects, plants, avifauna, mammals, and so on are to biocoenoses as cells to multicelled organisms and bees to hives. The whole Earth's living mantle might similarly be represented as a vast "comprehensive" organic being. In Wheeler's representation, each higher level of organization is "emergent"; thus, the whole cannot be reduced to the sum of the parts.

The original notion of an economy of nature was itself a metaphor which at the turn of the century ecologists began to unpack to construct an important theoretical model. Plant and animal associations might be studied as "biotic communities." Phytographers Eugenius Warming and Frederick Clements introduced the idea of competitive dynamic "succession" in plant communities which typically evolved from "pioneer" to "climax" stages.[21] In the 1920s and 1930s, zoologist Charles Elton developed the community analogy further, with a greater emphasis on structure than process. Each species occupies a "trophic *niche*" in the biotic community which is, as it were, a "profession" in the economy of nature.[22] There are three great guilds—producers (the green plants), first and second-order consumers (herbivorous and carnivorous animals respectively), and decomposers (fungi and bacteria). In biotic communities the myriad specialists in each great group are linked in "food chains" which when considered together constitute tangled "food webs." Certain common structures characterized all biotic communities, however different their component species, and peculiar professions. For example, the producers must be many times more numerous than the consumers and the prey many times more numerous than predators; nor might any two species share precisely the same ecological niche.

Oxford University ecologist Arthur Tansley coined the term *ecosystem* in 1935 deliberately to supplant the more metaphorical characterizations of biocoenoses as "communities" of plants and animals or as "super-organismic" entities.[23] Tansley's ecosystem model of biotic processes was intended to bring ecology as a science out of a qualitative, descriptive stage, with anthropomorphic and mystic overtones, and transform it into a value-free, exact quantitative science. Hence, Tansley suggested that measurable "energy" contained in food coursed through the ecosystem and was at the foundation of its structure.

The scientific exemplar to which Tansley looked was physics. Of the so-called "New Ecology," for which Tansley's ecosystem model was the critical ingredient, Donald Worster writes:

> It owed nothing to any of its forebears in the history of science. . . . It was born of entirely different parentage: that is, modern thermodynamic physics, not biology.[24]

Hence, it is no wonder that the New Physics and the New Ecology should be conceptually complementary and convergent. Tansley's exemplar for a new paradigm in ecology was, it turns out, the new paradigm emerging in physics. The ecosystem model was expressly designed to be the "field theory" of modern biology.

However, as Worster emphatically points out, the quantitative, thermodynamic, biophysical model of nature which is the hallmark of the New Ecology was immediately turned to economic advantage as a powerful new weapon in mankind's age-old campaign to conquer nature. With the quantitative precision of which Tansley's energy circuit model was capable, ecosystems could be made more "productive" and "efficient" so as to "yield" a higher caloric "crop." But just as the philosophical interpretation of the New Physics, the Copenhagen Interpretation and its variations and alternatives, is quite another thing from its economic and military applications—from TV to laser weaponry—so the philosophical interpretation of the new ecology is quite another thing from its agronomic and managerial applications—from Ducks Unlimited to the green revolution. As Worster prophetically remarks: "Organicism has a way of gaining a foothold on even the most unpromising surface."[25]

## IV

At mid-century ecologist and conservationist Aldo Leopold strove to erect a secular encironmental ethic on evolutionary and ecological foundations.[26] In his land ethic one finds traces of Burroughs' organic image of nature, although Leopold himself seems to have gotten the idea from the Russian philosopher P. D. Ouspensky.[27] And certainly crucial to the conceptual foundations of the land ethic is Elton's community concept. However, when Leopold turns more deliberately to the construction of a "mental image" of the natural environment in relation to which he urged new ethical sensibilities he sketches, in poetic terms, the physics-born ecosystem model. According to Leopold, *land*, his shorthand term for the natural environment

> is a fountain of energy flowing through a circuit of soils, plants, and animals. Food chains are the living channels conducting energy upward [*sc.*, to the apex of the trophic pyramid]. . . . The velocity and character of the upward flow of energy depend on the complex structure of the plant and animal community. . . . Without this complexity normal circulation would presumably not occur.[28]

Ecologist Paul Shepard, a decade or so later, developed more consciously the metaphysical overtones of this field theory of living nature adumbrated by Leopold. According to Shepard, from the modern classical perspective

> nature is epitomized by living objects rather than the complex flow patterns of which objects are temporary formations. . . . [T]he landscape [from the classical point of view] is a room-like collection of animated furniture. . . . [B]ut it should be noted that it is best describable in terms of events which constitute a field pattern.[29]

Shepard, thus, more abstractively than Leopold, suggested that an object ontology is inappropriate to an ecological description of the natural environment. Living natural objects should be regarded as ontologically subordinate to "events" and/or "flow patterns" and/or "field Patterns." As reflectively represented at mid-century, from the point of view of a mature ecological science, the biological reality seems to be, at the very least, more fluid and integrally patterned and less substantive and discrete than it had been previously represented.

In the early seventies Yale University biophysicist Harold Morowitz still more deliberately and emphatically set out the field ontology suggested by Leopold and Shepard as a more ecologically informed portrayal of the natural environment. According to Morowitz:

> Viewed from the point of view of modern [ecology], each living thing is a dissipative structure, that is, it does not endure in and of itself but only as a result of the continual flow of energy in the system.... From this point of view, the reality of individuals is problematic because they do not exist per se but only as local perturbations in this universal energy flow.... An example might be instructive. Consider a vortex in a stream of flowing water. The vortex is a structure made of an ever-changing group of water molecules. It does not exist as an entity in the classical Western sense; it exists only because of the flow of water through the stream. If the flow ceases the vortex disappears. In the same sense the structures out of which the biological entities are made are transcient, unstable entities with constantly changing molecules dependent on a constant flow of energy to maintain form and structure.[30]

Later in the same decade, Norwegian philosopher Arne Naess attempted to persuade the community of academic philosophers that ecology might have important and sweeping metaphysical implications. Naess entered a caveat similar to the one registered at the outset of this discussion, *viz.*, that metaphysical conclusions "are not derived from ecology by logic or induction."[31] Rather, according to Naess, ecology "suggests" or "inspires" a "relational total field image [in which] organisms [are] knots in the biospherical net of intrinsic relations."[32] Naess called this metaphysical dimension of ecology "Deep Ecology," and the nebula of normative and public policy tendencies associated with it, "the Deep Ecology movement."

Let me sum up and attempt to express more precisely the abstractive general concept of nature distilled from the New Ecology in the tradition of Leopold, Shepard, Morowitz, and Naess. First, in the "Organic" concept of nature implied by the New Ecology as in that implied by the New Physics, energy seems to be a more fundamental and primitive reality than material objects or discrete entities—elementary particles and organisms respectively.[33] An individual organism, like an elementary particle is, as it were, a momentary configuration, a local perturbation, in an energy flux or "field."

The metaphysical ecologists here quoted, however, if pressed, would seem hardly prepared to deny outright a primary reality to atomic and molecular matter per se in addition to energy and its "flow." Organisms, though conduits of and configured by energy, remain *composed* of molecules—soild material substances. Rather, ecological interactions, primarily and especially trophic relationships, constitute a macrocosmic network or pattern through which solar energy, fixed by photosynthesis, is transferred from organism to organism until it is dissipated. Organisms are moments in this network, knots in this web of life.

However, if we combine quantum theory with ecology, as well as compare them, and resolve the erstwhile solid and immutable atoms of matter which compose the molecules, which in turn compose the cells of organic bodies, into the ephemeral, energetic quanta, then we may say quite literally and un-ambiguously that organisms are, in their entire structure—from subatomic microcosm to ecosystemic macrocosm—patterns, perturbations, or configurations of energy.

Deep Ecology poet and philosopher Gary Snyder captured this vertical integration of metaphysical ideas in a poem: "Eating the living gems of grasses/Eating the ova of large birds . . . /Drawing on life of living/*clustered points of light spun/out of space/*hidden in the grape."[34] In these lines the "clustered points of light, spun out of space" apparently allude to the dynamic configurations of the microcosm, the patterns of energy in the subatomic world and, obviously, eating grains, eggs, and fruit, unambiguously calls attention—especially by the persistent use of progressive verb forms ("eating," "drawing")—to the dynamic, patterned energy flux at the core of ecological relationships as conceived on the ecosystem model.

Second, the concept of nature emergent from the New Ecology, as that emergent from the New Physics, is holistic. It is impossible to conceive of organisms—if they are, as it were, knots in the web of life, or temporary formations or perturbations in complex flow patterns—apart from the field, the matrix of which they are modes. Contrary to the object ontology of classical physics and biology in which it was possible to conceive of an entity in isolation from its milieu—hanging alone in the void or catalogued in a specimen museum—the conception of one thing in the New Physics and New Ecology necessarily involves the conception of others and so on, until the entire system is, in principle, implicated.

Naess points out another sense in which ecology implies a holistic conception of the organic world, the import of which only an academic philosopher would be likely to notice. He claims, in effect, that ecology revives the metaphysical doctrine of internal relations.[35] This suggestion, remarkably, had been advanced even earlier by comparative philosopher Eliot Deutsch who also connected it with the Vedantic concept of *karma*.[36]

The doctrine of internal relations is, of course, associated with nineteenth and early twentieth-century German and English idealism—with the philosophies of Hegel, Fichte, Bradley, Royce, and Bosanquet. The basic idea is that a thing's essence is exhaustively determined by its relationships, that it cannot be conceived apart from its relationships with other things. Whatever the motives of the idealists (coherency theories of truth, the omniscience and omnipresence of spirit or whatever) and notwithstanding the inevitable entanglement of the doctrine of internal relations with other concurrently fashionable topics by mid-century neo-scholastic, academic philosophers (with "bare particulars," nominalism, the analytic-synthetic distinction, and so on) internal relations are straigh forwardly implicated in ecology.

From the perspective of modern biology, species adapt to a *niche* in an ecosystem. Their actual relationships to other organisms (to predators, to prey, to parasites and disease organisms, etc.) and to physical and chemical conditions (to temperature, radiation, salinity, wind, soil and water pH, and so on) literally sculpt their outward forms, their metabolic, physiological, and reproductive processes, and even their psychological and mental capacities. A specimen is, in effect, a summation of its species' historical, adaptive relationship to the environment. This observation led Shepard to claim that "relationships of things are as real as the things."[37] Indeed, I would be inclined to go even further. To convey an anti-Aristotelian thought in an Aristotelian manner of speech one might say that from an ecological perspective, relations are "Prior" to the things related, and the systemic wholes woven from these relations are "prior" to their component parts. Ecosystemic wholes are "logically prior" to their component species because the nature of the part is determined by its relationship to the whole. That is, more simply and concretely expressed, a species has the particular characteristics that it has because those characteristics result from its adaptation to a *niche* in an *ecosystem.*

It is necessary to add immediately, however, that the holistic concept of nature implied, analogously, by the New Ecology and the New Physics at the macro- and micro-levels of organization is holism of a different stripe than that associated with classical Hindu metaphysics. Eliot Deutsch and Fritjoff Capra both have draw this, perhaps natural, but unfortunate comparison in either domain, respectively. According to Capra, in the "Eastern world view" (which he seems to regard as a monolithic body of wisdom encompassing the independent indigenous traditions of thought form the Indian subcontinent to Japan) and in "modern [quantum] physics, all things are seen as interdependent and inseparable parts of this cosmic whole; as different manifestations of the same ultimate reality. The Eastern traditions constantly refer to this ultimate, indivisible reality which manifests itself in things.... It is called *Brahman* in Hinduism...."[38] And according to Deutsch, "What does it mean to affirm continuity between man and the rest of life [as in ecology]? Vedanta would maintain that this means the recognition that fundamentally all life is one, that in essence everything is reality;...that *Brahman*, the oneness of reality, is the most fundamental ground of all existence."[39]

There is crucial conceptual distinction, however, in the very different ways in which things are thought to be one in classical Indian thought, on the one hand, and in contemporary ecology and quantum theory, on the other. In classical

Indian thought all things are one because all things are phenomenal and ultimately illusory manifestations or expressions of *Brahman*. The unity of things is, thus, substantive and essential and the experience of it homogeneous and oceanic. In both contemporary ecology and quantum theory at their respective levels of phenomena the oneness of nature is systemic and (internally) relational. No undifferentiated Being mysteriously "manifests" itself. Rather, nature is a *structured, differentiated* whole. The multiplicity of particles and living organisms, at either level of organization, retain, ultimately, their peculiar, if ephemeral, characters and identities. But they are systemically integrated and mutually defining. The wholes revealed by ecology and quantum theory are unified, not blankly unitary; they are one more as organisms are one, than one as an indivisible, homogeneous, quality-less substance is one.

## V

Ecology has rather signal implications for moral psychology which we may treat here for convenience as part of metaphysics. Since individual organisms, from an ecological point of view, are less discrete objects than modes of a continuous, albeit differentiated whole, the distinction between self and other is blurred. Hence, the central problem of modern classical moral philosophy as elegantly exposed by Kenneth Goodpaster in a recent discussion—the problem of either managing or overcoming egoism—is not solved by the moral psychology implicated in ecology so much as outflanked.[40]

Paul Shepard has remarked that

> In one aspect the self is an arrangement of organs, feelings, and thoughts—a "me"—surrounded by a hard body boundary: skin, clothes, and insular habits. ... The alternative [aspect] is a self as a center of organization, constantly drawing on and influencing the surroundings. ... Ecological thinking ... requires a kind of vision across boundaries. The epidermis of the skin is ecologically like a pond surface or a forest soil, not a shell so much as a delicate interpenetration. It reveals the self ennobled and extended ... as part of the landscape and the ecosystem.[41]

He then goes on to endorse a notion earlier crystallized by Alan Watts (whose inspiration came from oriental philosophies)—that "the world is your body."[42]

Environmental philosopher Holmes Rolston has alluded to and extended Shepard's notion of the "relational self" as implied by ecology. Meditating by the shores of a Rocky Mountain wilderness lake, Rolston asks:

> Does not my skin resemble this lake surface? Neither lake nor self has independent being. ... Inlet waters have crossed this interface

> and are now embodied within me. . . . The waters of North Inlet are part of my circulatory system; and the more literally we take this truth the more nearly we understand it. I incarnate the solar energies that flow through this lake. No one if free-living . . . *Bios* is intrinsically symbiosis.[43]

As one moves, in imagination, outwardly from the core of one's organism, it is impossible to find a clear demarcation between oneself and one's environment. The environing gases and fluids flow continuously in and out. The organisms outside (and inside!) one's osmotic envelope continually, albeit selectively, are transubstantiated into and through oneself. In the time-lapse cinematography of imagination one can see oneself arising from the earth, as it were, a pulsating structure in a vast sea of other patterns large and small—some of them mysteriously translating through oneself—finally to be transmuted into the others. The world is, indeed, one's extended body and one's body is the precipitation, the focus of the world in a particular space-time locale.

This idea is very old, even in the West, expressed abstractly and philosophically by Heraclitus in the Greek tradition and concretely and poetically—with the phrase, "for dust thou art, and unto dust shalt thou return"—by the author(s) of Genesis-J. In the West, however, there still lingers the image of the substantive *nephesh, psyche,* soul, or conscious mind—the more vulnerable and self-pitying, the more diaphanous and insubstantial its organic cladding is perceived to be. Paul Shepard, however, has pointed out that the relational concept of self extends to consciousness as well as organism, to mind as well as matter. According to Shepard:

> Internal complexity, as the mind of a primate, is an extension of natural complexity, measured by the variety of plants and animals and the variety of nerve cells—organic extensions of each other.
>
> The exuberance of kinds [is] the setting in which a good mind could evolve (to deal with a complex world). . . . The idea of natural complexity as a counterpart to human intricacy is essential to an ecology of man.[44]

In a subsequent discussion Shepard elaborates this insight.[45] The more primitive elements of animal consciousness—palpable hunger and thirst, fear and rage, pleasure and pain—are as clearly evolutionary adaptations to an ever more elaborate ecosystem as fur and feathers, toes and digits, eyes and ears. The distinctive mark of human consciousness and the material of human reason are the systems of concepts embodied by human languages. Shepard has suggested that conceptual thought evolved as the taxonomical array of animals and plants was mapped by the emergent consciousness of primate hunter-gatherers. In a very direct way, therefore, human consciousness, including abstract rational thought, is an extension of the environment, just as the "environment" becomes fully actual in the mind-body unity of the New Physics, only as it interacts with consciousness.[46]

Shepard has constructed on this basis an interesting argument for species conservation: if we simplify and impoverish the Earth's ecosystems, we risk rendering future generations of human beings mentally degenerate. Lacking a rich and complex natural environment to support—as correspondent, analogue, and stimulus—a rich and complex intelligence, human intelligence may atrophy.

The relational view of self—both self as bodily organism and self as conscious, thinking thing—transforms egoism into environmentalism, to borrow Kenneth Goodpaster's felicitous phrase. As I have elsewhere pointed out, egoism has been regarded as axiologically privileged.[47] The intrinsic value of oneself is taken as a given. How to account for the value of "others"—human others and now nonhuman natural others—has been the principal problematic of non-egoistic ethics.[48]

However, if the world is one's body and one's consciousness not only images in its specific content the world around, but the very structure of one's psyche and rational faculties are formed through adaptive interaction with the ecological organization of nature, then one's self, both physically and psychologically, gradiently merges form its central core outwardly to the environment. One cannot, thus, draw hard and fast boundaries between oneself, either physically or spiritually, and the environment.

For me this realization took concrete form, as I stood tow decades and an ecological education later, on the banks of the Mississippi River where I had roamed as a boy. As I gazed at the brown silt-choked waters absorbing a black plume of industrial and municipal sewage form Memphis and followed bits of some unknown beige froth floating continually down from Cincinnati, Louisville, or St. Louis, I experienced a palpable pain. It was not distinctly located in any of my extremities, nor was it like a headache or nausea. Still, it was very real. I had no plans to swim in the river, no need to drink from it, no intention of buying real estate on its shores. My narrowly personal interests were not affected, any yet somehow I was personally injured. It occurred to me then, in a flash of self-discovery, that the river was a part of me. And I recalled a line from Leopold's *Sand County Almanac*: "One of the penalties of an ecological education is that one lives alone in a world of wounds."[49]

Australian conservationist John Seed, musing on his efforts on behalf of rain forest preservation in Queensland, has come to a similar conclusion:

> [A]s the implications of evolution and ecology are internalized... there is an identification with all life.... Alienation subsides....

"I am protecting the rain forest" develops to "I am part of the rain forest protecting myself. I am that part of the rain forest recently emerged into thinking."[50]

Ecology, thus, gives new meaning as well as new substance to the phrase, "enlightended self-interest."

## Notes

1 See for example, E. A. Burtt, *The Metaphysical Foundations of Modern Science* (Garden City, N. Y.: Anchor Books, 1954) and Ernest Nagel, *The Structure of Science* (New York: Harcourt, Brace and World, 1961).

2 The term *New Ecology* was first used in H. G. Wells, with Julian Huxley and G. P. Wells, *The Science of Life* (New York: Garden City Publishing Co., 1939), p. 961, to characterize ecology after the quantifiable "ecosystem" model was developed by Arthur Tansley in 1935. See Warwick Fox, "Deep Ecology: A New Philosophy of Our Time?" *The Ecologist* 14 (1984): 194–200, and J. Baird Callicott, "Intrinsic Value, Quantum Theory, and Environmental Ethics," *Environmental Ethics* 7 (1985): 257–75, for a discussion of the convergence and complementary characteristics of the New Physics and New Ecology.

3 John Gribbin, *In Search of Schrodinger's Cat: Quantum Physics and Reality* (New York: Bantam, 1984). claims that while "Newton had [atomism] in mind in his work on physics and optics, atoms only really became a part of scientific thought in the latter part of the eighteenth century when the French chemist Antoine Lavoisier investigated why things burn" (p. 19). But, according to Thomas Kuhn, *The Copernican Revolution: Planetary Astronomy and the Development of Western Thought* (Cambridge, Mass.: Harvard University Press, 1957), whose historical point of view is somewhat broader than Gribbin's, "early in the seventeenth century atomism experienced an immense revival. . . . Atomism was firmly merged with Copernicanism as a fundamental tenet of the 'new philosophy' which directed the scientific imagination" (p. 237).

4 "Primary" and "secondary" qualities were terms given to the distinction between putative actual and nonactual qualities of the elements by John Locke, *Essay Concerning Human Understanding* (New York: E. P. Dutton and Co., 1961). Locke attempted to ground the distinction empirically rather than theoretically, the futility of which was subsequently demonstrated by Berkeley. The revealing terms, "the full" and "the empty," are attributed to the fifth-century atomists by Aristotle, *Metaphysics* 985b4.

5 See G. S. Kirk and J. E. Raven, *The Presocratic Philosophers: A Critical History with a Selection of Texts* (Cambridge: Cambridge University Press, 1962), and Burtt, *Metaphysical Foundations of Modern Science.*

6 Ibid.

7 See Ernst Nagel, *Structure of Science.*

8 Ibid.

9 For Democritus' materialistic psychology see W. K. C. Guthrie, *A History of Greek Philosophy* (Cambridge: Cambridge University Press, 1965), vol. 2; for Lucretius' position see Titus Lucretius Carus, *De Rerum Natura*, trans. Robert Latham (Harmondsworth: Penguin, 1951); and for Hobbes' views see Thomas Hobbes, *Leviathan* (New York: Collier Books, 1962). For Pythagoras' dualism see W. K. C. Guthrie, *A History of Greek Philosophy*, (Cambridge: Cambridge University Press,

1962), vol. 1: for Plato's see especially the *Phaedo;* and for Descartes' see *Meditations on First Philosophy*, in E. S. Haldane and G. R. T. Ross, trans., *The Philosophical Works of Descartes* (Cambridge: Cambridge University Press, 1911), vol. 1.

10 There is remarkable unanimity of thought on this head among Pythagoras, Plato, and Descartes, the West's most influential dualists.

11 See Hobbes, *Leviathan.*

12 See Immanuel Kant, *Foundations of the Metaphysics of Morals* (New York: Bobbs-Merrill Co., 1959).

13 See especially Plato, *Parmenides* and *Phaedo.*

14 See Aristotle, *De Partibus Animalium* and *Politicus.*

15 Ibid.

16 Anthony Quinton, "The Right Stuff," *New York Review of Books*, 5 December 1985, p. 52.

17 See Ernst Haeckel, *Generelle Morphologie der Organismen*, 2 vols. (Berlin: Reimer, 1966); and Carl Linnaeus, "Specimen Academicum de Oeconomia Naturae," *Amoenitates Academicae II: Holmae* (Lugdoni Batavorum: Apud Cornelium Haak, 1751).

18 Donald Worster, *Nature's Economy: The Roots of Ecology* (Garden City, N. Y.: Anchor Books, 1979).

19 See Gilbert White. *The Natural History of Selborne* (New York: Harper, 1842).

20 See John Burroughs, "The Noon of Science" in *The Writings of John Burroughs*, vol. 17: *The Summit of the Years* (Boston: Houghton Mifflin and Company, 1913), and William Morton Wheeler, *Essays in Philosophical Biology* (New York: Russell and Russell, 1939).

21 See Eugenius Warming, *The Oecology of Plants*, rev. ed. (Oxford: Clarendon Press, 1909); and Frederick Clements, *Plant Succession: An Analysis of the Development of Vegetation* (Washington, D.C.: Carnegie Institution of Washington, 1916).

22 See Charles Elton, *Animal Ecology* (New York: Macmillan, 1927).

23 See Arthur G. Tansley, "The Use and Abuse of Vegetational Concepts and Terms," *Ecology* 16 (1935): 292–303.

24 Worster, *Nature's Economy*, p. 303.

25 Ibid., p. 332.

26 See Aldo Leopold. *A Sand County Almanac and Sketches Here and There* (New York: Oxford University Press, 1949).

27 See Aldo Leopold, "Some Fundamentals of Conservation," *Environmental Ethics* 1 (1979): 131–48.

28 Aldo Leopold, *A Sand County Almanac*, p. 216.

29 Paul Shepard, "A Theory of the Value of Hunting," *Twenty-Fourth North American Wildlife Conference* (1957), pp. 505–06.

30 Harold J. Morowitz, "Biology as a Cosmological Science," *Main Currents in Modern Thought* 28 (1972): 156.

31 Arne Naess, "The Shallow and the Deep, Long-Range Ecology Movement. A Summary," *Inquiry* 16 (1973): 98.

32 Ibid., p. 95.

33 Werner Heisenberg, *Physics and Philosophy: The Revolution in Modern Science* (New York: Harper and Row, 1958), remarked "[W]e may say that all elementary particles consist of energy. This could be interpreted as defining energy as the primary substance of the world.... The elementary particles are certainly not etenal and indestructible units of matter, they can actually be transformed into each other.... Such events have been frequently observed and offer the best proof that all particles are made of the same substance: energy" (pp. 70–71).

34 Gary Snyder,"Song of the Taste," reprinted in Bill Devall and George Sessions, *Deep Ecology: Living as if Nature Mattered* (Salt Lake City: Peregrine Smith Books, 1985), p. 12.
35 See Naess, "The Shallow and the Deep, Long-Range Ecology Movement."
36 Eliot Deutsch, "Vedanta and Ecology," in T. M. P. Mehederan, ed., *Indian Philosophical Annual* 7 (Madras: Center for Advanced Study in Philosophy, 1970): 1–10.
37 Paul Shepard, "Ecology and Man: A Viewpoint" in Paul Shepard and Daniel McKinley, eds., *The Subversive Science*: *Essays Toward an Ecology of Man* (Boston: Houghton Mifflin, 1967), p. 3.
38 Fritjoff Capra, *The Tao of Physics: An exploration of the Parallels Between Modern Physics and Eastern Mysticism* (Boulder, Colo.: Shambala, 1975), pp. 30–31.
39 Eliot Deutsch, "Vedanta and Ecology," p. 4.
40 See Kenneth Goodpaster, "From Egoism to Environmentalism," in Kenneth Goodpaster and Kenneth Sayre, eds.; *Ethics and Problems of the 21st Century* (Notre Dame: University of Notre Dame Press, 1979), pp. 21–35.
41 Paul Shepard, "Ecology and Man: A Viewpoint," p. 2.
42 See Alan Watts, *The Book on the Taboo Against Knowing Who You Are* (New York: Pantheon Books, 1996).
43 Holmes Rolston, III, "Lake Solitude: The Individual in Wildness," *Main Currents in Modern Thought* 31 (1975): 122.
44 Paul Shepard, "Ecology and Man: A Viewpoint," p. 4.
45 See Paul Shepard, *Thinking Animals: Animals and the Development of Human Intelligence* (New York: Viking Press, 1978).
46 See Joanthan Powers, *Philosophy and the New Physics* (London: Methuen, 1982).
47 See Callicott, "Intrinsic Value, Quantum Theory and Environmental Ethics."
48 See Goodpaster, "From Egoism to Environmentalism."
49 Aldo Leopold, *A Sand County Almanac with Essays on Conservation from Round River* (New York: Ballantine Books, 1966), p. 197.
50 John Seed, "Anthropocentrism," Appendix E, in Bill Devall and George Sessions, *Deep Ecology*, p. 243.

101

# ECOSYSTEM ECOLOGY AND METAPHYSICAL ECOLOGY

## A case study

*Karen J. Warren and Jim Cheney**

Source: *Environmental Ethics*, 15 (1993), 99–116.

We critique the metaphysical ecology developed by J. Baird Callicott in "The Metaphysical Implications of Ecology" in light of what we take to be the most viable attempt to provide an inclusive theoretical framework for the wide variety of extant ecosystem analyses—namely, hierarchy theory. We argue that Callicott's metaphysical ecology is not consonant with hierarchy theory and is, therefore, an unsatisfactory foundation for the development of an environmental ethic.

### Introduction

Appeals to ecosystem ecology have figured importantly in much recent work in environmental ethics. Andrew Brennan, Holmes Rolston, III, and J. Baird Callicott, as well as deep ecologists and other, have emphasized the importance of a detailed understanding of the findings and theory of ecosystem ecology for the delineation of a satisfactory environmental ethic. We concur that it is important. These various philosophers, however, understand the details and implications of ecosystem ecology for environmental ethics in a variety of quite dissimilar ways. Particularly problematic is the tendency for some (but by no means *all*; e.g., Brennan and Rolston are important exceptions) of these philosophers to construct "metaphysical ecologies" inspired by scientific ecology and then to recommend that metaphysical ecology serve as the basis either for the construction of an environmental ethic or for the development of an "ecological consciousness." Although such projects must be looked at on a case-by-case basis,[1] we are concerned

* The authors thank Holmes Rolston, III and an anonymous referee for helpful comments.

that many philosophers have been too freewheeling in their appropriations of ecosystem ecology for the advancement of their metaphysical and ethical programs and projects. As a case study of this problem, we take a close look at the metaphysical ecology developed by J. Barid Callicott, particularly as it is articulated in "The Metaphysical Implications of Ecology."[2] We test Callicott's metaphysical ecology against hierarchy theory, which we take to be the most viable attempt to provide an inclusive theoretical framework for the wide variety of extant ecosystem analyses.[3]

## Hierarchy theory

Many ecological controversies in modern ecosystem ecology can be understood as arguments between two different approaches: the "population-community" approach and the "process-functional" approach. The population-community approach focuses on the growth of populations, the structure and composition of communities of organisms, and the interactions among individual organisms. It is grounded in Darwinian theory of natural selection. It "tends to view ecosystems as networks of interacting populations whereby the biota *are* the ecosystems and abiotic components such as soil or sediments are external influences."[4] The process-functional approach is based on a quantitative, thermodynamic, biophysical model, which emphasizes energy flows and nutrient cycling. It assumes that the fundamental unit of ecosystems includes both organisms and physical components.

Although discussions of ecology often present "the ecological perspective" as if there were only one perspective, debates arising from differences between the population-community and process-functional approaches reveal that there currently is no single model of ecosystem ecology. Recent work in ecological hierarchy theory suggests a new way to reconceive ecosystem ecology, one that reveals the limitations of the two traditional approaches while suggesting a way to combine their strengths into a third approach. Advocates of this approach view the population-community and process-functional approaches as complementary, rather than (as they are typically seen to be) contradictory, and try to develop an integrative approach to the study of ecosystems.

Central to the hierarchy theory approach is the notion of an *observation set*. It can be defined as follows:

> By an observation set we mean a particular way of viewing the natural world. It includes the phenomena of interest, the specific measurements taken, and the techniques used to analyze the data. The space-time scale is an important property of an observation set, determining the total number of measurements and the intervals between them.[5]

In order to develop an adequate conception of ecosystems and to make meaningful ecosystem comparisons, one must abandon attempts to develop one way of observing the natural world and, instead, address the complexity of the natural system by considering many observation sets. One way by which the problem of the complexity of ecosystems is overlooked or avoided is by focusing on a single observation set—for example, by focusing exclusively on either the population-community or the process-functional view of ecosystems.

An important characteristic of an observation set is its spatiotemporal scale, because this scale changes as the ecological problem changes and because "ecological principles often do not translate well across these scales." The space-time scale is important for understanding such concepts as *stability, equilibrium*, the *temporary*, the *enduring*, the *local*, and the *global*, since these concepts only have meaning relative to some particular scale. Depending on the space-time scale used in any given observation set, "ecosystems have been seen as static or dynamic, as steady-state or as fluctuating, as integrated systems or as collections of individuals."[6]

An ecosystem must be defined relative to the spatiotemporal scale of the problem being addressed. How one views a forest stand, for example, will depend on the spatiotemporal scale through which one is viewing the world. As the spatiotemporal scale changes, the forest will be viewed

> ... (1) as a dynamic entity in its own right, (2) as a constant (i.e., nondynamic) background within which an organism operates, or (3) as inconsequential noise in major geomorphological processes. Thus, it becomes impossible to designate *the* components of *the* ecosystem. The designations will change as the spatiotemporal scale changes."[7]

The population-community and process-functional approaches to studying ecosystems are based on different observation sets with different space-time scales. For instance, ecologists who study individual organisms use space-time scales which result in viewing organisms as sturdy, endurable, well-defined, and separate entities, while those who study energy flows view organisms as temporary reservoirs of energy, "momentary configurations," "local perturbations in an energy flux or field." To continue the forest stand example, the stand can be looked at from an organismic standpoint (e.g., as enduring, stable individual trees or populations of trees) *or* from an energy flow and nutrient cycling standpoint (e.g., as fluxes and flows of carbon and oxygen recycled through photosynthesis). The important thing to notice is that these two observations, depending on different observation sets, are space-time scale dependent.[8]

Since these two approaches focus on different observation sets, different concepts of the ecosystem are required in each case. A biotic emphasis on ecosystems, for instance, subdivides ecosystems into groupings of primary producers, consumers, and decomposers. "The ecosystem is conceptualized in terms of aggregates of biotic units."[9] Abiotic influences (e.g., soil) tend to be considered as external to the system, operating on it from outside. A biotic emphasis is designed to deal with problems concerning the contributions of species populations to ecosystem function rather than with function per se. In contrast, a functional emphasis on ecosystems subdivides ecosystems into energy capture, nutrient retention, and rate regulation. A tree, for example, does not belong to a single subsystem in a functional approach; leaves and boles are assigned to different subsystems (energy capture and nutrient retention, respectively). The functional view of ecosystems does not need a notion of ecosystems based on species or organisms.

From the perspective of hierarchy theory, "processes from predation to nutrient cycling occur across a broad spectrum of time and space scales and are associated with different levels of ecological organization."[10] Explicit consideration of multiple scales is essential for addressing the complexity of ecological phenomena. It is the failure to acknowledge the key role of observation sets with their spatiotemporal scalar properties which accounts for the failure of current ecological theory to deal adequately with both ecological complexity and the conceptualization of ecosystems.

Hierarchy theory provides a framework for understanding the relationship between organisms in a system and the functions of the system without reducing one to the other. The resolution of the problem "is simply that the two fields are looking at the natural world from different observation sets."[11]

> . . . the dilemma of population versus ecosystem ecology results from a false dichotomy. If we look across the range of observation sets that constitute the subject matter of ecology, we see the ecosystem is at least a dual entity. In one dimension, it is structured according to constraints involving organism interaction and natural selection. In another dimension, it is structured according to constraints that involve mass balance and thermodynamics.[12]

We have not yet touched on that aspect of hierarchy theory which gives the theory its name. the central notion is that ecosystems are organized into levels of organization ("hierachy structures") which result from differences in process frequency rates (with the natural world acting as the processor of energy and matter).[13] The structure imposed by differences in process rates is "sufficient to decompose a complex system into organizational levels and into discrete components within each level."[14] Rapid process rates occur at lower organizational levels and slower rates occur at higher organizational levels.

> For example, individual tree leaves respond rapidly to momentary changes in light intensity, $CO_2$ concentration, and the like. The

> growth of the tree responds more slowly and integrates these short-term changes. Changes in the species composition of the forest occurs even more slowly, requiring decades or even centuries.[15]

Differences in process rates not only isolate levels from each other, but also "help explain how the system responds to environmental fluctuations."[16] To continue with the tree example, the annual growth response of a tree reflects only the average integrated response of the leaves to sunlight; the oscillations at the lower level are dampened down and "the lower . . . level has communicated only its average response to the higher level."[17]

Frequency rates affect not only the relative disconnections *between* organizational levels, but also the connections *within* organizational levels:

> Within an organizational level, a hierarchical system can be further decomposed into subsystems or holons . . . on the basis of difference in rates. Within each holon [subsystem], components interact frequently (i.e., strongly) with each other but only infrequently (i.e., weakly) with components of other holons [subsystems].[18]

"Within-level structuring" helps explain why properties of components at various hierarchical levels are "averaged, filtered, and smoothed" as they become part of the *aggregated* output of a holon. Because of this filtering, the effect of lower-level behavior on higher-level behavior is less and less recognizable as the number of intervening levels increases.[19]

This last point has important implications for how one understands the oft-repeated remark that ecology demonstrates that everything is connected with everything else. In fact, a function of the hierarchical organization of an ecosystem is *not* to connect up everything with everything else, but rather to disconnect elements of the system. As O'Neill et al. point out,

> If all parts of a complex system interact directly and symmetrically, then there is a low probability that the system will endure [*precisely because*] . . . a disturbance anywhere in the system tends to affect all other parts of the system.[20]

Because process rates become slower at higher levels, the asymmetrical relationships resulting from the filtering and dampening effects of hierarchical organization produce systems in which "higher levels . . . are relatively unresponsive to changes in the lower level."[21]

> Hierarchy theory suggests that in a stable system, direct and symmetric connection should be isolated in well-defined holons. Interactions between components of different holons should be rare.

> The system should be further stabilized by constraints representing asymmetric relationships between levels in the hierarchy.[22]

When we put together the notion of the dual (population-community and process-functional) hierarchies with the notion we have just developed, that the relative disconnection between holons and levels is crucial to ecosystem stability, we can appreciate the further conclusion that O'Neill et al. draw: "It is quite feasible and even reasonable to maintain an individualistic (i.e., Gleasonian) concept of the community and a holistic concept of ecosystem function."[23] O'Neill et al. argue that functional and biotic hierarchies represent distinct hierarchies, each of which (at least) is required if we are to explain the full range of ecological phenomena; neither hierarchy can be reduced to the other.[24]

O'Neill et al. are mercifully restrained in their use of the overworked notion of non-reductionism. Although they argue that ecosystem hierarchies are not reducible to one another, they are quite clear that descriptions of organizational levels *within* hierarchies *can* be explained by reference to the properties and interactions of their component holons. As they put the matter,

> ... hierarchical descriptions help to manage complexity by isolating dynamics at a single level and ignoring details at lower organizational levels ... [W]e normally adopt hierarchical explanations when dealing with complex systems. We look to higher levels for significance and to lower levels for mechanisms. ...[25]

As Andrew Brennan has recently argued, there is no ontological significance in such explanations.[26] The mechanistic explanations within hierarchies are not intended to deny that the levels of organization explained by reference to the behavior of subsystems have properties in their own right, properties which are not properties of the component subsystems. The constraining functions of levels of organization are just such properties.

What O'Neill et al. object to is the naive reductionism which results from the assumption that there is only one way to subdivide an ecosystem; they object to the view that functional features of an ecosystem can be reduced to organisms, populations, or communities as their relevant subsystems and *vice versa*. The attempt to understand the stability-complexity hypothesis in terms of species diversity rather than in terms of the diversity and complexity of energy-flow pathways is just such an example of naive reductionism.[27]

> The real problem with [naive] reductionism is not that it advocates analysis into simpler subsystems. This [form of reductionism] is necessary to arrive at mechanistic explanations. The problem is that [naive] reductionism advocates, in advance, one set of subsystems, which it posits to be the only admissible set of components.[28]

The mode of explanation employed by hierarchy theorists is, thus, compatible with their view that several hierarchies may be discernible in ecosystem analysis—none of which is reducible to any other (biotic levels of organization, for example, do not reduce to process rate structures and vice versa)—*and* with their view that there are no "bottom line" components out of which ecosystems are constructed. Which analysis is appropriate varies from problem to problem depending upon the observation set and, more specifically, upon the spatiotemporal scale appropriate to the problem at hand. There is "no need to force the phenomena into a set of components that are observable and interesting at some other space and time scale."[29]

This discussion of reductionism leads to our final comment concerning hierarchy theory: it naturalizes ontology by privileging non-foundational, contextualist epistemology over foundational ontology. In hierarchy theory, certain notions which once figured in ecology as characterizations of ecosystems per se (e.g., equilibrium, stability, and the "balance of nature") are "contextualized" by fastening them to particular observation sets and spatiotemporal scales. Controversy over whether ecosystems or equilibrium or disequilibrium systems, or whether ecosystems are stable or unstable, for example, are not resolved by asking whether equilibrium or stability characterize ecosystems per se, but by contextualizing those notions to observation sets. According to O'Neill et al.,

> . . . the spatiotemporal scale of a problem will determine which dynamics are to be considered in equilibrium and which are to be considered as changing. Over long geologic time scales, the landscape is continuously changing due to uplift and erosion. On this time scale, the individual river must be considered as far from equilibrium. But if the observation set is much shorter, perhaps in years, then it is most useful to view the drainage system as being in equilibrium. Here attention focuses on the negative feedbacks and self-regulation processes that tend to keep the riverbed in relatively constant configuration.[30]

This contextual characterization of ecosystems suggests that the epistemological requirements of particular ecological problems dictate to ontology; an acceptable answer to an ecological problem must not be pressed into the shape of a preconceived, preferred ontology.

## Callicott's metaphysical ecology

We turn now to the work of J. Baird Callicott. In "The Metaphysical Implications of Ecology," Callicott spells out what he takes to be the metaphysical view suggested by scientific ecology.[31] The point of this project is to construct an environmental ethic based on those metaphysical implications, an ethic which will therefore be consonant with scientific ecology. For ease of reference, we number the claims Callicott makes.

1. Following Arne Naess and other deep ecologists, Callicott does not claim that the metaphysical views he teases from his reflections on the science of ecology (as well as from high-energy physics) are *logically* implied by ecology; he merely states that from the picture of the environment and our place in it supplied by contemporary ecology "we may *abstract* certain general conceptual notions." In other words, ecology *suggests* certain metaphysical views to Callicott.[32]

2. For Callicott, quantum theory aids in generating the metaphysical implications of ecology. "Ecology and contemporary physics," Callicott holds, "complement one another conceptually and converge toward the same metaphysical notions."[33]

3. Callicott endorses the view that an "object ontology is inappropriate to an ecological description of the natural environment. Living natural objects should be regarded as ontologically subordinate to 'events' and/or 'flow patterns' and/or 'field patterns.'" Stated differently, "energy seems to be a more fundamental and primitive reality than material object or discrete entities." Ecology establishes that organisms are *conduits* of, and *configured* by, energy. In accordance with point (2), it takes physical theory to establish that "organisms are, in their entire structure . . . patterns, perturbations, or configurations of energy."[34]

4. Appealing to "the doctrine of internal relations," i.e., the view (according to Callicott) that "a thing's essence is exhaustively determined by its relationships, that it cannot be conceived apart from its relationships with other things," Callicott holds that nature is essentially holistic:

> . . . the concept of nature emergent from the New Ecology, as that emergent from the New Physics, is holistic. It is impossible to conceive of organisms . . . apart from the field, the matrix of which they are modes. . . . [T]he conception of one thing in the New Physics and New Ecology necessarily involves the conception of others and so on, until the entire system is, principle, implicated.[35]

5. Going beyond the notion that relations are as real as the things related, Callicott holds that "relations are 'prior' [i.e., logically prior] to the things related, and the systemic wholes woven from these relations are 'prior' to their component parts." Thus, the holism Callicott invokes is one in which

> . . . the oneness of nature is systemic and (internally) relational. . . . The multiplicity of particles and living organisms . . . retain, ultimately, their peculiar, if ephemeral, characters and identities.[36]

6. Finally, for Callicott, not only is an object ontology obsolete, point (3), but also "Ecology undermines the concept of a separable ego or social atom and thus renders obsolete any ethics which involves the concepts of 'self' and 'other' as primitive terms." Otherwise put,

> As one moves, in imagination, outwardly from the core of one's organism, it is impossible to find a clear demarcation between oneself and one's environment.... The world is, indeed, one's extended body and one's body is the precipitation, the focus of the world in a particular space-time locale.[37]

For Callicott, this blurring of any distinction between self and other allegedly transforms egoism into environmentalism.[38]

With respect to point (1), there is something disingenuous about Callicott's disclaimer. Unlike Naess, who carefully details his view concerning the relationship between scientific ecology and deep ecological metaphysics,[39] Callicott's caveat seems a mere token. He proceeds to an account which reads for all the world as if the metaphysical view expressed *is* the one implied by a joint consideration of contemporary ecology and quantum physics. Nowhere does he discuss (or even hint at) the additional metaphysical and evaluative assumptions required to "abstract" his particular metaphysical views.

This point aside, one might well ask whether the metaphysical view developed by Callicott *is* in fact the position that is most in accord with state-of-the-art ecological theory, viz. hierarchy theory. We argue that it is not in the remainder of this paper.

With respect to Callicott's point (2)—the complementary nature of ecology and physics—we agree with Rolston when he argues that

> If quarks exist, the Yellowstone ecosystem is made of quarks, but the levels are so vastly different that we cannot generalize from observer dependent phenomena in quantum physics to the relation of the human self to the Yellowstone ecosystem. If there is an interdependence in an ecosystem that funds ethics, this will have to be shown without benefit of physics.[40]

Quantum physics is critically important for Callicott in establishing his "metaphysical implications of ecology," as point (3) makes clear. The holistic metaphysical cast of his "Implications," along with its downgrading of "object ontology," points (3) and (5), and its application of the doctrine of internal relations, point (4), is derived from Callicott's understanding of high-energy physics; it cannot be derived from scientific ecology alone. However, if hierarchy theory is on the right track, then object ontology-laden population-community observation sets *cannot* be reduced to process-functional observation sets. Organisms are *not* merely conduits of energy or functional elements in the flow of energy through the system. Yet, such a reductionism is precisely what is required in order to render organisms "ontologically subordinate" to events or flow patterns; it is the "naive reductionism"—resulting from the assumption that there is but one way to subdivide an ecosystem—to which O'Neill et al. object.[41]

We stated earlier that the methodology of hierarchy theory makes the epistemological requirements of particular problems (including requirements imposed by differing observation sets) dictate to ontology (not *vice versa*). With entirely different subsystems as foci of differing observation sets, it seems quixotic to strive for an articulation of *the* structure (even *the* hierarchiacal structure) of an ecosystem. Hierarchy theory is an epistemological/methodological means of investigating ecological problems. The ontology that emereges from any particular investigation is relative to the observation set which produces it. This relativity does not make that ontology subjective, but it does mean that "the ontological question is a moot point for scientist."[42]

The grand attempt to provide *a* metaphysics to ground morality puts the metaphysical/ontological cart before the epistemological/methodological horse. This is not say that the entities posited by ecosystem ecology are not there in the world; rather, it is to underscore the fact that the ontology embedded in both the explanation and phenomena being explained is a function of the appropriate observation set. We cannot generate an observation-set independent ontology which specifies *the* components of the universe to which various observation-set dependent ontologies must be reducible.[43] Consequently, Callicott's rejection of any object ontology, point (3), is unfounded.

Point (4) concerns Callicott's acceptance of a strong version of internal relations: the essence of a thing is *exhaustively* determined by its relationships and the conception of one thing *necessarily* involves the conception of others until, by reiteration, the entire system is drawn in. The image that emerges from Callicott's notion of internal relations is a natural world in which everything is connected to everything else in a densely interrelated system in such a way that a tug on any part of the system has an effect on every other part of system.[44] Yet, as we have seen, this image is one that O'Neill et al. (as well as Brennan) have been at pains to dispel. They argue that "the old imagery of the natural world as having everything connected to everything else is shortsighted. It is the relative disconnection that constitutes the organization of the system."[45] Critical to the stability of an ecosystem is the relative insulation of holons from one another and of levels of organization from the fluctuation in process rates of their subsystems. Overconnectedness in a system, where tugs on any part of the system produce effects on all parts of the system, is *un*stable.[46]

Callicott could respond that he does not intend to deny this point and, indeed, grants that organisms "retain, ultimately, their peculiar, if ephemeral, characters and identities." Perhaps it is a matter of emphasis: Callicott emphasizes the real connectedness that is there, while we emphasize the disconnectedness in the system. Perhaps. However, Callicott surely suggests and invokes the metaphysical notion of a densely interconnected universe when he claims that "The world is, indeed, one's extended body and one's body is the precipitation, the focus of the world in a particular space-time locale." This idea, in turn, leads to the notion that "one's self... *gradiently* merges from its central core outwardly to the environment," which then leads to the ethical notion that egoism is transformed into environmentalism: environmental protection is (quite literally and directly) *Self*-protection.[47]

In contrast, we suggest that, at the very least, an adequate metaphysical ecology must acknowledge that the world "strives," so to speak, to organize itself into discrete and relatively disconnected or autonomous holons and hierarchical levels of organization as a condition of its own stability. Relative disconnection and autonomy are at least as important a feature of our world as is its connectedness. Indeed, ecologically speaking, individuals come into their own with the same sterling ontological credentials as the energy flow patterns which emerge from process-functional analyses. Although everything may be tied to everything else in *some* sense, hierarchy theory suggests that these connections do not support Callicott's metaphysical ecology.

On our view, Callicott's claim that "a thing's essence is exhaustively determined by its relationships, that it cannot be conceived apart from its relationships with other things,... until the entire system is, in principle implicated,"[48] is best regarded as hyperbole from the point of view of informed ecological theory. We can agree with Brennan that "what we are and ought to be is partly determined by where we are" without buying into Callicott's metaphysical ecology. Certain ecological observation sets relevant to ethics yield an ontology of autonomous individual organisms interacting with one another. Other observation sets, of course, paint a holistic picture of ecosystem function. But there is no *a priori* reason (other than a misguided reductionism) to give ethical or metaphysical pride of place to the latter. Just as "It is quite feasible and even reasonable to maintain an individualistic (i.e., Gleasonian) concept of the community and a holistic concept of ecosystem function,"[49] so is it also quite feasible and reasonable to understand the moral community as consisting, in part, of autonomous agents with properties in their own right, while at the same time treating that community as in some respects holistic. We need to formulate our "ethical ontology" in light of an understanding of our epistemological relationship to the objects of moral concern. In terms of actual practice, we certainly can say things, significant and important things, about individuals without drawing the rest of the universe into our discussion.

Brennan takes this position in his discussion of the "duality" or "double aspect" of value, according to which an individual may be considered from the point of view both of its "basic" and its "supervenient" properties. As Brennan puts it, there is

> ...a certain truth is holism. In metaphysical terms, this truth involves the recognition of supervenient characteristics as significant, as part of the story of what a thing is. The supervenience, however, involves the dynamics of the particular item and its nesting within a larger environment of action.... For all forge, one way or another, their identities through processes of interaction with their surroundings... while maintaining their own separateness.[50]

Callicott's metaphysical holism is altogether too tight: it is highly problematic from the point of view of hierarchy theory and it is unnecessary for the construction of an ethic that recognizes communities and ecosystems as having properties in their own right—properties supervenient on individuals which have properties in *their* own right as well. The portrait or a densely interrelated universe probably has at its motivational core the view that if the universe, from the point of view of physical theory, is a field of energy, then organisms are merely perturbations in this field, intimately connected with all parts of the field. This picture not only ignores the important lesson learned from hierarchy theory that relative *dis*connection constitutes the organization of ecosystems, but also opts for a kind of metaphysical organicism in which relations are logically prior to the things related and in which the wholes woven from these relations are logically prior to their parts. Such a metaphysical organicism is simply untrue to ecosystem ecology, especially that aspect which focuses on population-community dynamics. Because Callicott mistakenly imposes a process functional framework on population-community accounts, he produces a simplistic metaphysical portrait of ecosystem organization. His position thus show that the temptations of the organismic metaphor often mask the particular and varied kinds of complexity actually manifested in the workings of ecosystems.[51]

With point (6), we confront the ethical implications that Callicott wishes to draw from his metaphysical ecology—namely, that the world is one's "extended body," rendering "obsolete any ethics which involves the concepts of 'self' and 'other' as primitive terms." With these claims, egoism becomes the foundation of environmentalism.

One phrase that Callicott uses is particularly revelatory of the way metaphysics figures in Callicott's ethical reflections. In the explication of his position he writes, "As one moves, in *imagination*, outwardly from the core of one's organism, it is impossible to find a clear demarcation between oneself and one's environment."[52] It may be true in *metaphysical* imagination, as we might justly amend the phrase, but is *not* true in experiential encounter. Ecosystem ecology provides an object lesson in just how faulty such metaphysical imagination can be.

What ethical mileage, then, can genuinely be made of Callicott's metaphysics of (systemic and internally relational) oneness? Granted that the objects of our ethical concern are often wholes or entities broader in scope than individual organisms (e.g., ecosystems, populations, species), we rarely if ever concern ourselves with the entire energy field which, for Callicott, is constitutive of the universe. We focus, rather, on chunks of this field—often, though not always, on individual organisms. When we do so, what are we to make, *ethically* speaking, of the claim that any ethic which employs *self* and *other* (or, for that matter, *anything* less than the whole fabric of existence) as "primitive" terms is obsolete? Is this claim supposed to imply that only the universe as a whole (which, on Callicott's view, *I am*) can be the direct object of moral regard?

We have argued at length that hierarchy theory implies that even if at some level of inquiry it is plausible to hold that the universe and everything in it is constituted of energy, that everything is perturbation in an all-encompassing energy field, such a theory does *not* imply that entities revealed through other observation sets are "primitive," that they are reducible to the ontology of some other observation set. We can get no ethical mileage from Callicott's view that *self* and *other* cannot be "primitive" terms in an ethic because that view is faulty. Even if (or, more appropriately, when and where) the view that we are all connected (because in some sense we are all perturbations in an all-encompassing energy field) is true, it says nothing whatsoever about whether concepts of self and other may or may not appropriately be "primitive" within that context of a particular set of ethical reflections.

Suppose, however, that we *were* to grant Callicott his metaphysical ecology. Would his view that egoism is transformed into environmentalism hold? That we are all systemically and relationally one could just as well be understood as implying that I am part of another's body. Why, that is to say, does Callicott take it that oneness implies that I am to understand the world as *my own* body rather than understand myself as part of the body of another?[53] If oneness were understood by me in this latter way, it would then be taken to enjoin concern and care for the way in which I am connected to the other for the sake of the other. In this way, one could just as well posit an ethic of self-sacrifice as an egoistic ethic. Then the emphasis of my concern would be on how I shape or affect the other, not on the other conceived as part of myself. While on both construals I am inextricably bound up with the other, in an ethic of self-sacrifice I must tend to my self primarily or only as a way of caring about the other, whose value and ethical standing is presumed. I am ethically enjoined to route my ethical reflection through the eyes of the other with whom I recognize myself as bound to as one. With the possibility of such an alternative "logic of oneness" in place, what is at issue here becomes clearer: Callicott's logic of oneness (his "egoism becomes environmentalism") is tied to an ethic of atomistic individualism just as the alternative logic we have sketched (but do not endorse) is tied to an ethic of self-sacrifice, each with historically gender-identified male and female associations, respectively. Neither seems to us an attractive candidate for a "metaphysics of morals."[54]

The truth buried in the view that egoism becomes environmentalism and that the world is my body is that a *contextual* understanding of individuals (egos and bodies as well as of populations and species) is important for ethical deliberation. Ecosystem ecology, understood through the lens of hierarchy theory, is of tremendous value in shaping such a contextual understanding. This contextual understanding is important for at least three reasons.

First, *what* a thing (person, community, population, species, river, etc.) is surely is partly a function of *where* it is, i.e., a function of the relationships in which it stands to other things and to its history, including (where applicable) its evolutionary history. As Brenman demonstrates, many of the ethical concerns of metaphysical holists such as Callicott and the deep ecologists can be addressed by reference to such a contextualism *without* any need to resort to a dubious metaphysical holism.

Second, a contextual understanding of individuals (populations, species) is important in order that we may assess the putatively universal claims of reason and ethical deliberation. Here we are speaking more specifically of a historicized understanding of human thought and reflection. We agree with Brennan when he argues that he does not

> ... deny the possibility of criticising a society and its form of life, either from the outside, or from within, by bringing forward considerations about rationality, dignity, respect and moral decency which are drawn from a fairly remote standpoint where human life has been considered in universal and highly general terms. What I do argue against is *ethical colonialism* whereby those situated in certain kinds of society, surrounded by certain kinds of goods and activities, declare that they have discovered universal ethical truths which do no more than reflect, in a suitably generalised way, their own local aspirations and ideals.[55]

Third, carrying Brennan's point still further, we stress the idiographic dimension of our ethical journeys and therefore of ethics itself. Holmes Rolston, III has been a strong advocate for the recognition of the idiographic aspect of ethical thought; his advocacy derives from his understanding that a thing is what it is, in part, because of where it is. As Rolston puts it,

> An ethics should be rational, but rationality inhabits a historical system. The place that is to be counted morally has a history; the ethics that befits such a place will take on historical form; the ethics will itself have a history. The place to be mapped ... will have twin

> foci. One focus will be nomothetic, recurrent; the other will be idiographic, uniquely particular . . . The rationality of the ethic, as well as the area to be mapped, will be historical. That is, logic will be mixed with story. The move from *is* to *ought* . . . is transformed into movement along a story line.[56]

We do not want to split off the idiographic as what ethics *permits*, provided that the *universal* demands of morality are met. Rather, the ethically idiographic is the very *center* of each individual's ethical life. It is not only the place from which we test the claims of the "universal" and the "rational," but also the place (not the *no*place) from which we construct the "universal" claims of "rationality." The "universal" and the "rational" are *always* in some manner or other inflected with historicity. The "universal" and "rational" are themselves moments in a story. It is this which makes the claims of otherness and an ethic of difference so important.

On our reading of Callicott's metaphysical ecology, such claims to otherness and difference, whether of individuals or ecosystems, are lost. As such, Callicott's metaphysical ecology is not an ecologically or ethically adequate basis for an environmental ethic.

## Notes

1 See, for example, Jim Cheney, "Eco-Feminism and Deep Ecology," *Environmental Ethics* 9 (1987): 115–45; "The Neo-Stoicism of Radical Environmentalism," *Enviromental Ethics* 11 (1989); 293–325; review of Arne Naess, *Ecology, Community and Lifestyle*, in Environmental Ethics 13 (1991): 263–73.

2 J. Baird Callicott, "The Metaphysical Implications of Ecology," *Environmental Ethics* 8 (1986): 301–16.

3 Our account of hierarchy theory is based upon R. V. O'Neill, D. L. DeAngelis, J. B. Waide, and T. F. H. Allen, *A Hierachical Concept of Ecosystems* (Princeton: Princeton University Press, 1986). See also R. V. O'Neill, "Perspectives in Hierarchy and Scale," in J. Roughgarden, R. M. May, and, S. A. Levin, eds., *Perspectives in Ecological Theory* (Princeton: Princeton University Press, 1988), pp. 140–56. Some of this material is drawn from Karen J. Warren and Jim Cheney, "Ecological Feminism and Ecosystem Ecology," *Hypatia* 6 (1991): 181–83. We do not here argue for our claim concerning the viability and philosophical sophistication of hierarchy theory. Nor do we examine the recent impact of the "science of chaos" on ecosystem ecology. For an interesting assessment of the latter, see Donald Worster, "The Ecology of Order and Chaos," *Environmental History Review* 14 (1990): 1–18.

4 O'Neill et al., *A Hierarchical Concept*, p. 8. Although an ecosystem ecologist who approaches ecosystem analysis from a population-community point of view may well count abiotic elements as part of the ecosystem, the population-community approach tends to treat abiotic components as "environment," the "backdrop or context within which biotic interactions occur," so that, in effect, "the biota *are* the ecosystem and abiotic components . . . are external influences."

5 Ibid., p. 7.

6 Ibid., p. 20.

7 Ibid., p. 83.

8 O'Neill et al. stress that they have exaggerated the differences between the biotic (population-community) and functional approaches, and that "few ecologists would hold to either extreme of the spectrum." The distinction between the two approaches is better viewed on a continuum, with the biotic and functional approaches at each end and ecologists "drawn in one direction or the other by the specific problems that interest them." Ibid., p.10.

9 Ibid., p. 11.

10 Ibid., p. 30.

11 Ibid., p. 186.

12 Ibid., p. 209.

13 Ibid., p. 75–76.

14 Ibid., p. 76.

15 Ibid.

16 Ibid., p. 78.

17 Ibid.

18 Ibid., p. 79. "Each holon can be defined in terms of a boundary or surface that encloses its components and separates them from components in the rest of the system. . . . " The boundaries of holons are not necessarily the recognizable boundaries of elements of the traditional hierarchy cell/organism/population/community. Boundaries, rather, are determined by difference in process rates.

19 Ibid., p. 80.

20 Ibid., p. 94.

21 Ibid., pp. 94–95. Instability, in fact, results from both underconnectedness and overconnectedness. Removal of a keystone predator, for example, results in an underconnected system setting off "a chain reaction in which populations are eliminated from the system and the constraint system is changed." Pests or infection in a dense population (e.g., a monoculture planting) result in over connectedness and its well-known effects. A tightly organized system doesn't *produce* stability; it in fact *requires* environmental stability to endure. This is why "the most tightly organized systems may exist in relatively stable environments such as the tropics." Ibid., pp, 126, 145.

22 Ibid., p. 127.

23 Ibid., p. 189.

24 O'Neill et al. point out that the process-functional and population-community hierarchies are but two of the observation sets that may be taken with respect to the natural world. They hint at a third: "At the present time, an awakening interest in spatial patterning . . . and landscape ecology . . . indicates that at least a third dimension must be investigated. In these studies, the interest is focused on the way ecological systems are arrayed in space. This perspective is essentially absent in the process-functional viewpoint and represents only a small proportion of population-community studies" (ibid., p. 211).

25 Ibid., pp. 55–56.

26 Andrew Brennan, *Thinking About Nature: An Investigation of Nature, Value and Ecology* (Athens: The University of Georgia Press, 1988), pp. 66–88.

27 O'Neill et al., *A Hierarchical Concept*, pp. 64–65. According to hierarchy theory, debate concerning the relationship between stability and diversity is an example of the conceptual confusion that can result from failure to distinguish between the population-community and process-functional points of view. The relationship was originally expressed as one between ecosystem stability and the *complexity* of energy or material flows in the ecosystem. "Later investigators," however, "focused almost entirely on *species diversity*, and stability defined as the persistence of

each species in the system." Functional *complexity* ended up being tested in the population paradigm employing the concept of species *diversity*. (Ibid., pp. 209, 191–93: our emphasis. For a discussion of problems with the *stability-diversity* hypothesis, see Mark Sagoff, "Ethics, Ecology, and the Environment: Integrating Science and Law," *Tennessee Law Review* 56 [1988]: 187–91).

28 Ibid., p. 70.

29 Ibid., pp. 68, 177.

30 Ibid., p. 203.

31 A different "metaphysics of morals" is developed by Callicott in his "Animal Liberation and Environmental Ethics: Back Together Again," in his *In Defense of the Land Ethic: Essays in Environmental Philosophy* (Albany: State University of New York Press, 1989): 49–59; and "The Case against Moral Pluralism," "*Environmental Ethics* 12 (1990): 99–124. This metaphysics of morals is critiqued in Jim Cheney, "Callicott's 'Metaphysics of Morals,'" *Environmental Ethics* 13 (1991): 311–25. It is unclear what the relationship, if any, is between Callicott's two metaphysics of morals, though our guess is that he would regard the one discussed in the present paper as the more fundamental.

32 Callicott, "Implications," p. 302.

33 Ibid.

34 Ibid., pp. 309–10.

35 Ibid., p. 311. *Holism* is defined by Sagoff as maintaining that "large ensembles have properties that cannot be derived analytically from those of their smallest parts." Sagoff, "Ethics, Ecology, and the Environment," p. 132. *Holism* is used in a stronger sense by Callicott. His brand of holism is defined by his points (3)-(6), none of which is entailed by Sagoff's definition. A minimal holism is endorsed by Brennan when he argues for the "supervenience of ecological role on more basic properties of organism and environment." Brennan, *Thinking about Nature*, pp. 118–19. Brennan explicitly rejects Callicott's metaphysical holism.

36 Callicott, "Implications, pp. 312–13. For Callicott, particles and organisms are not illusory, but, as we will argue, are ontologically parasitic upon (see note 41) and, in that sense, reducible to energy.

37 Ibid., p. 314.

38 Ibid., pp. 301, 314–15. Cf. J. Baird Callicott, "Intrinsic Value, Quantum Theory, and Environmental Ethics," *Environmental Ethics* 7 (1985): 257–75.

39 See Cheney, review of Naess's *Ecology*.

40 Holmes Rolston, III, review of Andrew Brennan, *Thinking About Nature*, in *Environmental Ethics* 11 (1989): 260.

41 One reviewer of this paper thought it odd that we call Callicott a reductionist, given that Callicott is a self-identified holist. Must he not be one or the other, but not both? We think not. As Brennan has persuasively argued (*Thinking about Nature*, pp. 66–88), the *terminological* reductions typical of the sciences have no ontological implications. Presumably, the real concern of our reader is the *direction* of Callicott's reduction. Ontological reductionism is usually thought of as the reduction of putative wholes to the properties and behavior of their component parts. Antireductionism, by contrast, holds that these wholes "have properties that cannot be derived analytically from those of their smallest parts" (see n. 35). Callicott's holism is certainly antireductionist in *this* sense (what we call a Gestaltish sense). He is nonetheless reductionist in a different sense (what we call a primitive sense); the *direction* of the reduction is merely reversed. Brennan calls the type of reduction effected by Callicott *ontological parasitism* or *ontological primitivism* (ibid., pp. 76–88). To say that one kind of entity is ontologically parasitic upon (or primitive with respect to) another is to say that

the one is nothing but the other (as when a table is said to be nothing but a swarm of molecules). The parasitic entity is then said either not to exist at all—reference to such entities is a mere convenience or convention—or to have less reality than the entity upon which it is parasitic. The reduction of a whole to its parts is achieved by arguing that the whole is parasitic upon its parts. It is quite possible to reverse the reduction by reversing the parasitic relationship, arguing that the *parts* are ontologically parasitic upon the *whole*, less real than the whole, or dependent for their existence upon the whole. This is what Callicott does: "energy seems to be a more fundamental and primitive reality than material objects or discrete entities"; organisms are "configured" by energy, they are "in their entire structure . . . patterns, perturbations, or configurations of energy"; "a thing's essence is exhaustively determined by its relationships"; "relations are 'prior' to the things related, and the systemic wholes woven from these relations are 'prior' to their component parts." Callicott's reductionism (primitive sense) is especially clear in the following argument: "ecologists . . . would seem hardly prepared to deny outright a primary reality to atomic and molecular matter per se in addition to energy and its 'flow.' Organisms, though conduits of and configured by energy, remain *composed* of molecules. . . . [E]cological interactions . . . constitute of macrocosmic network or pattern through which solar energy . . . is transferred from organism to organism until it is dissipated. . . . [Now the reduction:] However, if we . . . resolve the erstwhile solid and immutable atoms of matter . . . into the ephemeral, energy quanta, then we may say quite literally and unambiguously that organisms are, in their entire structure—from subatomic microcosm to ecosystem macrocosm—patterns, perturbations, or configurations of energy" ("Implications," p. 310). Not all endorsements of holism are reductive in this primitive sense. Brennan's minimal holism, for example, is not (see n.35). It is the primitive reductionism to which O' Neill et al. strenuously object.

42 O'Neill, "Perspectives," p. 145.

43 The position we attribute to O'Neill et al. is sometimes called *antirealist*. This term is misleading, however. The antirealist position must be understood in contrast to the *particular* realist position it rejects. As Sagoff puts it: "Realists believe that scientific knowledge has a metaphysical foundation, insofar as true hypotheses go beyond the phenomena studied and correspond to the world as *given* or *as it exists in itself*, independently of our concepts, experience, and beliefs" (Sagoff, "Ethics, Ecology, and the Environment," p. 94). Clearly, to reject this particular form of realism is not equivalent to holding that the entities and processes described by ecosystem ecology are not real in *some* reasonably robust sense or that they are merely subjective. The "antirealist" position of some biologists is described in this way by Sagoff: "These biologists hold that the world is correctly described in many ways; these descriptions depend upon the concepts that are available, the data that are accessible, the other beliefs that we accept, and the purposes which we seek to serve. Under this approach, no set of biological concepts is metaphysically privileged; instead, a vocabulary is to be preferred only in the context of a particular purpose and because it serves that purpose better than another vocabulary." Ibid., pp. 102–03. "Antirealists" also point out that facts are theory-laden and that theories are, in turn, value-laden. From this perspective, we must understand our ontological commitments (e.g., what it means to say that ecosystems really exist) in light of our epistemic relationship to the world. To say that knowledge is socially constructed, that it is contextual, that it is value-laden, etc., helps define what it means to say that a theory is true or that it correctly describes the word.

44 The things so connected, it should be remembered, are reduced by Callicott to nodes of energy which, in turn, are conceived as a field, all parts of which are

connected to every other part. Consequently, organisms (as nodes in an energy field) are all connected to one another. The ecologist may talk in terms of organism and niche rather than in terms of energy flow patterns. Still, according to Callicott, a reduction is always possible: population-community studies can be conceived of in process-functional terms because, energy being the more fundamental or primitive reality, organisms, properly conceived, are mere perturbations of energy. This view, however, is precisely what hierarchy theory denies.

45 O'Neill et al., *A Hierarchical Concept*, p. 86.

46 O'Neill et al. state: "Hierarchically structured systems will not be fully connected, and, due to differences in rate processes, some relationships will be asymmetric: *X* will affect *Y*, but *Y* cannot affect *X*. This asymmetry becomes a fundamental organizing principle" (ibid., pp. 86, 94).

47 Callicott, "Implications," pp. 313–15 (emphasis added).

48 Ibid., p. 311.

49 O'Neill et al., *A Hierarchical Concept*, p. 189.

50 Brennan, *Thinking about Nature*, pp. 216–17.

51 See O'Neill et al., *A Hierarchical Concept*, p. 14.

52 Callicott, "Implications," p. 314 (emphasis added). This is not to say that Callicott's account is without merit. The extension of the relation concept of the self to consciousness, which Callicott borrows from Paul Shepard, is extremely provocative. Our position is simply that this relational view of the self is completely detachable from Callicott's metaphysical ecology.

53 Callicott's reply would be that "egoism has been regarded as axiomatically given," that the problem is "How to account for the value of 'others'—human others and now nonhuman natural others." Callicott, "Implications," p. 315. But this response merely invites the following questions: first, is it axiomatically given *for ethical theory?* Second, is not this supposed givenness, if such it be, itself problematic? We are simply looking at what the implications of Callicott's metaphysical ecology might be if the axiom were denied, if ethical reflection began with the givenness of the value or ethical considerability of *others* and proceeded from *there*.

54 See Cheney, "Eco-Feminism." For an account of the respects in which eco-feminism and hierarchy theory are engaged in complementary, mutually supportive projects, see Warren and Cheney, "Ecological Feminism and Ecosystem Ecology."

55 Brennan, *Thinking about Nature*, p. 176.

56 Holmes Rolston, III, *Environmental Ethics: Duties to and Values in the Natural World* (Philadelphia: Temple University Press, 1988), pp. 341–2. According to Donna Haraway, "the life and social sciences . . . are composed through complex, historically specific storytelling practices. Facts are theory-laden; theories are value-laden; values are story-laden. Therefore, facts are meaningful within stories." Donna Haraway, "Primatology is Politics by Other Means," in Ruth Bleier, ed., *Feminist Approaches to Science* (New York: Pergamon Press, 1986), p. 79. See also Jim Cheney, "Postmodern Environmental Ethics: Ethics as Bioregional Narrative," *Environmental Ethics* 11 (1989): 117–34.

# 102

# THE SHIFTING PARADIGM IN ECOLOGY

*S. T. A. Pickett and Richard S. Ostfeld*

Source: Richard Knight and Sarah Bates (eds), *A New Century for Natural Resources Management*, Washington: Island Press, 1995, pp. 261–78.

> To the ecological mind, balance of nature has merits and also defects. Its merits are that it conceives of a collective total, that it imputes some utility to all species, and that it implies oscillations when balance is disturbed. Its defects are that there is only one point at which balance occurs and that balance is normally static.
>
> (Aldo Leopold *The River of the Mother of God and Other Essays*)

This chapter summarizes how the science of ecology has changed over the last few decades, and explores the implications of those changes for natural resource managers. The changes we describe are the most general ones a science can exhibit: changes in its paradigm. A paradigm is the viewpoint a science takes of the world, which consists of the often unspoken background assumptions of the science, and the way the science approaches and answers questions. These two components are often summarized as a world view or belief system, and the exemplars for problem solving. Because a paradigm is so broad in scope, it is an excellent way to encapsulate the more detailed changes in data, techniques, and theories that a science might experience.

We use the synthetic tool of the paradigm to suggest the management implications of some of the vast changes that ecology has undergone. Many of these changes reinforce changes in management strategy already being practiced, while others invite continued evaluation of management goals and techniques.

We first present elements of the classical paradigm in ecology. We then indicate the implications of the classical paradigm for resource management. The failings of the classical strategy are indicated in general terms. The relationship of the classical paradigm and management to the everyday metaphor of "the balance of nature" will be explored. Because the classical paradigm has failed, we outline the new paradigm in ecology, and point out its significance for management. Along the way, we describe a metaphor, the "flux of nature," which is consistent with the new paradigm, and discuss important cautions that managers must practice when employing the new paradigm and its common sense label. The new paradigm calls for new kinds of management, such as process management, ecosystem management, or adaptive management. But it also calls for a firm ecological foundation, and a broad and careful vision in its application.

## The classical paradigm in ecology

The classical ecological paradigm comprises six key points that from a network of closely related background assumptions (1). Historically, ecological systems were considered to be primarily closed, self-regulating, and subject to a single stable equilibrium. Furthermore, any changes in communities or ecosystems through time were thought to occur by successions that must always pass through the same phases. Any disturbances that might affect natural systems were considered to be exceptional events, and humans were excluded from the roster of normal ecological factors. What does each of these points mean in real ecological systems? Before giving examples, we must emphasize that the points refer to a whole array of ecological units and entities, including individuals, populations, communities, landscapes, and ecosystems. Throughout this chapter, the term *system* is used to mean any of these units of ecological interest.

Considering ecological systems to be closed means that the important structures and interactions occur within the boundaries set for studying them. Even under this assumption, of course, thermodynamics calls for energetic openness of ecological systems, and ecologists have uniformly recognized external physical constraints or inputs imposed on ecological systems by climate. Thus, assuming ecological systems to be closed is paradigmatic only when applied to the local factors that ecologists invoke to explain the mechanistic workings of their objects of study. For example, the nutrient capital of an ecosystem would be derived from weathering of the local bedrock or parent material; and the organisms involved in important mutualisms would be local residents only.

The paradigmatic assumption that ecological systems are self-regulating follows, in part, from the assumption that systems are closed. If systems are indeed self-contained, then they must be internally regulated if they are to persist. For example, the local cycling of nutrients would govern productivity; within-community interactions would account for species coexistence; internal interactions within populations would regulate the density of the population.

Classically, ecologists focused on single, stable equilibria as the defining points of reference for their systems. But such end points were taken as much more than useful abstract points of reference. They were for a long time considered concrete states that must exist in nature. Such a focus on equilibrium points limits the explanatory mechanisms that can be applied to answer ecological questions. Examples of equilibrium points include the single "climax" state for a community on a coarse scale, the fixed carrying capacity of a population and the saturation number of species fixed by resource partitioning in a community.

When the classical paradigm allowed change in ecological systems, it was by fixed pathways. For example, successions were deterministic in that they required that invading plants appear before later stages; overshoot of a population from its carrying capacity would necessarily decline again to that level, perhaps with orderly fluctuation around that constant; community membership would be reproduced after a perturbation.

Disturbance was thought to be an exceptional occurrence under the classical ecological paradigm. This meant that the primary goal of ecological studies could be the understanding of undisturbed systems. As a result, ecologists emphasized pristine and apparently "natural" systems. Alternatively, they could focus on the return of systems from a disturbed state to the single point of equilibrium. Ecologists largely ignored such systems that were burned or blown down, populations apparently not regulated internally, managed landscapes, and organisms living beyond their usual ranges.

Humans were not considered by ecologists to be part of the classical paradigm because humans violated, either accidentally or intentionally, many of the other assumptions of that paradigm. Humans were often purposely left out of ecology because they introduced multiple states to systems, acted as disturbance agents, transported materials and organisms beyond their usual distributions, acted as external regulators of ecological systems, and prevented orderly, deterministic successions.

### *Relationship to resource management*

The classical paradigm has clear parallels in classical management. Not all resource managers have accepted all of these assumptions, however, as there have always been independent and visionary practitioners (2). Indeed, rejection of these assumptions is increasingly seen in the management of natural areas. Laying out the assumptions in their bald form, as we do here, almost immediately illuminates their barrenness.

#### *Systems as closed*

If managers accept this assumption, they would consider any unit of the natural world to be manageable as a separate entity. National parks would be cordoned off, toxic waste sites would be mitigated as isolates, an endangered species could be saved as an entity.

*Systems as self-regulating*

The internal dynamics of natural resource systems would adjust to environmental changes. Such systems could simply be left alone if they were self-regulating. For example, a mammal population would be regulated by internal behavioral and reproductive feedbacks to maintain a relatively stable density through time. Likewise, a species distributed patchily throughout a landscape would maintain its density and distribution by adjusting rates of immigration and emigration among patches. Therefore, the dominant ingredient in management of many systems would be benign neglect.

*Equilibrium as a point*

Systems possess a single point at which their composition or function is in equilibrium with the environment. The implication is that ecological systems are at or close to equilibrium. For example, there is a single stable state for vegetation set by a particular climate, soil, elevation, and exposure. Likewise, the "carrying capacity" for a population is a fixed constant. If a single stable point equilibrium is the dominant kind of stability in nature, then managers can simply observe how nature is at any one time, and do whatever is necessary to maintain that state. A second bias about equilibria found in the classical paradigm is that they were the most desirable state of nature (3). These assumptions about equilibria as fixed points link with the previous one to underwrite management by benign neglect.

*Succession as fixed*

This assumption suggests that systems subjected to a disturbance will recover their previous state through an obligatory succession. Managers may usefully help the rate of succession along, but the pattern of states through time is fixed, and it inexorably leads to the expected end point, which was often assumed to be the most desirable state of the system. For example, although the post-agricultural succession in tallgrass prairie was long considered to be the necessary sequence of annual weeds, annual grasses, bunch grasses, and climax prairie, that sequence is not often found (4).

*Disturbance as exceptional*

If disturbance, the disruption by a sudden event of the structure, resource availability, or environmental controls of a system, is seen as very uncommon or restricted in space, then managers could make and execute their plans without taking it into account. This assumption would lead managers to be unprepared for disturbance when it finally did occur. The assumption could even lead them to attempt to prevent or to compensate inappropriately for the effects of disturbance that should in fact be a part of the system. For example, the assumption that fire did not belong in presumably equilibrium systems led to its exclusion from management strategies in many systems. A related error is made by those who believe that a large patch of wind-thrown trees in a moist, broadleaved forest is an unfortunate scar rather than an event that is part of the dynamics of such forests.

### *Humans excluded*

The assumption that humans are not a component of systems to be managed could lead to management plans that neglected historic or contemporary human influences on a particular system. If such influences were actually important structuring variables in the system, the management strategy would be flawed (5). For example, forgetting the effect of Native Americans on vegetation and animal communities could result in neglecting important ecological controls on systems. Neglecting human effects is unfortunately easy, because many such effects are subtle, or originate at a distance from the site of interest.

In pointing out the general ways on which the assumptions of the classical paradigm could translate into management decisions, we have suggested how they can fail. In idealizing and simplifying the ecological world, the tenets of the classical paradigm have blinded ecologists and managers to critical factors and events that can govern systems. The assumptions have also caused scientists and managers to neglect important dynamical pathways and states, and to disregard important connections among different systems (6).

## ***Balance of nature and the classical paradigm***

While it would be unwise to either blame or credit ecology for too much of the practice of natural resource management, at the very least, it is clear that the two areas have shared many of the same assumptions in the past. This situation may reflect the cultural origins of some of the largest assumptions any science makes. Ecology and resource management may both have adopted certain key components of their classical outlooks from the larger society (7). Such a situation is suggested by the pervasiveness and apparently comfortable acceptance of the idea of the "balance of nature."

The balance of nature is a poorly articulated idea that is a cultural metaphor rather than an exact scientific concept. The metaphor implies that natural systems exhibit tight control and adjustment to an equilibrium state. It has extremely deep historical roots (8), is held in high esteem, and often is misused as an apparently unassailable bastion in discussion. Because the idea is not scientific, it is unclear just what its assumptions are, where and when it might apply, what mechanisms might lead to it, and how one might test it.

Yet the balance-of-nature metaphor can stand for some valid scientific ideas. The fundamental truth about the natural world that the idea may relate to is the fact that natural systems persist, and they do so by differential response of various components. The idea also points toward the ecological principle that there are limitations in natural systems. No component of a natural ecological system, at whatever level of organization, grows without limit. Furthermore, the idea also indicates the frequently observed trends of ecological systems toward equilibrium states, given a specified environment and a sufficiently long time. Examples are density-dependent processes (i.e., the tendency of populations to grow when small and shrink when large) and the existence of successional trajectories by which communities tend to approach a stable "climax" state. All these principles are powerful ecological generalizations. But, ecologists themselves have sometimes relied on the venerable, yet vague, metaphor of the balance of nature to represent these principles in a public dialogue about conservation and management. This is inappropriate for two reasons.

First, the balance-of-nature metaphor has serious limitations even as a vessel for valid ecological generalizations. It can lull people into accepting equilibrium points as persistent, and into assuming that trajectories toward presumed equilibrium will necessarily succeed. Second, the metaphor has been closely associated with the classical paradigm in ecology, with its emphasis on equilibrium, containment, regulation, and so on. This is a serious liability now that the basic assumptions of the classical paradigm in ecology have failed. The failure of the classical paradigm is most cogently indicated by how rarely a stable equilibrium point is actually achieved (9). Constraints on equilibrium-seeking tendencies are common and effective, as illustrated by the burgeoning study of natural disturbance (10). The examples of the six assumptions of the classical paradigm we gave earlier in the chapter indicate the pervasive problems with them. We elaborate the failings and replacement paradigm below.

## The flux of nature and a new paradigm

The pitfalls of the classical paradigm calls for a new paradigm in ecology. The nonequilibrium paradigm can be connoted in the metaphor, the "flux of nature" (11). The term *flux* highlights variation, fluidity, and change in natural systems, rather than stasis, which is implied by the term *balance*. Although this metaphor does not deny the existence of stable points in nature, it focuses our attention on the fact that natural systems, which certainly do persist, do so as a result of a variety of fluxes. These include not only the obvious exchanges of energy and matter that are the currencies for trophic relationships of species, but also fluxes that range from the relative migration of various phenotypes and genotypes, through the shifts of dominance within communities, to the movements of patch types in landscapes. Many other kinds of fluxes could be cited. Some may exhibit equilibrium distributions while others may exhibit erratic or steady trajectories.

The new paradigm also must accommodate the contradiction of the six background assumptions of the classical paradigm. Ecological systems are never closed, but rather experience inputs such as light, water, nutrients, pollution, migrating genotypes, and migrating species. Note that the novel emphasis of the modern paradigm is not that ecological systems are self-regulating, but rather experience important limits from external sources. For example, successional rates can be controlled by herbivores from adjacent communities, and populations can rely on mutualists that reside elsewhere. Stable point equilibria are rare, although some systems of sufficient size and duration may exhibit stable frequency distributions of states. For example, a landscape may be a shifting mosaic of patches or community types, and in some cases, the number of young and old communities can remain constant, even though specific spots change as a result of disturbance and succession. Successions are rarely deterministic, but are affected by specific histories, local seed sources, herbivores, predators, and diseases. Disturbance is a common component of ecological systems, even though some sorts of disturbance are not frequent on the scale of human lifetimes. Indeed, many biotic interactions have the same sort of impacts as physical disturbances, like wind and fire, in altering species composition, interaction between species, and availability of resources in systems. And finally, landscapes that have not experienced important human influences have been the exception for hundreds if not thousands of years (12).

These points together constitute the contemporary paradigm in ecology. In a sense, this contradicts the classical paradigm, since it shows that the classical background assumptions are not necessarily true. But although the classical assumptions are not universally true, on certain temporal or spatial scales and under certain conditions, those assumptions may apply. For example, equilibrium states can be a special case under the new paradigm. The new paradigm does not deny the empirical existence or theoretical utility of equilibrium states as special cases or points of reference, but it does not assume that equilibria are the dominant or controlling states in the real world.

So far, we have treated the new paradigm as both a contradiction of the old and an expansion containing the good points of the classical paradigm. To completely appreciate the new paradigm, we must examine several other modern tools for understanding the natural world.

The entities that ecologists study are parts of networks. This, of course, emphasizes the openness of systems, and the potential that regulation resides outside the system of interest. For example, changes in populations other than the one an ecologist is trying to understand may change or constrain its density (13). Interacting populations may be inconspicuous, such as parasitic lung worms relative to their elk hosts. Or the interacting populations can live separately from the focal population for a time. For instance, the butterflies whose larvae attack passion flower vines rely as adults on the flowers of cucumber vines that live in entirely different habitats (14). Changes in number of a prey population may result from asynchronous changes in the dynamics of a predator population that indirectly affects its numbers (15). Similarly, there may be lags between the time a population receives a signal from other populations, and the time when it responds reproductively or behaviorally (16). Finally, internal changes in the population of interest, whether genetic or phenotypic, may affect how that population responds to other populations and physical signals (17). In a sense, for the population level of organization, the general message is that population ecology must in fact be community ecology (18).

In studying networks or communities in ecology, relating change and stability become important tasks. Pimm (19) has defined four kinds of change in networks. Resilience is the rate of recovery to a specified former state. Persistence is the period over which a state exists before a new state is reached. Resistance is the change in a variable following change in interacting variables. For example, an ecological community is resistant if its species composition changes little after the invasion of an exotic species Finally, variability is the degree of change over time. Variability indicates that fluctuations are accounted for in Pimm's (20) conceptual system.

The new paradigm recognizes that understanding and working with ecological systems depends on their spatial and temporal extent relative to the extent and duration of processes and structures in those systems. *Scale* is the term used to describe the relationships between two measurements such as extent over which a process occurs and spatial extent of a system. Scaling determines what is internal or external to the system. Whether some event is considered a disturbance or not, what is considered normal variation, what is self-regulation, and in fact whether the system is in equilibrium, all depend on the scale the system takes. At some scales, events that are considered external disturbances become incorporated into the system. For example, observing biomass of a prairie over a few years detects a fire as a disturbance to the biomass value. In contrast, observing species composition of that same prairie over decades accommodates a regime of fires as a regular part of the system (21). The contemporary paradigm equips us to deal with the structure and function of systems regardless of what scale they are observed on, and whether that scale incorporates all disturbances or regulatory factors within the system. Of course, the lives of both scientists and managers become simpler if we are assigned complete systems that internalize all important factors and exist at scales over which equilibrium can be attained. But neither the natural world nor administrators are usually so kind.

A final spur to adopt the new paradigm is the growing awareness of chaos (22). Chaos technically refers to system behavior that seems totally unpredictable and random, but is driven by strictly necessary (deterministic) relationships without any random elements. Although exactly how chaos might be expressed in ecological systems, and whether our data sets are long or detailed enough to allow that detection are open questions (23), the possibility of chaos in ecology casts further doubt on the utility of the equilibrium paradigm with its focus on linear, deterministic relationships (24).

## Opportunities for new management strategies

We have already outlined some of the ways traditional management approaches have been limited by the classical ecological paradigm. Assuming systems to be closed and self-regulating, and to possess a stable point equilibrium reached by deterministic successions after rare disturbance events unaffected by humans, has reinforced a particular philosophy of management. Additional kinds of limitations have grown out of this classical paradigm. Management has been limited by its focus on individual components versus management of systems (25).

Here we want to emphasize the organizational implications of the term *system.* A system is any collection of entities linked by functional relationships. Managing an individual entity may neglect important relationships or contexts that are in fact critical to the persistence of the focal entity. Management of commodities, for example, may neglect the remainder of the system, especially inconspicuous or infrequent interactions, on which the commodity depends. Managing an individual population, such as a fishery, is also an example. Modeling specific fisheries as isolated populations has often failed to predict the collapse of the stock. While obvious requirements such as habitat and food may be provided for, the structuring of the population by historical events, or its periodic control by fluctuations in mutualists or consumers may be harder to detect without an explicit system approach. For example, the decline in the black-footed ferret resulted from both a decline in availability of its main prey (prairie dogs) and the introduction of canine distemper resulting from close proximity to humans and their pets. An additional danger of focusing on isolates of systems is that desirable attributes of the entire system may be damaged in the effort to optimize a single entity (26). This has been the case with isolated guilds, such as edge-requiring wildlife, which have been managed to the detriment of more sensitive forest-interior species (27). The increase in white-tailed deer and the brown-headed cowbird are examples of untoward results of management of a single guild or species. Deer inhibit forest regeneration, while the parasitic cowbirds reduce the nesting success of forest interior birds.

The contemporary paradigm suggests that the fluctuations in the natural world are not simply oddities and dismissable excursions from nature's "real" state. Rather, fluctuations in the form of physical disturbances, climate shifts, population cycles, range shifts and the like are significant normal parts of the real world. Management cannot proceed as though such fluctuations did not occur or were not important. Managers must proceed as though fluctuations mattered.

Fortunately, just as we are now in a scientific window of opportunity for developing and applying new management strategies, we may well be in a social window as well. There are enough well-publicized and debated cases of the failure of classical management, or of the failure of policy that did not account for the flux of nature, that the time may be ripe for establishing a new foundation for management. Management within a world of open systems that are subject to internal and externally driven fluxes can be labeled "ecosystem management" or "process management." Whatever it is called, and whether its focus is on some spatial unit or on some dynamic process, management must be able to adapt to natural change.

Fortunately, there is an emerging framework for the accommodation of natural flux, that of adaptive management (28). Under this strategy, the goals of the management plan are articulated using an explicit model of the system that should encompass the components, interactions, and likey fluctuations. Then, appropriate techniques are applied to the system, the results are monitored, and the tactics or even the goals are modified according to what is learned from the response of the system. In other words, management can, like science, be used to test hypotheses. The difference is that in testing management hypotheses, the purpose of the management can be changed if it is found to be unrealistic, ineffective, or counter to the persistence of the system (29).

For example, the management by neglect practiced in many old-growth oak forests had been found to permit those forests to change composition and structure in undesirable ways. Many are shifting to a predominance of maples with a concomitant decline in understory diversity. One major way in which adaptive management can be useful is in refining the basic model of the system. The failure of a management strategy may indicate that the model of the system is incomplete, perhaps because it does not incorporate some important flux or connection that actually functions in the system. Because conceptual models of systems, whether they be production landscapes or nature reserves, are always tentative, such a strategy is important to adopt widely.

## Process management and its constraints

We have said that management for maintaining processes that structure the system of interest is especially appropriate to the new paradigm in ecology. The processes include the relationships to environmental resources and regulators, interactions among populations, and system dynamics or succession over time. However, the concept of "process" should not be taken too narrowly. A narrow meaning might focus solely on biogeochemical flows without concern for the diversity of biological entities that interact with biogeochemical transformations. Ecosystem models are often reduced to such a form, consisting of boxes for the pools of chemicals and arrows showing the chemical transformations within the ecosystem. However, ecosystem scientists recognize the community and population context of the biogeochemical pools and transformations encapsulated in such an abstract model (30).

But we must recognize additional kinds of processes in management. A complete framework for processes of concern in management would include the following: (1) disturbance regimes; (2) movement of materials, energy, and organisms; (3) succession; and (4) species interactions. It may be necessary to prioritize processes and manage to optimize certain ones, depending on the goals. However, even if only one or a few processes are to be optimized, their relationships to other processes must be known in order to avoid management surprises.

A disturbance regime is the spatial and temporal pattern of disturbance in some area. It consists of the agents, intensities, sizes, distributions, and frequencies of disturbances that affect a system. The term *disturbance regime* does not mean that the identity, temporal, or spatial patterns of disturbance are fixed and regular, only that such distributions can be characterized and understood as system attributes. They can be modified and compensated for by management.

Movement of materials and entities is a second focus of process management. The things moving can be either abiotic or biotic. Abiotic fluxes include the familiar ecosystem fluxes of energy and nutrients. In human-modified landscapes, which is beginning to mean most of the Earth, the materials would also include pollutants and plumes of thermally altered air and water, which can interact with natural ecosystems, communities, and populations.

Biotic fluxes include the movement of organisms. The specific processes include natural dispersal, migration, and the transport of exotics. But biotic fluxes also involve larger scales. Shifting mosaics of communities in landscapes is an example of a coarse-scale biotic flux (31).

One of the most universal of ecological processes is succession, the change in community composition or structure through time. Succession must therefore be a concept in the toolkits of all managers (32). It is driven by characteristics of sites determined by natural or human disturbances, by the differential availability of organisms, and by the differential performance of organisms at those sites. Managing a system without information on its successional status or rate is risky, and is likely to produce unexpected results (33).

Species interactions are among the most well-known interactions that must be managed. Although interactions are often labeled by their "net effect" on the populations involved, such that competition is a joint negative effect and mutualism is a joint positive effect, it is important to know just how such interactions proceed. This is because there can be hidden, indirect chains of effects, or interactions mediated by third parties. Thus, competition between two species may be only "apparent" (34), and the outcome driven instead by the stimulatory effects of one of the species in the "competitive pair" on an enemy of the second. Voles and shrews are affected by the same predators, and change in one of the prey species may cause the predator to change its impact on the other prey (35). If just the small mammals are studied, the dynamic would appear as competition. Indirect effects such as apparent competition appear to be legion in natural communities.

How does a process approach to management relate to "ecosystem management"? In reality they can be considered the same thing, because a good management plan will be based on a comprehensive model of an ecosystem. The model will indicate the spatial extent of interest, the inputs and outputs of resources and populations, and the direct and indirect interactions of populations, as well as the interactions between the populations and the physical substrate and fluids in the ecosystem. So ecosystem management must include the processes that occur within that ecosystem and which connect it with others. The phrase *ecosystem management* is valuable because of the spatial implications of the term *ecosystem.* Ecosystem management reminds us to consider the landscape or regional connections of the situation to be managed (36). And the metaphor, flux of nature, is a reminder that the nature of the system and its context can both change through time, sometimes episodically.

## Caveats for using the flux of nature

The metaphor, the flux of nature, is a powerful tool for communicating the new ecological paradigm. It reminds us that the basic substance of nature, the networks of species interacting with one another and exchanging matter and energy with their surroundings, do persist over long periods. But the persistence is the result of dynamism and flux on different scales (37). The assemblages of species change in identity with evolution on the long term, and with natural disturbance and succession on the short term. The assemblages shift in local composition and architecture on the local scale, while the arrangement of patterns in landscapes shift under disturbance regimes and climate change. The richness that evolution has produced is a response to the opportunities produced by disturbance, natural extinction, climate shifts, and the arrival of new immigrants. On a more local scale, even though the time frame may be long for an individual human life, the fall of a mature tree in an old-growth forest generates opportunities for establishment, growth, or reproduction of species that had been excluded from or suppressed on the forest floor. Various animals, microbes, and plants may use the new site, while others avoid it. But over the long term and large areas, the variety of opportunities, resources, and environmental signals is answered by a richness of organisms. That is the beauty of the flux of nature.

Most spasms of nature become opportunities, and if the accidents of evolution and history have made organisms available, then they will appear to fill the stage. Subsequent changes in the environment, the species pool, or the nature of interactions may occur. They are all part of the flux of nature, none more intrinsically desirable than another. This is the shifting stage and shifting play that ecologists must understand and managers must deal with.

For all its scientific intrigue and poetic beauty, the flux of nature is a dangerous metaphor. The metaphor and the underlying ecological paradigm may suggest to the thoughtless or the greedy that since flux is a fundamental part of the natural world, any human-caused flux is justifiable. Such an inference is wrong because the flux in the natural world has severe limits.

Knowing what these limits of natural flux are is a critical step in management. In general terms, these limits are functional, historical, and evolutionary. Functional limits are the physiological constraints that organisms operate under. For example, certain molecules cannot be detoxified, leaves must remain open to pollutants if they are to take up $CO_2$ for photosynthesis, and high temperatures denature proteins. Such limits cannot be overcome due to the mechanics and chemistry of cells and organs. An anthropogenic flux outside those limits is an insult to natural flux.

Historical limits are those that result from the suites of species that have become established through time at a site, the accumulation of resources and organic matter, and the kinds and patterns of disturbances that have operated there over time. For example, if the disturbance regime of a site has not included massive stand-opening events that leave the roots and soil intact, then there are not likely to be species present that can respond rapidly to a novel anthropogenic disturbance of that type.

Evolutionary limits are imposed by the amount of genetic variation available for selection to act on, the intensity of selection, and time over which selection may have acted. Evolution is also constrained by the physical structures elaborated in the past. If human-induced changes reduce the capacity of lineages to evolve by reducing genetic variation, elevating levels of gene flow, or altering population mating structures, then their effect on the flux of nature will be damaging, perhaps permanently.

Problematic human changes of fluxes are those that are beyond the limits of physiology to tolerate, history to be prepared for, or evolution to react to. Two characteristics of a human-induced flux would suggest that it would be excessive: fast rate and large spatial extent. Natural systems respond to, even exploit flux, if the rates of flux are within the capacity of physiology to adjust, new assemblages to accumulate, and evolution to generate new diversity. Where rates of human-generated flux exceed the natural rates of response, policy makers and managers should be alerted that undesired changes may result. Managers may be in a position to compensate for inappropriate rates of human-caused changes.

The second problem of current human-generated flux is its great spatial extent. Natural flux takes space for the shifting mosaics of landscapes, the sources of genetic novelty, and the refuges from disturbance, predation, and disease. Human-generated flux that subtracts from the area available for natural flux must be carefully evaluated and compensated for.

## Conclusion

We have examined the paradigm shift that has been taking place in ecology over the past several decades. The classical paradigm, the equilibrium world view of the discipline, assumed that ecological systems were closed, self-regulating, possessed of single equilibrium points reached by deterministic dynamics, rarely disturbed naturally, and separate from humans. Classical management strategies were cast as though they followed these same assumptions. However, the failures of management reflecting this paradigm became apparent, and along with the accumulation of new data and scales of observation in ecology, contributed to the demise of the classical, or equilibrium, paradigm. With the demise of the equilibrium paradigm, the inappropriateness of the formerly comfortable metaphor of the balance of nature became apparent.

A new ecological paradigm has emerged that recognizes ecological systems to be open, regulated by events arising outside of their boundaries, lacking or prevented from attaining a stable point equilibrium, affected by natural disturbance, and incorporating humans and their effects. A new metaphor of the flux of nature symbolizes the new, or nonequilibrium, paradigm effectively. The new paradigm suggests that the management approaches being developed are much more appropriate to the variability, contingency, and openness of ecological systems. The intimate linkage of opportunity and response that is embedded within the flux of nature takes long times and large spaces to work. Ecosystem management, carried out in an adaptive way in which management plans are viewed as hypotheses based on inclusive models of the systems to be managed, is likely to exploit the contemporary understanding of ecological systems.

Neither process nor ecosystem management should be considered to mean that nutrient or energy flows should be managed without attention to the biodiversity with which they interact. In addition, management for the flux of nature must respect the physiological, historical, and evolutionary limits inherent in natural systems if management is to be successful and sustainable.

## Acknowledgments

We thank Dan Binkley and Rick Knight for reviews that improved the structure and clarity of our presentation. This is a contribution to the program of the Institute of Ecosystem Studies with partial support from the Mary Flagler Cary Charitable Trust.

## Notes

1 Pickett, S.T.A., and P.S. White. 1985. The ecology of natural disturbance and patch dynamics. Orlando, Florida: Academic Press.
2 Holt, S.J., and L.M. Talbot. 1978. New principles for the conservation of wild living resources. Louisville, Kentucky: Wildlife Society.
3 Jefferson, R.G., and M.B. Usher. 1986. Ecological succession and the evaluation of non-climax communities. In Wildlife conservation evaluation. M.B. Usher, ed. 69–91. London, England: Chapman and Hall.
4 Collins, S.L., and D.E. Adams. 1983. Succession in grasslands: thirty-two years of change in a central Oklahoma tallgrass prairie. Vegetation 51:181–190.
5 Wagner, F.H., and C.E. Kay. 1993. "Natural" or "healthy" ecosystems: are U.S. national parks providing them? In Humans as components of ecosystems: the ecology of subtle human effects and populated areas. M.J. McDonnell and S.T.A. Pickett, eds. 257–270. New York: Springer-Verlag.
6 Botkin, D.B. 1990. Discordant harmonies: a new ecology for the twenty-first century. New York: Oxford University Press.
7 Oelschlaeger, M. 1991. The idea of wilderness from prehistory to the age of ecology. New Haven, Connecticut: Yale University Press.
8 Egerton, F.N. 1973. Changing concepts of the balance of nature. Quarterly Review of Biology 48:322–350.
9 Botkin 1990 (6).
10 Pickett and White 1985 (1).
11 Pickett, S.T.A., V.T. Parker, and P. Fiedler. 1992. The new paradigm in ecology: Implications for conservation biology above the species level. In Conservation biology: the theory and practice of nature conservation, preservation, and management. P. Fiedler and S. Jain, eds. 65–88. New York: Chapman and Hall.
12 McDonnell, M.J., and S.T.A. Pickett, eds. 1993. Humans as components of ecosystems: the ecology of subtle human effects and populated areas. New York: Springer-Verlag.
13 Pimm, S.L. 1991. The balance of nature? Ecological issues in the conservation of species and communities. Chicago, Illinois: University of Chicago Press.
14 Gilbert, L.E. 1980. Food web organization and the conservation of neotropical diversity. In Conservation biology: an evolutionary-ecological perspective. M.E. Soule and B.A. Wilcox, eds. 11–13. Sunderland, Massachusetts: Sinauer Associates.
15 Hansson L., and H. Henttonen. 1988. Rodent dynamics as community processes. Trends in Ecology and Evolution 3:195–200.
16 Hanski, I., P. Turchin, E. Korpimäki, and H. Henttonen. 1993. Population oscillations of boreal rodents: regulation by mustelid predators leads to chaos. Nature 364:232–235.
17 Ostfeld, R.S., C.D. Canham, and S.R. Pugh. 1993. Intrinsic density dependent regulation of vole populations, Nature 366:259–261.
18 Pimm 1991 (13).
19 Pimm 1991 (13).
20 Pimm 1991 (13).
21 Allen, T.F.H., and E.P. Wyleto. 1983. A hierarchical model for the complexity of plant communities. Journal of Theoretical Biology 101:529–540.

22 Schaffer, W.M., and M. Kot. 1985. Nearly one-dimensional dynamics in a simple epidemic. Journal of Theoretical Biology 112:403–427.
23 Sugihara, G., and R.M. May. 1990. Nonlinear forecasting as a way of distinguishing chaos from measurement error in time series. Nature 344:734–741.
24 Allen, J.C., W.M. Schaffer, and D. Rosko. 1993. Chaos reduces species extinction by amplifying local population noise. Nature 364:229–232.
25 Franklin, J.F. 1993. Preserving biodiversity: species ecosystems, or landscapes. Ecological Applications 3:202–205.
26 Committee on Scientific and Technical Criteria for Federal Acquisition of Lands for Conservation. 1993. Setting priorities for land conservation. Washington, D.C.: National Research Council.
27 Harris, L.D. 1988. Edge effects and conservation of biotic diversity. Conservation Biology 2:330–332.
28 Barrett, G.W. 1985. A problem-solving approach to resource management. BioScience 35:423–427; Schroeder, R.L., and M.E. Keller. 1990. Setting objectives: a prerequisite of ecosystem management. In Ecosystem management: rare species and significant habitats. R.S. Mitchell, C.J. Sheviak, and D.J. Leopold, eds. 1–4. Albany, New York: New York State Museum.
29 Irwin, L.L., and T.B. Wigley. 1993. Toward an experimental basis for protecting forest wildlife. Ecological Applications 3:213–217.
30 Jones, C.G., and J.H. Lawton. 1995. Linking species and ecosystems. New York: Chapman and Hall.
31 Pickett and White 1985 (1).
32 Niering, W.A. 1987. Vegetation dynamics (succession and climax) in relation to plant community management. Conservation Biology 1:287–295.
33 Luken, J.O. 1990. Directing ecological succession. New York: Chapman and Hall.
34 Holt, R.D. 1997. Predation, apparent competition and the structure of prey communities, Theoretical Population Biology 12:197–229.
35 Hansson and Henttonen 1988 (15).
36 Risser, P.G. 1985. Toward a holistic management perspective. Bio-Science 35:414–418.
37 Allen, T.F.H., and T.W. Hoekstra. 1992. Toward a unified ecology. New York: Columbia University Press.

# INDEX